Lecture Notes in Mathematics

Edited by A. Dold and B. Eckmann

1354

A. Gómez F. Guerra M. A. Jiménez
G. López (Eds.)

Approximation and Optimization

Proceedings of the International Seminar
held in Havana, Cuba, Jan. 12–16, 1987

Springer-Verlag
Berlin Heidelberg New York London Paris Tokyo

Editors

Juan Alfredo Gómez-Fernandez
Institute of Mathematics, Cuban Academy of Sciences
Calle 0 #8, Vedado, Havana 4, Cuba

Francisco Guerra-Vázquez
Guillermo López-Lagomasino
Faculty of Mathematics, University of Havana
Havana 4, Cuba

Miguel A. Jiménez-Pozo
Cuban Mathematical Society, University of Havana
Havana 4, Cuba

Mathematics Subject Classification (1980): 30-06, 41-06, 49-06, 65-06, 42-06, 42C05, 90C05, 93C05

ISBN 3-540-50443-5 Springer-Verlag Berlin Heidelberg New York
ISBN 0-387-50443-5 Springer-Verlag New York Berlin Heidelberg

Printing and binding: Druckhaus Beltz, Hemsbach/Bergstr.
2146/3140-543210

PREFACE

This volume contains the proceedings of the Seminar on Approximation and Optimization, which took place in January 12-16, 1987 at the University of Havana, Havana, Cuba. The seminar was jointly organized by the University of Havana, the Cuban Academy of Science and the Cuban Mathematical Society to promote scientific contacts between specialists of two very closely related branches of mathematics, namely approximation theory and optimization theory.

We wish to thank the International Mathematical Union and the International Council of Scientific Unions for sponsoring the seminar: their financial support was decisive in obtaining a considerable participation from mathematicians of Western Europe, North America and Latin America. The Third World Academy of Sciences also made a financial support.

The contributions to this volume include original research papers as well as a few survey articles. All these papers were refereed. We have divided the contents into three sections: the first one contains the papers submitted by some of the invited speakers; in the last two, the rest of the papers are classified according to their contents.

Alfredo Gomez — ICIMAF, Academia de Ciencias de Cuba, 0 #8, Habana 4, Cuba

Francisco Guerra — Fac. de Mat. y Cib., Univ. de La Habana, Habana 4, Cuba

Miguel Jiménez — Sociedad Cubana de Matemáticas, Habana 4, Cuba

Guillermo López — Fac. de Mat. y Cib., Univ. de La Habana, Habana 4, Cuba

TABLE OF CONTENTS

Nonparametric Polynomial Density Estimation in the L^p Norm

Z. CIESIELSKI

Abstract. *A simple construction of polynomial estimators for densities and distributions on the unit interval is presented. For densities from certain Lipschitz classes the error for the mean L^p deviation is characterized. The Casteljeau algorithm for calculating the values of the estimators is applied.*

1. Introduction. The space of all real polynomials of degree not exceeding m is denoted by Π_m. In Π_m we have the Bernstein basis i.e.

$$\Pi_m = \mathrm{span}[N_{i,m}, i = 0, \ldots, m],$$

where

$$N_{i,m}(x) = \binom{m}{i} x^i (1-x)^{m-i}, \qquad i = 0, \ldots, m.$$

The Casteljeau algorithm is based on the identity

$$N_{i,m}(x) = (1-x)N_{i,m-1}(x) + xN_{i-1,m-1}(x). \tag{1.1}$$

For given $w \in \Pi_m$

$$w(x) = \sum_{i=0}^{m} w_i N_{i,m}(x), \tag{1.2}$$

where the coefficients w_i are unique. Using (1.1) we find that for $0 \leq k \leq m$

$$w(x) = \sum_{i=0}^{m-k} w_i^{(k)}(x) N_{i,m-k}(x), \tag{1.3}$$

where $w_i^{(k)} \in \Pi_k$, and for $0 \leq k < m$ we have

$$w_i^{(k+1)}(x) = (1-x)w_i^{(k)}(x) + xw_{i+1}^{(k)}(x), \qquad i = 0, \ldots, m-k-1. \tag{1.4}$$

In particular, $w(x) = w_0^{(m)}(x) = const.$

Some more properties of the Bernstein polynomials will be needed. Our attention will be restricted to the interval $I = [0,1]$ and the following notation will be used

$$(f,g) = \int_I f(x)g(x)\,dx,$$

$$\|f\|_p = \Big(\int_I |f|^p\Big)^{\frac{1}{p}}.$$

It is convenient to use simultaneously with $N_{i,m}$ the polynomials

$$M_{i,m} = (m+1)N_{i,m}.$$

The following elementary properties of the polynomials $N_{i,m}$ and $M_{i,m}$ will be used:

1°. $N_{i,m}(x) \geq 0$ for $x \in I, i = 0, \ldots, m$.

2°.

$$\sum_{i=0}^{m} N_{i,m} = 1.$$

3°. $(M_{i,m}, 1) = 1$, for $i = 0, \ldots, m$.

4°. For w as in (1.2) we have

$$Dw = m\sum_{i=0}^{m-1} \Delta w_i N_{i,m-1}$$

$$= \sum_{i=0}^{m-1} \Delta w_i M_{i,m-1},$$

where $\Delta w_i = w_{i+1} - w_i$ and $Dw = dw/dx$.

5°. For $i = 0, \ldots, m$

$$DN_{i,m} = M_{i-1,m-1} - M_{i,m-1}$$

with $M_{j,m} = 0$ whenever $j < 0$ or $j > m$.

2. Polynomial operators. A linear operator in a function space with range contained in Π_m for some m is called a *polynomial operator* . The space of all real functions of bounded variation on I which are left continuous is denoted by $BV(I)$ and it is equipped with the norm

$$\|F\|_{BV(I)} = |F(0)| + var(F).$$

Moreover, define

$$\mathbf{D}(I) = \{F \in BV(I) : F \textit{ is nondecreasing on } I,\ F(0) = 0,\ F(1) = 1\}$$

The polynomial operator T_m is now defined for $F \in BV(I)$ by the formula

$$T_m F(x) = \sum_{i=0}^{m} \int_I M_{i,m}\,dF \int_0^x N_{i,m}(y)\,dy. \tag{2.1}$$

It then follows that

$$T_m : BV(I) \to \Pi_{m+1}, \tag{2.2}$$

and

$$T_m : \mathbf{D}(I) \to \Pi_{m+1} \cap \mathbf{D}(I). \tag{2.3}$$

The polynomial operators corresponding to the densities are going to be defined naturally by means of the kernel

$$R_m(x,y) = \sum_{i=0}^{m} M_{i,m}(x) N_{i,m}(y). \tag{2.4}$$

It follows by the definitions and properties of $M_{i,m}$ and $N_{i,m}$ that

$$R_m(x,y) = R_m(y,x), \quad 0 \le R_m(x,y) \le m+1 \quad for \quad x,y \in I. \tag{2.5}$$

Define

$$R_m f(x) = \int_I R_m(x,y) f(y)\, dy$$

Clearly, $R_m : L^p \to \Pi_m$ and since by (2.4)

$$R_m f = \sum_{i=0}^{m} (M_{i,m}, f) N_{i,n}, \tag{2.6}$$

it takes by 2° and 3° densities into densities.

It is worth to notice that for F being absolutely continuous (2.1) gives

$$DT_m F = R_m DF. \tag{2.7}$$

PROPOSITION 2.8. *For F in $BV(I)$ we have*

$$T_m F(x) = F(0)(1-x)^{m+1} + F(1)x^{m+1} + \sum_{i=1}^{m} (F, M_{i-1,m-1}) N_{i,m+1}(x).$$

PROOF. Direct computation gives

$$\int_I M_{i,m}\, dF = (m+1)\{\delta_{i,m}F(1) - \delta_{i,0}F(0) + (F, M_{i,m-1}) - (F, M_{i-1,m-1})\},$$

and therefore by 5°

$$
\begin{aligned}
T_mF(x) &= F(0) + \sum_{i=0}^{m} \int_I M_{i,m}\, dF \int_0^x N_{i,m}(x)\, dx \\
&= F(0)(1-x)^{m+1} + F(1+)x^{m+1} \\
&+ \sum_{i=0}^{m-1} (F, M_{i,m-1}) \int_0^x (m+1)(N_{i,m}(y) - N_{i+1,m}(y))\, dy \\
&= F(0)(1-x)^{m+1} + F(1+)x^{m+1} \\
&+ \sum_{i=0}^{m-1} (F, M_{i,m-1}) \int_0^x DN_{i+1,m+1}(y)\, dy.
\end{aligned}
$$

3. Approximation properties of the polynomial operators. In this section we state the necessary results on approximation by the operators T_m and R_m. The following is a consequence of Proposition 2.8

COROLLARY 3.1. *For $m = 0, 1, \dots,$ and $F \in BV(I)$ we have*

$$\|T_mF\|_\infty \le 3\|F\|_\infty, \tag{3.2}$$

and for $F, G \in \mathbf{D}(I)$

$$\|T_mF - T_mG\|_\infty \le \|F - G\|_\infty. \tag{3.3}$$

PROPOSITION 3.4. *For $f \in L^p(I)$ we have*

$$\|R_mf\|_p \le \|f\|_p, \quad m = 0, 1, \dots, \tag{3.5}$$

and if $f \in L^p(I)$, then

$$\|f - R_mf\|_p \to 0 \ \ as \ \ m \to \infty. \tag{3.6}$$

For the proof we refer to [1].

PROPOSITION 3.7. *Let $F \in C(I)$. Then $\|F - T_mF\|_\infty \to 0$ as $m \to \infty$.*

PROOF: Since (3.2) takes place it is sufficient to check the statement for absolutely continuous F. However, in this case (2.6) implies for $f = DF$

$$|F(x) - T_mF(x)| \le \|DF - DT_mF\|_1 = \|f - R_mf\|_1,$$

and the last term by (3.6) tends to 0 as $m \to \infty$.

In order to define the proper Lipschitz classes following [6] we need the step-weight function

$$\phi(x) = \sqrt{x(1-x)}, \ x \in I$$

and the symmetric difference of the second order

$$\Delta_h^2 f(x) = f(x+h) - 2f(x) + f(x-h).$$

Now, the modulus of smoothness with step-weight ϕ is given as follows

$$\omega_{2,\phi,p}(f;\delta) = \sup_{0<h<\delta} \|\Delta^2_{h\phi(x)} f(x)\|_2,$$

where $\Delta^2_{h\phi(x)}$ is zero whenever either $x + h\phi(x)$ or $x - h\phi(x)$ is not in I. Now, we can formulate the important for us auxiliary result (see [5], Theorm 3.4).

PROPOSITION 3.8. *Let α, and f be given such that $0 < \alpha < 1$, $f \in L^p(I)$, $1 \le p \le \infty$. Then*

$$\|f - R_m f\|_p = O\left(\frac{1}{m^\alpha}\right) \quad as \quad m \to \infty \quad \Longleftrightarrow \quad \omega_{2,\phi,p}(f;\delta) = O\left(\delta^{2\alpha}\right) \quad as \quad \delta \to 0_+.$$

4. The estimators. Let us start with a simple sample of sise $n : X_1, \ldots X_n$. It is assumed that the common distribution function F of these i.i.d. random variables has its support in I. For the given sample let us introduce

$$f_{m,n}(x) = \frac{1}{n}\sum_{j=0}^{n} R_m(X_j, x), \quad x \in I. \tag{4.1}$$

Clearly $f_{m,n}$ is a polynomial of degree m which, by (2.6), is a density on I. Let now F_n be the empirical distribution i.e. $F_n = |\{i : X_i < x\}|/n$ and let

$$F_{m,n} = T_m F_n. \tag{4.2}$$

It follows by (2.1) that

$$DF_{m,n} = f_{m,n}. \tag{4.3}$$

PROPOSITION 4.4. *Let F and $X_1, X_2, \ldots$ be given as above. Then*

$$P\{\, F_{m,n} \Longrightarrow F \;\; as \;\; m,n \to \infty\} = 1,$$

where $\Longrightarrow$ means the weak convergence of probability distribution functions.

PROOF: Let us start with following identity

$$F - F_{m,n} = (F - T_m F) + T_m(F - F_n). \tag{4.5}$$

It will be shown at first that $T_m F$ converges weakly to F as $m \to \infty$ for each $F \in \mathbf{D}(I)$. For ϕ continuous on $(-\infty, \infty)$ and with compact support according to (2.1) and (3.6) we obtain

$$\int_{-\infty}^{\infty} \phi \, dT_m F = \int_0^1 R_m(\phi|_I)\, dF \to \int_{-\infty}^{\infty} \phi \, dF, \quad as \quad m \to \infty.$$

For the second part of (4.5) we obtain by (3.3) that

$$\|T_m(F - F_n)\|_\infty \le \|F - F_n\|_\infty,$$

but by Glivenko's theorem (see [8])

$$P\{\|F - F_n\|_\infty \to 0 \quad as \quad n \to \infty\} = 1.$$

Thus, with probability one $T_m(F - F_n)$ tends uniformly on I to 0 as $m, n \to \infty$. Since $F(x) - T_m F(x)$ tends to zero as $m \to \infty$ at each continuity point of F it follows by (4.5) that with probability one $F_{m,n}(x) \to F(x)$ with at all such points.

PROPOSITION 4.6. *Let $F \in \mathbf{D}(I) \cap C(I)$. Then*

$$P\{\|F - F_{m,n}\|_\infty \to 0 \text{ as } m, n \to \infty\} = 1.$$

This follows from the proof of Proposition 4.4 and by Proposition 3.7. We need the following inequality from [7].

LEMMA 4.7. *For continuous probability distribution F on $(-\infty, \infty)$ there are constants C, γ such that $0 < \gamma \leq 2$, $0 < C < \infty$, and*

$$Pr\{\|F - F_n\|_\infty > \frac{\lambda}{\sqrt{n}}\} \leq Ce^{-\gamma\lambda^2} \qquad for \qquad \lambda > 0,\ n = 1, 2, \ldots$$

For later convenience let us introduce the set of all densities on I

$$\mathbf{P}(I) = \{f \in L^1(I) : \int_I f = 1,\ f \geq 0\}.$$

LEMMA 4.8. *Let $f \in L^p(I) \cap \mathbf{P}(I)$ for some p, $1 \leq p \leq \infty$. Then*

$$|\ \|f - f_{m,n}\|_p - \|f - R_m f\|_p| \leq \|R_m f - f_{m,n}\|_p \leq 2(m+1)\|F - F_n\|_\infty$$

Moreover, for each finite p there is finite C such that

$$(E\|R_m f - f_{m,n}\|_p^p)^{\frac{1}{p}} \leq C\,\frac{m+1}{\sqrt{n}} \qquad for \quad n, m+1 = 1, 2, \ldots$$

PROOF: The (2.1), 2° and 5° of Section 1 give for fixed $x \in I$

$$|R_m f(x) - f_{m,n}(x)| = |DT_m(F - F_n)(x)|$$

$$= |\sum_{i=0}^{m} \int_I M_{i,m}\, d(F - F_n)\ N_{i,m}(x)| = |\sum_{i=0}^{m} \int_I (F - F_n)(y) DM_{i,m}(y)\, dy\ N_{i,m}(x)|$$

$$\leq \|F - F_n\|_\infty \sum_{i=0}^{m} \|DM_{i,m}\|_1 N_{i,m}(x) \leq 2(m+1)\|F - F_n\|_\infty,$$

whence $\|R_m f - f_{m,n}\|_p \leq 2(m+1)\|F - F_n\|_\infty$, and this completes the first part of the proof, which in combination with Lemma 4.7 gives the second part.

THEOREM 4.9. *Let either $f \in L^p(I) \cap \mathbf{P}(I)$ for some p, $1 \le p < \infty$, or $f \in C(I) \cap \mathbf{P}(I)$, and let $m = [n^\beta]$ for some $\beta > 0$. Then, for $0 < \beta < \frac{1}{2}$*

$$Pr\{\|f - f_{m,n}\|_p = o(1) \quad as \quad n \to \infty\} = 1.$$

Moreover, if $0 < \alpha < 1$, $0 < \beta < \frac{1}{2}\frac{1}{1+\alpha}$, then the following conditions are equivalent:

(i)
$$\omega_{2,\phi,p}(f;\delta) = O(\delta^{2\alpha}) \qquad as \qquad \delta \to 0_+,$$

(ii)
$$Pr\{\|f - f_{m,n}\|_p = O(\frac{1}{n^{\alpha\beta}}) \;\; as \;\; n \to \infty\} = 1.$$

PROOF: We know from [1] that $\|f - R_m f\|_p = o(1)$. On the other hand Lemma 4.7 gives for $\epsilon > 0$

$$Pr\{(m+1)\|F - F_n\|_\infty > \epsilon\} \le C\, e^{-\gamma\epsilon^2\frac{n}{m^2}} \le C\, e^{-\gamma\epsilon^2 n^{1-2\beta}}. \tag{4.10}$$

This implies

$$Pr\{(m+1)\|F - F_n\|_\infty = o(1) \quad as \quad n \to \infty\} = 1.$$

To complete the first part of the proof it is sufficient now to apply Lemma 4.8. Substituting in (4.10) $\epsilon = \frac{c}{n^{\alpha\beta}}$ we get

$$Pr\{(m+1)\|F - f_n\|_\infty > \frac{c}{n^{\alpha\beta}}\} \le C\, e^{-\gamma c^2 n^{1-2\beta}}, \tag{4.11}$$

which implies

$$Pr\{(m+1)\|F - F_n\|\infty = O(\frac{1}{n^{\alpha\beta}}) \quad as \quad n \to \infty\} = 1.$$

Now the equivalence of (i) and (ii) follows by Lemma 4.8.

Next Theorem concerns the order of the mean L^p deviations for the estimators $f_{m,n}$. To this end we need the following auxiliary inequalities. The first is elementary and it is well known.

PROPOSITION 4.12. *Let $J = < -a, a >$, $a > 0$, $R = (-\infty, \infty)$. Then,*

$$0 \le |x+h|^p + |x-h|^p - 2|x|^p \le p(p-1)a^{p-2}|h|^2 \quad for \quad p > 2,\ x+h, x-h \in J.$$

To formulate the second inequality we recall the definition of the customary second order modulus of smoothness i.e for $g \in C(J)$ define

$$\omega_{2,\infty}(g;\delta)_J = \sup_{x_1,x_2 \in J, |x_1 - x_2| \le 2\delta} |g(\frac{x_1+x_2}{2}) - \frac{g(x_1)+g(x_2)}{2}|, \qquad 0 < \delta \le \frac{1}{2}. \tag{4.13}$$

The following useful estimate we find in [9].

PROPOSITIOPN 4.14. *Let X be a random variable with values in J, J being finite or infinite interval, $EX^2 < \infty$ and let $g \in C(J)$. Then*

$$|g(EX) - Eg(X)| \le 15\,\omega_{2,\infty}\left(g; \frac{1}{2}\sqrt{E(X-EX)^2}\right)_J .$$

LEMMA 4.15. *Let $f \in L^p(I) \cap \mathbf{P}(I)$, $1 \le p < \infty$. Then, for the sample $X_1, \dots, X_n$ corresponding to the density f we have*

$$\left(E\|R_m f - f_{m,n}\|_p^p\right)^{\frac{1}{p}} \le C\,\frac{(m+1)^{\frac{1}{q\wedge 2}}}{n^{\frac{1}{p\vee 2}}},$$

where $\frac{1}{p} + \frac{1}{q} = 1$, $a \vee b = max(a,b)$, $a \wedge b = min(a,b)$.

PROOF: The case $1 \le p \le 2$ is easy. As in [2] we have

$$\left(E\|R_m f - f_{m,n}\|_p^p\right)^{\frac{1}{p}} \le \left(E\|R_m f - f_{m,n}\|_2^2\right)^{\frac{1}{2}} \le \left(\frac{m+1}{n}\right)^{\frac{1}{2}} .$$

Let now $p > 2$ and let for $i = 0, \dots, m$,

$$X^{(i)} = \frac{1}{n}\sum_{j=1}^{n}\left(M_{i,m}(X_j) - EM_{i,m}(X_j)\right).$$

It follows that the values of $X^{(i)}$ are in $J = (-m-1, m+1)$. Since $EX^{(i)} = 0$, Proposition 4.14 applied to $X^{(i)}$ and to $g(x) = |x|^p$ gives

$$|Eg(X^{(i)})| \le 15\,\omega_{2,\infty}\left(g; \frac{1}{2}\sqrt{\frac{1}{n}\int_I M_{i,m}^2 f}\right)_J ,$$

whence by Proposition 4.12

$$E|X^{(i)}|^p \le C^p (m+1)^{p-2}\frac{1}{n}\int_I M_{i,m}^2 f \le C^p \frac{(m+1)^p}{n}\int_I N_{i,m} f.$$

Now, by Jensen's inequality

$$E\|R_m f - f_{m,n}\|_p^p = \int_I E\left|\sum_{i=0}^{m} X^{(i)} N_{i,m}(y)\right|^p dy$$

$$\le \int_I \sum_{i=0}^{m} E|X^{(i)}|^p N_{i,m}(y)\,dy = \frac{1}{m+1}\sum_{i=0}^{m} E|X^{(i)}|^p \le C^p \frac{(m+1)^{p-1}}{n}.$$

To formulate the last result we introduce for $0 < \alpha < 1$ and $1 \le p < \infty$

$$\psi(\alpha,p) = \begin{cases} \frac{1}{2\alpha+1}, & \text{if } 1 \le p < 2; \\ \frac{1}{p\alpha+p-1}, & \text{if } 2 \le p < 2 + \frac{1}{\alpha+1}; \\ \frac{1}{2\alpha+2}, & \text{if } p \ge 2 + \frac{1}{\alpha+1}. \end{cases}$$

THEOREM 4.16. *Let $1 \le p < \infty$ and let $f \in L^p(I) \cap \mathbf{P}(I)$. Let α, $0 < \alpha < 1$, and β, $0 < \beta \le \psi(\alpha,p)$ be given. Moreover, let $m = [n^\beta]$. Then the following conditions are equivalent:*

(i)
$$\omega_{2,\phi,p}(f;\delta) = O(\delta^{2\alpha}) \quad as \quad \delta \to 0_+,$$

(ii)
$$(E\|f - f_{m,n}\|_p^p)^{\frac{1}{p}} = O(\frac{1}{n^{\alpha\beta}}) \quad as \quad n \to \infty.$$

For the proof we apply Lemmas 4.15, 4.8 and Proposition 3.8.

COROLLARY 4.17. *Under the assumptions of Theorem 4.16 the best choice of β with respect to (ii) is given by formula $\beta = \psi(\alpha,p)$.*

EXAMPLE: Using the examples on page 228 of [6] we find that for $1 \le p < 2$ the arcsin law density i.e. for

$$f_0(x) = \frac{1}{\pi}\frac{1}{\sqrt{x(1-x)}}, \qquad x \in <0,1>$$

we have

$$\omega_{2,\phi,p}(f_0;\delta) \sim \delta^{\frac{2}{p}-1} \quad as \quad \delta \to 0_+.$$

Thus, for this density $\alpha = \frac{1}{p} - \frac{1}{2}$ and the optimal choice for β is $\beta = \frac{p}{2}$.

5. Algorithm for computing the density and distribution estimators. Let $X_1, \ldots, X_n$ be given as in the previous section. Since

$$f_{m,n}(x) = \frac{1}{n}\sum_{j=1}^{n} R_m(X_j, x),$$

to compute $f_{n,m}(x)$ for fixed x we need to compute $R_m(X_j, x)$ for $j = 1, \ldots, n$. However,

$$R_m(X_j, x) = \sum_{i=0}^{m} M_{i,m}(X_j) N_{i,m}(x)$$

and therefore we use the Casteljeau algorithm for the first time to compute $M_{i,m}(X_j)$ and for the second time to calculate $R_m(X_j, x)$. Now, the density $f_{m,n}$ has also the following representation

$$f_{m,n}(x) = \sum_{i=0}^{m} a_i N_{i,m}(x),$$

where

$$a_i = \frac{1}{n}\sum_{j=1}^{n} M_{i,m}(X_j), \qquad i = 0,\dots,m.$$

Thus, at almost no cost the following coefficients

$$b_0 = 0,\ b_1 = 1,\ b_j = \frac{a_0 + \dots + a_{j-1}}{m+1}, \qquad j = 1,\dots,m+1$$

can be computed. To compute $F_{m,n}(x)$ one applies once more the Casteljeau algorithm to the following formula

$$F_{m,n}(x) = \int_0^x f_{m,n}(y)\,dy = \sum_{j=1}^{m+1} b_j M_{j,m+1}(x).$$

6. Comments. This note is related to [3] and [2] but the tools used here are different. This made it possible to extend the results from [2]. The author is indebted to G. Krzykowski who has brought to our attention Lemma 4.7.

References

[1] Z.Ciesielski, *Approximation by polynomials and extension of Parseval's identity for Legendre polynomials to the L_p case,* Acta Scient. Mathematicarum **48**(1985),65-70.

[2] Z.Ciesielski, *Nonparametric Polynomial Density Estimation,* Prob. and Math. Statistics (to appear).

[3] Z.Ciesielksi, *Haar System and Nonparametric Estimation in Several Variables,* Math. Inst. Polish Acad. of Sciences 1987, 12 pages (preprint).

[4] Z.Ciesielski and J.Domsta, *The degenerate B-splines as basis in the space of algebraic polynomials,* Ann. Polon. Math. **46**(1985),71-79.

[5] Z.Ditzian and K.Ivanov, *Bernstein type operators and their derivatives,* Journal of Approximation Theory (to appear).

[6] Z.Ditzian and V.Totik, *Moduli of smoothness,* Dept. of Math., Univ. of Alberta, Edmonton (1986), 477 pages (preprint).

[7] A.Dvoretzki, A.Kiefer and J.Wolfowitz, *Asymptotic minimax character of the sample distribution function and of the classical multinomial estimator,* Ann. Math. Stat. **27.3**(1956), 642-669.

[8] A.Renyi, *Wahrscheinlichkeitsrechnung,* Berlin 1962.

[9] L.I.Strukov and A.F.Timan, *Mathematical expectation of a function of a random variable , smoothness and deviation,* Sibirskii Mat. J.**18.3**(1977),658-664 (in Russian).

Instytut Matematyczny PAN
ul. Abrahama 18
81-825 Sopot, POLAND

Local Spline Interpolation Schemes in One and Several Variables

W. Dahmen *
Universität Bielefeld
Fakultät für Mathematik
48 Bielefeld, West Germany

T. N. T. Goodman
Department of Mathematics
University of Dundee
Dundee, Scotland

Charles A. Micchelli *
Mathematical Sciences Department
IBM Thomas J. Watson Research Center
Yorktown Heights, New York 10598

Abstract: In the first part of this paper we briefly review some recent results pertaining to the construction of compactly supported fundamental functions for univariate Lagrange interpolation by splines. In the second part of the paper we discuss several possible extensions of these results to a multivariate setting.*

* Partially supported by NATO Travel Grant DJRG 639/84.

1. Introduction

The importance of splines for the numerical solution of interpolation problems is a well-established fact. Nevertheless, computationally spline interpolation requires the solution of large sparse linear systems whose order is roughly equal to the number of data being interpolated. This is reflected in the fact that generally the interpolant at any point depends on all the data. Equivalently, if we let $\{L_i\}_{i\in\mathbb{Z}}$ be the fundamental Lagrange splines for interpolation on the sequence $\{x_i\}_{i\in\mathbb{Z}}$, that is,

(1.1) $$L_i(x_j) = \delta_{ij},\ i,j \in \mathbb{Z}$$

then each L_i is supported on all of $\mathbb{R}$. It seems desirable to have fundamental functions of *compact support*. This can prove useful when updating of the interpolant is desired as new data are available or when solving several smaller linear systems to determine the Lagrange splines is preferred over solving one large set of equations. The use of compactly supported fundamental functions has already proved useful in numerical grid generation [7] as well as in computer aided design, [1].

An efficient method for constructing Lagrange splines of compact support is the addition of knots beyond those chosen at the data locations. The problem then is how to use these degrees of freedom in such a way that either shape control and/or high accuracy is achieved. Various such questions have been systematically analyzed in [6]. Some of these results are briefly reviewed in Section 2. In Section 3 we propose several extensions of these results to multivariate interpolation problems. Due to the wider variety of possibilities in the multivariate case, these results only provide an initial investigation into a problem that has important applications for practical data fitting in several variables.

2. Univariate Compactly Supported Fundamental Functions

Let $X = \{x_i\}_{i\in\mathbb{Z}}$ be a strictly increasing sequence of real numbers. As usual, the B-splines of order k on X are defined by

$$N_{i,k,X}(x) = (x_{i+k} - x_i)\,[x_i, \dots, x_{i+k}](\bullet - x)_+^{k-1}$$

where $[x_i, \dots, x_{i+k}]f$ denotes the k-th order divided difference of f and

$$x_+^\ell = \begin{cases} x^\ell, & x \geq 0, \\ 0, & x < 0. \end{cases}$$

For any fixed integer $1 \leq q \leq k-1$ the function

(2.1) $$L_{i+q}(x) = \left(\prod_{\substack{j=1 \\ j\neq q}}^{k-1} \frac{x - x_{i+j}}{x_{i+q} - x_{i+j}} \right) N_{i,k,X}(x)/N_{i,k,X}(x_{i+q})$$

is a piecewise polynomial of degree $2k-3$ having $k-2$ continuous derivatives. Moreover, as is clearly apparent

(2.2) $$L_i(x_j) = \delta_{ij},\ i,j \in \mathbb{Z},$$

and

$$\operatorname{supp} L_i = [x_{i-q}, x_{i-q+k}] = \operatorname{supp} N_{i-q,k,X}.$$

The Catmull Rom method used in the design of free form curves and surfaces (cf. [1]) is based on the interpolation operator

$$(2.3) \qquad (If)(x) = \sum_{i \in \mathbb{Z}} f(x_i)\, L_i(x).$$

The fact that the smoothness of the interpolant is relatively low compared to its degree may be viewed as a disadvantage. Moreover, the accuracy of the method turns out to be restricted as well.

Theorem 2.1. [6] Let $X = \mathbb{Z}$. The interpolation operator given by (2.3) reproduces all linear functions P, that is,

$$IP = P$$

but not all quadratics.

The construction of the Catmull Rom interpolant may be viewed as inserting additional knots into the set X, namely, each x_i becomes a (k-1)-fold knot. As we will now explain below much higher accuracy as well as smoothness can be obtained by inserting fewer knots. To be precise, for an arbitrary strictly increasing sequence X as above, let

$$T = X \cup Y$$

where $Y = \{y_i\}_{i \in \mathbb{Z}}$ satisfies

$$(2.4) \qquad x_i < y_i < x_{i+1}, \quad i \in \mathbb{Z}.$$

Theorem 2.2. [6] Let $k \geq 3, -1 \leq \ell \leq k-1$ and $1 \leq \nu \leq k-2+\ell$ be fixed integers. Then there exists a unique sequence $\{L_i\}_{i \in \mathbb{Z}} \subset \mathscr{S}_k(T) = \text{span}\,\{N_{i,k,T} : i \in \mathbb{Z}\}$ such that

$$(2.5) \qquad \text{supp}\, L_i = [x_{i-\nu}, x_{i-\nu+k+\ell-1}]$$

$$(2.6) \qquad L_i(x_j) = \delta_{ij}, \ i, j \in \mathbb{Z}$$

such that the corresponding operator

$$(If)(x) = \sum_{i \in \mathbb{Z}} f(x_i)\, L_i(x)$$

is exact of degree ℓ.

The special case $X = \mathbb{Z}$, $Y = \mathbb{Z} + 1/2$, $\nu = \ell = k - 1$ already appeared in [8] and another proof is given in [6]. The proof of Theorem 2.2 requires a completely different approach which we outline next. First, we show that the conditions

$$(2.7) \qquad \begin{aligned} &L_i(x_j) = \delta_{ij}, \ j = i-\nu+1, \dots, i-\nu+\ell+k-2 \\ &L'_i(x_{q_j}) = a_{ij}, \ j = 1, \dots, \ell+1 \end{aligned}$$

uniquely determine for any choice of $\{q_1, \dots, q_{\ell+1}\} \subseteq \{i-\nu+1, \dots, i-\nu+\ell+k-2\}$ and any $a_{ij}, j = 1, \dots, \ell+1$ an element L_i in $\mathscr{S}_k(T)$ satisfying (2.5). The parameters a_{ij} are then uniquely determined by the exactness requirement through the following systems of equations

$$(2.8) \qquad \sum_{i \in \mathbb{Z}} x_i^r L'_i(x_j) = r\, x_j^{r-1}, \ r = 0, \dots, \ell, \ j \in \mathbb{Z},$$

$$(2.9)\quad \sum_{i\in\mathbb{Z}} x_i^r L_i^{(m)}(x_0) = r(r-1)\dots(r-m+1)x_0^{r-m},\ r=0,\dots,\ell,\ m=2,\dots,k-3.$$

For $k>3$, the corresponding equations in the unknowns a_{ij} are coupled. However, a way of decoupling them, i.e., to determine the a_{ij} by small linear systems whose order depends only on k and not on the number of interpolation conditions is described in [6].

The situation is much simpler though in the quadratic case $k=3$. In fact, choosing for $\nu=\ell=2$ and

$$(2.10)\qquad a_{ij} = L'_i(x_j) = Q_{ij}(x_j),\quad j=i-1,\ i,\ i+1$$

where

$$Q_{ij}(x) = \prod_{\substack{r=j-1\\ r\neq i}}^{j+1} \frac{x-x_r}{x_i-x_r},\quad i=j-1,\ j,\ j+1$$

gives rise to an interpolant of maximal order two. This scheme is easily seen to be equivalent to Hermite interpolation on X from the space $\mathcal{S}_3(T)$ where the derivative data are obtained by taking corresponding derivatives of local interpolating quadratics.

Similarly, a completely local scheme for higher degree is easily obtained at the expense of adding an appropriate number of additional knots. Suppose $Y=\{y_{ij}\}_{j=1,\,i\in\mathbb{Z}}^{k-2}$ satisfies

$$x_i < y_{i,1} \le \dots \le y_{i,k-2} < x_{i+1},\quad i\in\mathbb{Z},$$

and let $T = X \cup Y$.

Theorem 2.3. [6] Let ν, $1\le\nu\le k$ be fixed. For each $i\in\mathbb{Z}$ there exists a unique function $L_i\in\mathcal{S}_k(T)$ satisfying

$$(2.11)\qquad L_i(x_j)=\delta_{ij},\quad \operatorname{supp} L_i = [x_{i-\nu}, x_{i-\nu+k+1}],\ i,j\in\mathbb{Z}$$

as well as

$$(2.12)\qquad L_i^{(m)}(x_j) = Q_{ij}^{(m)}(x_j),\ m=0,\dots,k-2$$

where

$$Q_{ij}(x) = \prod_{\substack{r=j+\nu-k\\ r\neq i}}^{j+\nu-1} \frac{x-x_r}{x_i-x_r},\quad i=j+\nu-k,\dots,j+\nu-1.$$

Moreover, the corresponding interpolation operator is exact of degree $k-1$.

Remark 2.1. Decreasing the support of the fundamental splines will result in correspondingly lower order of exactness, see [6] for further details.

3. Multivariate Schemes

Whenever the data are located on a rectangular grid tensor products of the above interpolation schemes readily lead to multivariate procedures which are also based on compactly supported fundamental functions. In this section we will focus, however, on spline interpolants for regular triangular grids in $\mathbb{R}^2$ or even more general partitions of $\mathbb{R}^s$, $s \geq 2$ and correspondingly chosen data points.

We will describe first a bivariate analog of Theorem 2.3 using cubic polynomials. To this end, let Δ denote the regular triangulation of the plane which is induced by the integer translates of the lines $x=0$, $y=0$, and $x-y=0$. Connecting the centroid of each triangle $T \in \Delta$ with its vertices induces the so-called Clough-Tocher split shown in Figure 3.1.

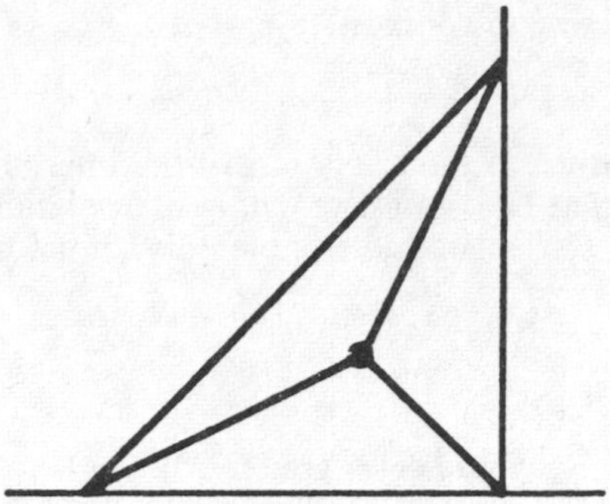

Figure 3.1

Let us denote by Δ_C the corresponding refinement of Δ. It is well-known that every element $S \in S_3^1(\Delta_C)$, the space of all C^1 piecewise cubics on Δ_C, is uniquely determined by its function values and gradients at the vertices of Δ and by the normal derivatives at the midpoints of the edges of Δ.

The fundamental function which we are going to construct now has its support Σ as shown below in either Figure 3.2(a) or (b).

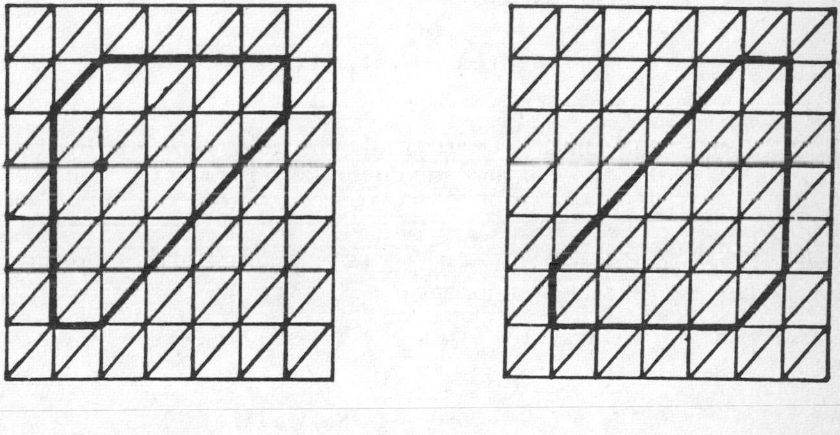

(a) (b)

Figure 3.2

We may choose any lattice point in the interior of the support as the origin. For definiteness, we fix the position of Σ relative to the origin as indicated in Figure 3.2(a) and let

$$\Omega = \operatorname{int} \Sigma \cap \mathbb{Z}^2.$$

According to the above remarks any function $L \in S_3^1(\Delta_C)$ with

$$\text{(3.1)} \qquad \begin{aligned} &\operatorname{supp} L = \Sigma, \\ &L(\alpha) = \delta_{0\alpha}, \quad \alpha \in \mathbb{Z}^2 \end{aligned}$$

is uniquely determined by fixing in addition,

$$\operatorname{grad} L(x)\,|_{x=\nu}, \qquad \nu \in \Omega$$

as well as the normal derivatives at the midpoints of the interior edges in Σ. To specify these derivatives, we note that Ω is unisolvent for interpolation from the space of all cubics on $\mathbb{R}^2$. Hence for $\nu \in \Omega$ we can find a cubic polynomial Q_ν satisfying

$$\text{(3.2)} \qquad Q_\nu(\gamma) = \delta_{-\nu,\gamma}, \quad \gamma \in -\Omega.$$

Then for $\nu \in \Omega$ we set

$$\text{(3.3)} \qquad \operatorname{grad} L(\nu) = \operatorname{grad} Q_\nu(0).$$

Moreover, for $\varepsilon_i \in \{\pm 1/2\}$, $i = 1,2,3$ and $\nu \in \Omega$ define

$$\text{(3.4)} \qquad \begin{aligned} \frac{\partial}{\partial x_2} L(\nu + (\varepsilon_1, 0)) &= \frac{\partial}{\partial x_2} Q_\nu(\varepsilon_1, 0), \\ \frac{\partial}{\partial x_1} L(\nu + (0, \varepsilon_2)) &= \frac{\partial}{\partial x_1} Q_\nu(0, \varepsilon_2), \\ \left(\frac{\partial}{\partial x_1} - \frac{\partial}{\partial x_2}\right) L(\nu + (\varepsilon_3, \varepsilon_3)) &= \left(\frac{\partial}{\partial x_1} - \frac{\partial}{\partial x_2}\right) Q_\nu(\varepsilon_3, \varepsilon_3), \end{aligned}$$

while at the midpoints of all remaining edges the corresponding normal derivatives of L are set equal to zero.

Theorem 3.1. The interpolation operator

$$\text{(3.5)} \qquad (If)(x) = \sum_{\alpha \in \mathbb{Z}^2} f(\alpha)\, L(x - \alpha)$$

reproduces cubic polynomials. Consequently, the scaled version of I given by $E_h I E_{h^{-1}}$ where $(E_h f)(x) = f(hx)$ approximates smooth functions to order $O(h^4)$, $h \to 0^+$.

Proof: By construction, If interpolates f at lattice points, that is, $(If)(\alpha) = f(\alpha)$, $\alpha \in \mathbb{Z}^2$. Moreover, for any $\beta \in \mathbb{Z}^2$ (3.3) yields

$$\begin{aligned} \operatorname{grad} (IP)(\beta) &= \sum_{\beta - \alpha \in \Gamma} P(\alpha) \operatorname{grad} L(\beta - \alpha) \\ &= \sum_{\beta - \alpha \in \Gamma} P(\alpha) \operatorname{grad} Q_{\beta - \alpha}(0). \end{aligned}$$

This suggests that we consider the cubic polynomial

$$H(x) = \sum_{\beta-\alpha\in\Gamma} P(\alpha)\, Q_{\beta-\alpha}(x)$$

and observe that for $\gamma - \beta \in -\Omega$

$$H(\gamma-\beta) = \sum_{\beta-\alpha\in\Gamma} P(\alpha)\, Q_{\beta-\alpha}(\gamma-\beta) = P(\gamma).$$

When P is a cubic polynomial it follows that

$$H(x) = P(x+\beta)$$

which yields the equation

$$\text{grad}\left(\sum_{\beta-\alpha\in\Gamma} P(\alpha)\,\text{grad}\, Q_{\beta-\alpha}(x)\right)\Big|_{x=0} = \text{grad}\, P(\beta)$$

and confirms that

$$\text{grad}\,(IP)\,(\beta) = \text{grad}\, P(\beta),\ \ \beta \in \mathbb{Z}^2.$$

Similarly by (3.4)

$$\frac{\partial}{\partial x_2}(IP)(\beta + (\varepsilon_1, 0)) = \frac{\partial}{\partial x_2}\left(\sum_{\beta-\alpha\in\Gamma} P(\alpha)\, Q_{\beta-\alpha}(x)\right)\Big|_{x=(\varepsilon_1,0)}$$

$$= P(\beta + (\varepsilon_1, 0))$$

and likewise

$$\frac{\partial}{\partial x_1}(IP)\,(\beta + (0, \varepsilon_2)) = \frac{\partial}{\partial x_1} P(\beta + (0, \varepsilon_2)),$$

$$\left(\frac{\partial}{\partial x_1} - \frac{\partial}{\partial x_2}\right)(IP)\,(\beta + (\varepsilon_3, \varepsilon_3)) = \left(\frac{\partial}{\partial x_1} - \frac{\partial}{\partial x_2}\right)P(\beta + (\varepsilon_3, \varepsilon_3)),\ \ \beta \in \mathbb{Z}^2$$

finishing the proof.

If only exactness for linear or quadratic functions is required, one can shrink the size of the support of L. Figure 3.3 shows typical supports allowing for the reproduction of linear (a) and quadratic (b) functions

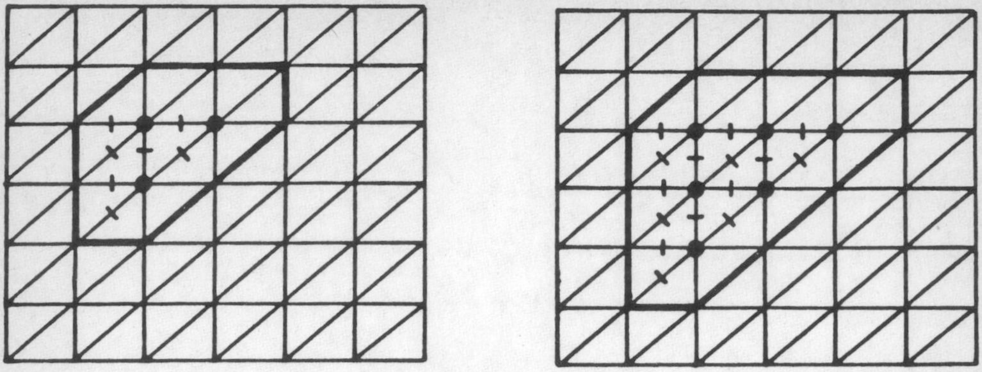

(a) (b)

Figure 3.3

Since the corresponding sets of interior lattice points in (a), (b) are again unisolvent for interpolation from quadratics and linear functions, respectively, the same construction as before works. The marks on the midpoints of interior edges in Figure 3.3 (a),(b) indicate where conditions analogous to (3.4) could be imposed. A similar approach is conceivable for irregular triangulations since the regularity of Δ is used here only to assure that certain sets of vertices are unisolvent for polynomial interpolation.

Next we turn to another extension of Theorem 2.3 based on the concept of box spline. To explain this, let $X = \{x^1, \dots, x^n\} \subseteq \mathbb{R}^s \setminus \{0\}$ be any set of not necessarily distinct vectors. For the sake of convenience, the matrix with columns x^i, $i = 1, \dots, n$ will also be denoted by X. The box spline $B(\bullet \mid X)$ is then defined by requiring that

$$\int_{\mathbb{R}^s} f(x)\, B(x \mid X) dx = \int_{[0,1]^n} f(Xu) du$$

for any $f \in C(\mathbb{R}^s)$. It is well known [2-5] that when

$$\operatorname{span}\{X\} = \langle X \rangle = \mathbb{R}^s$$

$B(\bullet \mid X)$ is a piecewise polynomial of degree $\leq$ n-s with support

$$Z(X) = \{Xu : u \in [0,1]^n\}.$$

Moreover,

$$B(\bullet \mid X) \subseteq C^{d-1}(\mathbb{R}^s)$$

if

(3.6) $$\langle X \backslash V \rangle = \mathbb{R}^s, \quad \forall\, V \subseteq X, \ |V| = d.$$

When $X \subseteq \mathbb{Z}^s$ one considers linear combinations of translates of $B(\bullet \mid X)$

$$S(x) = \sum_{\alpha \in \mathbb{Z}^s} c_\alpha B(x - \alpha \mid X) \tag{3.7}$$

and for present purposes we will assume that

$$|\det Y| = 1\,, \quad Y \in \mathcal{B}(X) \tag{3.8}$$

where $\mathcal{B}(X) = \{Y \subseteq X : |Y| = \dim \langle Y \rangle = s\}$ and $|Y|$ denotes the cardinality of Y. This requirement is equivalent to saying that the translates $B(\bullet - \alpha \mid X)$, $\alpha \in \mathbb{Z}^s$, are locally linearly independent, a result proved in [4].

The polynomial pieces of the spline S given by (3.7) are generally separated by the X-planes, the totality of which are given by the set

$$c(X) = \cup \{\alpha + \langle Y \rangle : Y \subset X,\ |Y| = \dim \langle Y \rangle = s - 1,\ \alpha \in \mathbb{Z}^s\}.$$

For any y not on an X-plane, that is, $y \in \mathbb{R}^s \backslash c(X)$ let

$$b(y \mid X) = \{\alpha \in \mathbb{Z}^s : B(y - \alpha \mid X) \neq 0\}.$$

It was shown in [4], that for any $X \subseteq \mathbb{Z}^s \backslash \{0\}$

$$|b(y \mid X)| = \mathrm{vol}_s(Z(X)) = \sum_{Y \in \mathcal{B}(X)} |\det Y|\,. \tag{3.9}$$

Hence, when (3.8) holds, one has $|b(y \mid X)| = |\mathcal{B}(X)|$. The last identity in (3.9) reflects the fact that $Z(X)$ can be decomposed into translates of parallelepipeds spanned by all the bases in $\mathcal{B}(X)$, [4].

Since $B(\bullet \mid x^1, \ldots, x^n) = B(\bullet - x^1 \mid -x^1, \ldots, x^n)$ we may suppose, by replacing the box spline by a suitable integer translate of itself, that zero is not contained in the convex hull of X. This condition implies that $\sum_{j=1}^{n} x^j = w$ is an extreme point of $Z(X)$.

Any simply connected maximal domain on which $B(\bullet \mid X)$ is a polynomial is called an X-region. Let Γ be any X-region whose closure contains w. Choose $Y \in \mathcal{B}(X)$ such that

$$\left(\sum_{v \in X \backslash Y} v\right) + y \in \partial Z(X), \quad y \in Y$$

and that

$$\Gamma \subset \left(\sum_{v \in X \backslash Y} v\right) + Z(Y).$$

Then for any $u \in \Gamma$ we have

$$0 \in b(u \mid X) \subseteq Z(X). \tag{3.10}$$

Since $\{\alpha \in \mathbb{Z}^s,\ B(\alpha \mid X) \neq 0\} \subset b(u \mid X)$ we conclude that for some $z \in -Z(Y)$ and N sufficiently large the set of lattice points

$$A_N = z + b(u \mid X)/N$$

satisfies

(3.11) $$(\beta + \operatorname{int} Z(X)) \cap \mathbb{Z}^s \subseteq b(u \mid X), \quad \beta \in A_N.$$

The following result will be of central importance.

Proposition 3.1. Suppose (3.8) holds. Then the linear system

(3.12) $$\sum_{\beta \in A_N} c_\beta B(\alpha - z - \beta \mid X) = \delta_{\alpha,\gamma}, \quad \alpha \in b(u \mid X)$$

has for every $\gamma \in b(u \mid X)$ a unique solution.

Proof: Suppose that $\{c^*_\beta\}_{\beta \in A_N}$ is a vector satisfying the homogeneous linear system

$$\sum_{\beta \in A_N} c^*_\beta B(\alpha - z - \beta \mid X) = 0, \quad \alpha \in b(u \mid X),$$

and consider the function

(3.13) $$G(x) = \sum_{\beta \in A_N} c^*_\beta B(x - z - \beta \mid X).$$

We introduce the notation $D_v f = \sum_{j=1}^{s} v_j \frac{\partial f}{\partial x_j}$, for the directional derivative of f in the direction of $v \in \mathbb{R}^s$.
We also write $D_V f = \left(\prod_{v \in V} D_v\right) f$ for any set $V \subseteq X$ and define

$$D(X) = \{f \in \mathscr{D}'(\mathbb{R}^s) : D_V f = 0, \ \forall V \subseteq X, \ < X \backslash Y > \neq \mathbb{R}^s\}.$$

It was shown in [4] that D(X) is a finite dimensional space of polynomials of degree at most n-s and that

(3.14) $$\dim D(X) = |\mathscr{B}(X)|.$$

For any fixed polynomial $P \in D(X)$, consider the function

(3.15) $$F(x) = \sum_{\nu \in \mathbb{Z}^s} P(\nu) G(x - \nu).$$

By (3.13), we may rewrite F(x) as

(3.16) $$F(x) = \sum_{\beta \in A_N} \sum_{\nu \in \mathbb{Z}^s} c^*_\beta P(\nu) B(x - \nu - \beta - z \mid X) = \sum_{\beta \in A_N} c^*_\beta Q(x - \beta)$$

where $Q(x) = \sum_{\nu \in \mathbb{Z}^s} P(\nu) B(x - \nu - z \mid X)$. We recall the following result from [2,3].

Proposition 3.2. The mapping $T : f \to \sum_{\alpha \in \mathbb{Z}^s} f(\alpha) B(x - \alpha \mid X)$ takes D(X) one-to-one and onto D(X).

Hence both Q and F are polynomials in D(X) and since $Q(x - \beta)$ agrees on $\mathbb{Z}^s$ with the function

$$\sum_{\nu \in \mathbb{Z}^s} P(\nu - \beta)\, B(x - \nu - z \mid X)$$

they are everywhere equal. Consequently, (3.16) may be rewritten as

$$\begin{aligned} F(x) &= \sum_{\beta \in A_N} \sum_{\nu \in \mathbb{Z}^s} \overset{*}{c}_\beta P(\nu - \beta)\, B(x - z - \nu \mid X) \\ &= \sum_{\nu \in \mathbb{Z}^s} \left(\sum_{\beta \in A_N} \overset{*}{c}_\beta P(\nu - \beta) \right) B(x - z - \nu \mid X) \qquad (3.17) \\ &= \sum_{\nu \in \mathbb{Z}^s} P_N(\nu)\, B(x - z - \nu \mid X), \end{aligned}$$

where

$$P_N(x) = \sum_{\beta \in A_N} \overset{*}{c}_\beta P(x - \beta)$$

and so P_N is a polynomial in D(X). By assumption, $G(\alpha) = \sum_{\beta \in A_N} \overset{*}{c}_\beta B(\alpha - z - \beta \mid X) = 0$, $\alpha \in b(u \mid X)$ and because (3.11) gives (int supp G) $\cap\, \mathbb{Z}^s = b(u \mid X)$ we conclude that

$$G(\alpha) = 0, \quad \alpha \in \mathbb{Z}^s.$$

Hence the polynomial F given by (3.15) also vanishes on all lattice points which means $F \equiv 0$. Proposition 3.3 can be invoked again to conclude

$$P_N(\nu) = 0, \quad \nu \in \mathbb{Z}^s.$$

Since this equation holds for every $P \in D(X)$ we may now use a result from [4] that states under the assumption (3.8) the set A_N is unisolvent for interpolation from D(X). Hence for any $\beta' \in A_N$ there exists $P_{\beta'} \in D(X)$ such that

$$P_{\beta'}(-\beta) = \delta_{\beta,\beta'}, \quad \beta \in A_N.$$

Thus

$$0 = P_N(0) = \sum_{\beta \in A_N} \overset{*}{c}_\beta P_{\beta'}(-\beta) = \overset{*}{c}_{\beta'},$$

which proves the assertion.

We make use of Proposition 3.1 by letting $\{c_\beta\}_{\beta \in A_N}$ be the unique solution of (3.12). We introduce the functions

$$G_\gamma(x) = \sum_{\beta \in A_N} c_\beta B(x - z - \beta \mid X)$$

and

$$L(x) = G_\gamma(x - \gamma)$$

so that

$$L(\alpha - \nu) = \delta_{\alpha,\nu}, \quad \alpha, \quad \nu \in \mathbb{Z}^s.$$

Therefore the operator

$$(3.18) \qquad (I f)(x) = \sum_{\nu \in \mathbb{Z}^s} f(\nu) L(x - \nu)$$

is an interpolant of f whose smoothness is controlled by the choice of X. Furthermore, since $IP \in D(X)$ for all $P \in D(X)$ we have

Theorem 3.2. The interpolation operator I defined by (3.18) reproduces D(X). Since D(X) contains all polynomials of degree *d* (see (3.6)) the scaled version of I given by $E_h I E_{h^{-1}}$ approximate smooth functions with order $d + 1$.

We recall from [5] that linear combinations of box splines may be efficiently evaluated by means of subdivision algorithms. This offers a means for computing the fundamental spline L constructed above and therefore the interpolant (3.18).

As our last example we consider a variant of the Catmull Rom concept applied to bivariate box splines. We restrict ourselves to sets X of the type

$$X_m = \{\underbrace{(1,0), \dots, (1,0)}_{m_1}, \underbrace{(0,1), \dots, (0,1)}_{m_2}, \underbrace{(1,1), \dots, (1,1)}_{m_3}\}, \quad m = (m_1, m_2, m_3).$$

When $m_3 = 0$, $B(\bullet \mid X_m)$ reduces to the familiar tensor product B-spline supported on $[0, m_1] \times [0, m_2]$. For $m_i \neq 0$, $i = 1, 2, 3$, $c(X_m)$ induces the triangulation Δ considered at the beginning of this section. It is not hard to see that when $m_i > 1$

$$\Lambda = \{\alpha \in \mathbb{Z}^2 \colon \alpha \in \text{int } Z(X_m)\} \subseteq b(u \mid X_{m-e^i})$$

where $e^i = (\delta_{ij})_{j=1}^3$ for some $u \in \mathbb{R}^2 \backslash c(X_m)$. Fixing some $\alpha' \in \Lambda$ let $P \in D(X_{m-e^i})$ be given by (cf. [4])

$$P(\alpha) = \delta_{\alpha',\alpha}, \quad \alpha \in b(u \mid X_{m-e^i})$$

so that

$$L_{\alpha'}(x) = P(x) B(x \mid X) / B(\alpha' \mid X)$$

satisfies

$$L_{\alpha'}(\alpha) = \delta_{\alpha',\alpha}, \quad \alpha \in \mathbb{Z}^2.$$

Hence setting

$$L(x) = L_{\alpha'}(x + \alpha')$$

the operator

$$(3.19) \qquad (I f)(x) = \sum_{\alpha \in \mathbb{Z}^2} f(\alpha) L(x - \alpha)$$

produces a piecewise polynomial on Δ of degree at most $2(m_1 + m_2 + m_3) - 5$ which interpolates f on $\mathbb{Z}^2$. More precisely,

$$I f \in C^{d-1}(\mathbb{R}^2)$$

where $d = \min\{m_i + m_j,\ i \neq j\}$.

Let us briefly discuss now the degree of exactness for these methods. When $m_3 = 0$ the univariate result [6] readily assures that all bilinear functions are reproduced. For the general situation assume that Q is some homogeneous linear function and consider

$$\sum_{\alpha \in \mathbb{Z}^2} Q(\alpha)L(x-\alpha) = \frac{1}{K} \sum_{\alpha \in \mathbb{Z}^2} Q(\alpha)P(x+\alpha'-\alpha)B(x+\alpha'-\alpha \mid X_m)$$

where $K = B(\alpha' \mid X_m)$. Setting $x + \alpha' = y$ we obtain

$$\begin{aligned}(I\,Q)(x) &= \frac{1}{K} \sum_{\alpha \in \mathbb{Z}^2} Q(\alpha)P(y-\alpha)B(y-\alpha \mid X_m) \\ &= \frac{1}{K} \sum_{\alpha \in \mathbb{Z}^2} Q(-(y-\alpha))P(y-\alpha)B(y-\alpha \mid X_m) \qquad (3.20) \\ &+ \frac{Q(y)}{K} \sum_{\alpha \in \mathbb{Z}^2} P(y-\alpha)B(y-\alpha \mid X_m).\end{aligned}$$

Let $n = m_1 + m_2 + m_3$ and expand P in the last summand in a Maclaurin series,

$$\sum_{|\nu| \le n-1} \frac{y^\nu}{\nu!} \sum_{\alpha \in \mathbb{Z}^2} (D^\nu P(-\alpha))B(y-\alpha \mid X_m)$$

Proposition 3.2 implies that $\sum_{\alpha \in \mathbb{Z}^2} P(y-\alpha)B(y-\alpha \mid X_m)$ is a polynomial since $P \in D(X_{m-e_i}) \subset D(X_m)$. But for $y = \beta \in \mathbb{Z}^2$ we get

$$\sum_{\alpha \in \mathbb{Z}^2} P(\beta-\alpha)B(\beta-\alpha \mid X_m) = \sum_{\alpha \in \mathbb{Z}^2} P(\alpha)B(\alpha \mid X_m) = B(\alpha' \mid X_m) = K$$

by the definition of P. Hence the second summand on the right hand side of (3.20) reduces to $Q(y) = Q(x) + Q(\alpha')$. If now

$$Q(-x)P(x) \in D(X_m) \qquad (3.21)$$

the same reasoning gives for the first summand on the right hand side of (3.20)

$$\sum_{\alpha \in \mathbb{Z}^2} Q(-(y-\alpha))P(y-\alpha)B(y-\alpha \mid X_m) = Q(-\alpha')B(\alpha' \mid X_m)$$

which shows that $I\,Q = Q$ when (3.21) holds. In general (3.21) will not hold for every linear function Q as the following example shows (thus not even exactness of degree one may be expected). For $m_1 = 2$, $m_2 = m_3 = 1$ one may check that

$$D(X_m) = \operatorname{span}\{1, x, y, x^2, 2xy - y^2\}$$

while

$$D(X_{m-e^1}) = \operatorname{span}\{1, x, y\}.$$

Choosing for $b(u \mid X_{m-e^1})$ the set indicated by the dots in Figure 3.4 below

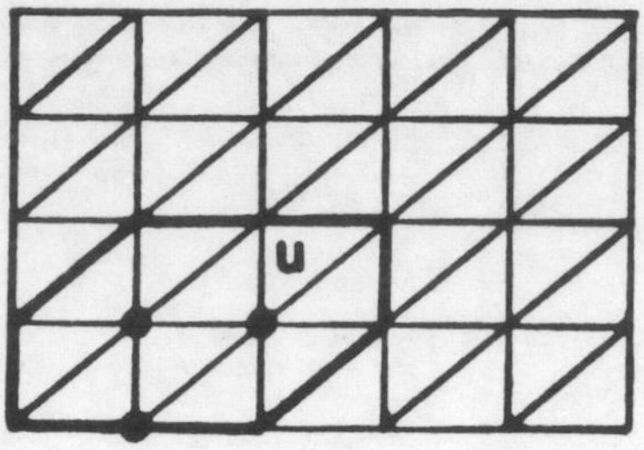

Figure 3.4

and choosing P as $y - x + 1$, i.e. $P(1, 1) = 1$, $\alpha' = (1, 1)$, we have

$$xP(x, y) = -x^2 + xy + x \notin D(X_m).$$

Even if we chose for $b(u \mid X_{m-e^3}) = \{(1,1), (2,1)\}$ giving $P(x, y) = 2 - x$ we obtain $yP(x, y) = -xy + 2y \notin D(X_m)$ while now, however, $xP(x, y) = 2x - x^2 \in D(X_m)$.

Remark 3.1. It can be shown even when (3.21) holds for all linear functions exactness of degree one is the most one can expect, as in the univariate case.

References

[1] R.H. Bartels, J.C. Beatty, B.A. Barsky, An Introduction to the use of splines in Computer Graphics, SIGGRAPH Lecture Notes, San Francisco, 1985.

[2] C. de Boor, K. Höllig, B-splines from parallelepipeds, J. d'Analyse Math. 42 (1982/83), 99-115.

[3] W. Dahmen, C.A. Micchelli, Translates of multivariate splines, Linear Algebra Appl. 52/53 (1983), 217-234.

[4] W. Dahmen, C.A. Micchelli, On the local linear independence of translates of a box spline, Studia Mathematica, 82 (1985), 243-263.

[5] W. Dahmen, C.A. Micchelli, Subdivision algorithms for the generation of box spline surfaces, Computer Aided Geometric Design, 1 (1984), 115-129.

[6] W. Dahmen, T.N.T. Goodman, C.A. Micchelli, Compactly supported fundamental functions for spline interpolation, IBM Research Report, June 1987, to appear Numerishe Mathematik.

[7] P.R. Eiseman, High Level Continuity for Coordinate Generation with Precise Controls, Journal of Computational Physics, 47 (1982), 352-374.

[8] D.X. Qi, A class of local explicit many knot spline interpolation schemes, University of Wisconsin, MRC Technical Summary Report #2238, 1981.

On the rate of rational approximation of analytic functions

A.A.Gončar and E.A.Rakhmanov
Math. Inst. of the USSR Acad. of Science

Introduction. In the past few years, the problems related with the constructive rational approximation of analytic functions have obtained a relevant development. In particular, this allowed to obtain the solution of a series of problems on the rate of rational approximation of analytic functions. In the present paper, a general theorem in this direction (theorem 1, section 2) is introduced. This result has a direct application to the known question of the rate of rational approximation of the exponential function on the half-line (theorem 2, section 3). In section 1, the problem of the equilibrium distribution of a charge on the plates of a condenser is considered, under the condition that an "exterior field" is acting on one of the plates; in terms connected with this equilibrium problem, a concept of "symmetry" (S-property) is introduced, which plays a fundamental role in the following. The statements of the results of the present paper were published in survey [1] (a lecture of the first author at the International Congress of Mathematicians, Berkeley, August 3-12, 1986).

I. Equilibrium. S-property. To start, let us introduce the notations which we will maintain in the following. Let K be a compact set in the extended complex plane $\overline{\mathbb{C}}$. A measure (charge) on K is a positive (correspondingly, real) Borel measure ν, whose support $s(\nu)$ is contained in K; M(K) is the set of all unit measures on K which satisfy the condition

$$\int_{|t|>1} \log |t| \, d\nu(t) < +\infty .$$

the logarithmic potential of a charge (in particular, measure) ν will be denoted by V^{ν}:

$$V^{\nu}(z)= \int \log |t-z|^{-1} d\nu(t) ;$$

if $\nu\in M(K)$, then $V^{\nu}(z) > -\infty$ for all $z\in\mathbb{C}$.

Now, let us consider the regular condenser (E,F) in $\overline{\mathbb{C}}$. Suppose that on the plate F of this condenser a continuous function (exterior field) $\varphi:F\longrightarrow \mathbb{R}$ is given. The condenser (E,F) with the given field φ on F will be denoted (E,F,φ).

Let $M(E,F)$ be the set of all charges of type $\mu= \mu_F-\mu_E$, where $\mu_E\in M(E)$ and $\mu_F\in M(F)$. The (doubled) energy of the charge μ in the field φ will be denoted by $I_{\varphi}(\mu)$:

$$I_{\varphi}(\mu) = \iint \log |t-z|^{-1} d\mu(t)d\mu(z) + 2\int \varphi(t) d\mu_F(t).$$

Known methods of potential theory (see, for example, [2] chap.2) allow to prove the following proposition:

<u>There exists a unique charge</u> $\lambda\in M(E,F)$, <u>which minimizes the energy</u> I_{φ} <u>(in this class)</u>:

$$I_{\varphi}(\lambda) = \min \{I_{\varphi}(\mu): \mu\in M(E,F)\}.$$

<u>The charge</u> λ <u>and only this charge</u> (<u>in the class</u> $M(E,F)$) <u>satisfies the following equilibrium condition</u>

$$\begin{aligned} &V^{\lambda}(z) \equiv w_E , \quad z\in E; \\ &(V^{\lambda}+\varphi)(z) \equiv \min_F (V^{\lambda}+\varphi) = w_F , \quad z\in s(\lambda_F). \end{aligned} \tag{1}$$

The charge $\lambda = \lambda_F-\lambda_E$ which satisfies relations (1), will be called the <u>equilibrium charge</u> with respect to the equipped condenser (E,F,φ). The assertion about the <u>uniqueness</u> of the equilibrium charge λ plays a fundamental role in our considerations. Put $w=w_F-w_E$ ($w=w(E,F,\varphi)$).

On more general equilibrium problems under the influence of exterior fields see [3].

Let us now define the concept of "symmetry". By Γ_0 we will denote the set of all points ζ of the compact Γ which satisfy the condition that there exists a neighborhood of the point ζ whose intersection with Γ is a simple analytic arc. We will say that the the compact Γ is admissible if $\Gamma_0 \neq \emptyset$ and cap $(\Gamma \setminus \Gamma_0) = 0$ (cap$(\cdot)$ denotes the logarithmic capacity).

We will say that the equipped condenser (E,F,φ) satisfies the S-property and write $(E,F,\varphi,)\in S$, if the following conditions are satisfied:

i) φ is a harmonic function on a certain neighborhood Ω of the plate F;

ii) the support $\Gamma = s(\lambda_F)$ of the F-component of the equilibrium charge λ is an admissible compact;

iii) $$\frac{\partial(V^\lambda + \varphi)}{\partial n_+}(\zeta) = \frac{\partial(V^\lambda + \varphi)}{\partial n_-}(\zeta), \quad \zeta\in\Gamma_0,$$

($\partial/\partial n\pm$ denote the normal derivatives to Γ in opposite directions).

It would be more exact to say that (under the imposed conditions) the plate F of the condenser (E,F,φ) satisfies the S-condition; nevertheless, in the present paper, we consider only this alternative of symmetry and this precision is not necessary.

2. The main result. Let f be a continuous function on the compact E. $\mathcal{R}_n$ the class of all the rational functions on z of order not greater than n. By $\rho_n(f,E)$ we denote the distance of f to $\mathcal{R}_n$ in the uniform metrics on E:

$$\rho_n(f,E) = \inf_{r\in\mathcal{R}_n} \| f - r \|_E,$$

where $\|\cdot\|_E$ is the sup norm on E.

Theorem 1. Let E be a continuum on $\overline{\mathbb{C}}$, F a rectifiable Jordan curve lying in $\mathbb{C}\setminus E$, and $\{\phi_n\}$ a sequence of holomorphic functions on a neighborhood Ω of the compact F. If

$$\frac{1}{2n}\log\frac{1}{|\phi_n(z)|} \rightrightarrows \varphi(z), \quad z\in\Omega,$$

and $(E,F,\varphi)\in S$ then the sequence of functions

$$f_n(z) = \int_F \frac{\phi_n(t)\,dt}{z - t}, \quad z \in E, \tag{2}$$

satisfies relation

$$\lim_n \rho_n(f_n, E)^{1/n} = e^{-2w}, \quad w = w(E,F,\varphi). \tag{3}$$

We formulated theorem 1 for the simplest geometric configuration; it is also true in the case when E is the union of a finite number of continuums and F is a contour consisting of a finite number of rectifiable Jordan arcs or curves. Relation (3) is also true for sequences of functions of type

$$f_n(z) = \int_\gamma \frac{\phi_n(t)\psi(t)\,dt}{z - t}, \quad z \in E$$

where ψ is a holomorphic function on $\Omega \backslash F$ which has a sufficiently regular jump on Γ_0 (i.e. if it is continuous and different from zero), and γ is an arbitrary contour contained in $\Omega \backslash F$ which separates E from F.

The scheme of the proof of the theorem is based on the construction of multipoint Padé approximants (interpolating rational functions with free poles); in a simpler situation, this scheme was used in problems on the rate of rational approximation in paper [4] of A.A.Gončar (see also [5]). The problems connected with the limit distribution of the zeros of "complex" orthogonal polynomials and the convergence of Padé approximants obtained an essential development in the recent papers of H.Stahl [6]. Stahl's method (in which the S-property plays an essential role) allows to study the asymptotic behavior of multipoint Padé approximants also in the case considered by us; new here is the fact that the "complex" relations of orthogonality (for fixed n) contain a complementary factor ϕ_n and the limit behavior of the multipoint Padé approximants can be described in terms of the equilibrium problem for a condenser under the presence of an exterior field φ (in Stahl's papers $\varphi \equiv 0$). The corresponding result constitutes the contents introduced in lemma 1 section 2.2; this lemma allowed to substantially widen the scope of applications of the indicated scheme. For the real case, on this observation see [7].

Let us sketch the proof of theorem 1. Without loosing generality we may suppose that E is a bounded continuum and the complement of E is connected.

2.1. Suppose that $\{\alpha_{n,k}\}$, $k=1,\ldots,n$, $n=1,2,\ldots$ is a triangular table of points (interpolation nodes) contained in the compact set E. Let us consider the sequence of polynomials

$$\omega_n(z) = \prod_{k=1}^{n} (z - \alpha_{n,k}), \qquad n=1,2,\ldots$$

and its associated sequence of unit measures

$$\mu(\omega_n) = \frac{1}{n} \sum_{k=1}^{n} \delta_{\alpha_{n,k}}, \qquad n=1,2,\ldots$$

(δ_ζ denotes Dirac's measure at point ζ). We choose the table $\{\alpha_{n,k}\}$ so that the following relation takes place

$$\mu(\omega_n) \longrightarrow \lambda_E , \tag{4}$$

where λ_E is the E-component of the equilibrium charge λ corresponding to (E,F,φ) (the symbol $\longrightarrow$ applied to a sequence of measures stands for weak convergence).

Let us fix a natural number n. Let us consider a rational function R_n of type P_n/Q_n, where P_n, Q_n are arbitrary polynomials that satisfy the relations

$$\deg P_n \le n-1, \ \deg Q_n \le n \ (Q_n \not\equiv 0), \quad \frac{Q_n f_n - P_n}{\omega_{2n}} \in H(E) \tag{5}$$

(H(E) denotes the class of all the holomorphic functions on E). The last requirement means that $Q_n f_n - P_n = 0$ at the zeros of ω_{2n}. The polynomials P_n and Q_n which satisfy relations (5) exist for any $f_n \in H(E)$; their ratio determines a unique rational function R_n. This rational function is called the multipoint Padé approximant of type [n-1,n] of function f_n, corresponding to the interpolation nodes $\alpha_{2n,1},\ldots,\alpha_{2n,2n}$. In the following, we will suppose that the leading coefficient of the polynomial Q_n is equal to one.

The rational function R_n not necessarily interpolates f_n at all the zeros of the polynomial ω_{2n}. Nevertheless, the loss of d_n interpolation conditions is accompanied by a reduction of the degrees of the numerator and the denominator of R_n by the same value. More precisely, if the number of zeros of the polynomial ω_{2n}, at which $f_n-R_n=0$, is equal to $2n-d_n$ and $R_n=P_n^*/Q_n^*$, $(P_n^*,Q_n^*)=1$, then $\deg P_n^*\le n-1-d_n$, $\deg Q_n^*\le n-d_n$ (we wish to underline that the zeros and the interpolation points are always considered according to their multiplicity).

We have

$$\left(\frac{Q_nf_n-P_n}{\omega_{2n}}\right)(z) = O\left(\frac{1}{z^{n+1}}\right),\quad z\longrightarrow\infty$$

Hence, using that $(Q_nf_n-P_n)/\omega_{2n}$ is holomorphic in the region $\overline{\mathbb{C}}\setminus F$ and the expression (2), it is not difficult to obtain orthogonality relations for the polynomials Q_n:

$$\int_F Q_n(t)t^k\,\frac{\phi_n(t)\,dt}{\omega_{2n}(t)} = 0,\qquad k=0,1,\dots,n-1 \tag{6}$$

and a formula for the error:

$$(f_n-R_n)(z) = \left(\frac{\omega_{2n}}{Q_nQ}\right)(z)\int_F\frac{(Q_nQ\phi_n)(t)dt}{\omega_{2n}(t)(z-t)},\qquad z\in D,$$

where Q is an arbitrary polynomial of degree not greater than n and D is the complement to F. In particular, we have

$$(f_n-R_n)(z) = \left(\frac{\omega_{2n}}{Q_n^2}\right)(z)\int_F\frac{(Q_n^2\phi_n)(t)dt}{\omega_{2n}(t)(z-t)},\qquad z\in D, \tag{7}$$

2.2. For any natural number n, let us fix a pair of polynomials (P_n,Q_n) which satisfy conditions (5). Put

$$\mu(Q_n) = \frac{1}{n}\sum_n \delta_{\beta_{n,k}},\quad n=1,2,\dots,$$

where the sum is taken over all the zeros $\beta_{n,k}$ of the polynomial Q_n. With the aid of the orthogonality relations the following can be proved:

Lemma 1. Suppose that the conditions of theorem 1 are satisfied and the interpolation nodes are chosen so that relation (4) takes place. Then

(i) under a convenient normalization of the polynomials Q_n we have $\mu(Q_n) \longrightarrow \lambda_F$, where λ_F is the F-component of the equilibrium charge λ;

(ii) $$\left| \int_F \frac{(Q_n^2 \phi_n)(t)dt}{\omega_{2n}(t)(z-t)} \right|^{1/n} \longrightarrow \exp(-2w_F)$$

in capacity on each compact subset of D.

As was said above, the proof of lemma 1 is carried out with the aid of a suitable modification of Stahl's method [6]; the presence of the factor ϕ_n under the integral sign in (6) and (7) manifests itself in that an exterior field apppears in the corresponding condenser problem. We note that the proof of lemma 1 is technicallly the most difficult step in the proof of theorem 1. From lemma 1 and formula (7) follows:

Lemma 2. Under the conditions of lemma 1 we have:

$$|f_n - R_n|^{1/n} \longrightarrow \exp 2(V^\lambda - w_F)$$

in capacity on each compact subset of D.

The fact that above we are talking about convergence in capacity is connected with the essence of the problem; although most part of the poles of R_n accumulate on F (see assertion (i) of lemma 1), outside of a fixed neighborhood of F can lay o(n) poles of these rational functions (as $n \longrightarrow \infty$).

The proof of relation (3) is based on lemma 2; essentially we repeat below the arguments of paper [4]. These arguments are carried out in a fixed (small) neighborhood of the continuum E. We observe that in the case considered ((E,F) is a regular condenser) the potential V^λ of the equilibrium charge λ is continuous in $\overline{\mathbb{C}}$. Suppose that $g(z,\zeta)$ is Green's function for the complement G of the continuum E. Let us fix an arbitrarily small $\varepsilon>0$ and then a constant $\theta_o>0$ so that in the region $g(z,\infty)<\theta_o$ takes place the inequality

$$|V^\lambda - w_E| < \varepsilon. \tag{8}$$

Let us also introduce the notations

$$\gamma(\theta) = \{z: g(z,\infty) = \theta\},\ 0<\theta\le\theta_o$$

$$\sigma(\theta',\theta'') = \{z: \theta'< g(z,\infty)< \theta''\},\ 0\le\theta'<\theta''\le\theta_o;$$

the curves $\gamma(\theta)$ are analytic and the sets $\sigma(\theta',\theta'')$ are annular regions which surround the continuum E.

Let us fix an arbitrary region $\sigma(\theta,\theta_o)$, $\theta>0$. From lemma 2 follows that the inequality

$$\overline{\lim_n} \| f_n - R_n \|_{\gamma_n}^{1/n} \le \exp(-2w+\epsilon) \tag{9}$$

holds for a certain sequence of curves $\gamma_n=\gamma(\theta_n)$, $\theta<\theta_n<\theta_o$ (a set of sufficiently small capacity cannot intersect all the curves $\gamma(t)$, $\theta<t<\theta_o$); the lenghts of the curves γ_n are uniformly bounded. For z surrounded by $\gamma(\theta)$ put

$$r_n(z) = \frac{1}{2\pi i}\int_{\gamma_n}\frac{R_n(t)\,dt}{t-z}\quad;\quad r_n\in\mathcal{R}_n .$$

From the formula

$$(f_n - r_n)(z) = \frac{1}{2\pi i}\int_{\gamma_n}\frac{(f_n-R_n)(t)\,dt}{t-z}\ ,\quad z\in E\ ,$$

and inequality (9) we obtain

$$\overline{\lim_n} \| f_n - r_n \|_E^{1/n} \le \exp(-2w+\epsilon),$$

in view of the arbitrariness of $\varepsilon>0$ follows the estimate

$$\overline{\lim_n}\, \rho_n(f_n,E)^{1/n} \le e^{-2w}.$$

Let us prove the corresponding lower estimate:

$$\underline{\lim_n}\, \rho_n(f_n,E)^{1/n} \ge e^{-2w}. \tag{10}$$

If this inequality doesn't take place, then there exists an $\eta>0$ and a sequence of rational functions r_n, $n\in\Lambda\subset\mathbb{N}$ such that

$$r_n= p_n/q_n,\ \deg p_n\le n-1,\ \deg q_n\le n$$

$$\| f_n - R_n \|_E < \exp[-(2w+3\eta)n], \qquad n\in\Lambda \tag{11}$$

Let us show that this assumption leads to a contradiction. Put

$$\psi_n(z) = \log |(f_n-R_n)(z)| - \sum_k g(z,\zeta_{n,k}),$$

where $\zeta_{n,k}$ denote the poles of the rational function r_n. The constant $\theta_o>0$ which appears above will be selected so that the inequality (8) takes place for $\varepsilon=\eta$. Function ψ_n is subharmonic in the region $\sigma(0,\theta_o)$. From (11) we obtain

$$\psi_n(z) < -(2w + 3\eta)n, \quad z\in \partial E, \tag{12}$$

for $n\in\Lambda$ (further we consider only such n's). From the maximum principle for subharmonic functions follows the inequality

$$\log |r_n(z)| - \sum_k g(z,\zeta_{n,k}) \le \log \|r_n\|_E, \quad z\in G.$$

Using this inequality and the obvious estimates for f_n (under the conditions of theorem 1), we obtain

$$\psi_n(z) < Cn, \quad z\in\gamma(\theta_o) \tag{13}$$

(C is independent of n). Suppose that $\omega(z)$, $z\in\sigma_o$ is the harmonic measure of ∂E with respect to $\sigma_o=\sigma(0,\theta_o)$. From (12),(13), and the two constants theorem follows the estimate

$$\psi_n(z) < -(2w + 3\eta)n\omega(z) + Cn(1 - \omega(z)), \quad z\in\sigma_o.$$

Hence, it follows that if $\theta''\in(0,\theta_o)$ is sufficiently small we have

$$\psi_n(z) < -2(w + \eta)n, \quad z\in\sigma(0,\theta'')$$

Let us fix an arbitrary θ'', for which this inequality holds; hence,

$$|(f_n - r_n)(z)| < \exp[-2(w+\eta)n + \sum_k g(z,\zeta_{n,k})], \quad z\in\sigma(0,\theta'') \tag{14}$$

Let us now choose the parameter $\theta'\in(0,\theta'')$; this selection is based on the following lemmas.

<u>Lemma</u> 3. <u>If</u> $F_t= \{z: \sum_k g(z,\zeta_{n,k})\ge nt\}$, $t>0$, <u>then</u> $\mathrm{cap}(E,F_t)\le 1/t$.

Here, cap(·) denotes the capacity of the indicated condenser.

<u>Lemmma</u> 4. <u>Let</u> $A>0$ <u>be</u> <u>an</u> (<u>arbitrarily</u> <u>large</u>) <u>number</u>. <u>Then</u> <u>there</u> <u>exists</u> $\theta'\in(0,\theta'')$ <u>such</u> <u>that</u> <u>for</u> <u>any</u> <u>continuum</u> $K\subset\bar\sigma(\theta',\theta'')$ <u>which</u> <u>has</u>

<u>non-empty intersection with $\gamma(\theta')$ and $\gamma(\theta'')$ holds the inequality</u> cap(E,K)>A.

Let us fix the parameter $\theta'\in(0,\theta'')$ so that for any continuum K figuring in lemma 4 takes place the inequality cap(E,K)>$2/\eta$. Then, using (14) and lemmas 2 and 3 (the last with $t=\eta$) we can show that for any sufficiently large $n\in\Lambda$ there exists a curve $\gamma_n\subset \sigma(\theta',\theta'')$ enveloping E on which simultaneously hold the following inequalities:

$$\min_{z\in\gamma_n} |(f_n - R_n)(z)| > e^{-(2w+\eta)n}, \tag{15}$$

$$\max_{z\in\gamma_n} |(f_n - r_n)(z)| < e^{-(2w+\eta)n}, \tag{16}$$

Now we can conclude the proof. Suppose that the rational function R_n interpolates function f_n at $2n-d_n$ zeros of the polynomial ω_{2n}. Let us represent the function $R_n - r_n$ in the form

$$R_n - r_n = \frac{P_n^* q_n - p_n Q_n^*}{Q_n^* q_n} .$$

From (15), (16) and Rouche's theorem it follows that the polynomial

$$P_n^* q_n - p_n Q_n^*$$

has at least $2n-d_n$ zeros; at the same time, the degree of this polynomial is not greater that $2n-d_n-1$ (see section 2.1). Consequently, $r_n \equiv R_n$, which contradicts (15)-(16). This proves relation (10) and hence (3).

3. <u>Rate of rational approximation of the exponential on the half-line.</u>

Put $\rho_n = \rho_n(e^{-x},[0,+\infty])$. A detailed survey on results of the form

$$0 < c_1 \le \underline{\lim}_n \rho_n^{1/n} \le \overline{\lim}_n \rho_n^{1/n} \le c_2 < 1$$

and the conjectures connected with the (existence and the) limit value of

$$v = \lim_n \rho_n^{1/n} \tag{17}$$

can be found in R.Varga's book [8]; see also the subsequent publications [9,10,11]. In the paper of Opitz and Scherer [9] the inequality $c_2 < 1/9.037$ was proved (disproof of the conjecture "v=1/9"). In the papers of Trefethen and Gutknecht [10] and Carpenter, Ruttan and Varga [11], on the basis of computational analysis of the problem, approximate values of the constant v were obtained (the question of the existence of the limit (17) remained open). So, in [11] the following value was introduced

$$v = 1/9.289\ 025\ 491\ 920\ 81... \qquad (18)$$

and the coefficients of the numerator and the denominator of the rational functions of best approximation were approximately calculated for $n \leq 30$. Judging from [11], a great computational work was carried out; as it follows from our theorem (see below), all the digits in (18) are correct!

Recently, A.Magnus in [12] determined the value of the constant v with the aid of Caratheodory-Fejer's method, adjusted for rational approximation of real functions on the segment [-1,1] as appears in [10] (using this method in [10] the value $v \approx 1/9.289\ 03$ was obtained). Magnus found the correct answer; that is

$$v = \exp(-\pi K'/K),$$

where K' and K are the complete elliptic integrals of the first kind for the modules k and $k' = (1-k^2)^{1/2}$, and module k satisfies the equation $K(k) = 2E(k)$ ($E(\cdot)$ is the complete elliptic integral of the second kind). On the basis of this characterization he calculated 30 digits of the constant v. However, the technique described in [12] for determining the value of v has a heuristic character in essential points (and from the mathematical point of view, Magnus' result had the character of a conjecture). At present it is not clear, in particular, how to rigorously justify the transition from the n-multiple integrals to the equations which describe the limit distributions (as $n \longrightarrow \infty$). J.Nuttal used such type of transitions in the anaysis of the asymptotic properties of Hermite-Padé polynomials and in other problems (saddle point method); the justification of the corresponding technique in the "complex" case would serve to solve a series of problems in the theory of rational approximation which still remain open.

With the aid of theorem 1 we can <u>prove</u> that limit (17) exists and describe the value v in terms related with a problem of equilibrium for a condenser of the form (E,F), $E=[0,+\infty]$, with the condition that the plate F is under the influence of the exterior field $\varphi(z)=\frac{1}{2}\operatorname{Re} z$, moreover, $(E,F,\varphi)\in S$. This equilibrium problem can be solved explicitly in terms of elliptic functions and integrals. There are different possible ways (in form) of expressing constant v; in the following theorem, we formulate the answer in a form which seems to us the most interesting (other forms of the answer will be introduced below).

<u>Theorem</u> 2. <u>There exists the limit</u> $v = \lim_n \rho_n^{1/n}$, <u>moreover</u> v <u>is the (unique) positive root of the equation</u>

$$\sum_{n=1}^{\infty} a_n v^n = 1/8 \quad , \qquad a_n = \Big|\sum_{d|n} (-1)^d \cdot d\Big| . \tag{19}$$

The computation of constant v on the basis of (19) constitutes no difficulty.

The reduction of theorem 2 to theorem 1 (more precisely, to the corresponding theoretical-potential problem) is based on the following obvious observation. We have

$$\rho_n = \rho_n(e^{-x},E) = \rho_n(e^{-nx},E); \tag{20}$$

here and in the following $E=[0,+\infty]$. Let us fix the point $b=3+i\beta$, $\beta<0$ (β will be given later). Let F be an arbitrary rectifiable curve, lying in the region $G=\mathbb{C}\setminus E$ which connects the points b and $\bar{b}$. γ is an unbounded contour consisting of F and two rays (a.e. parallel to

$\mathbb{R}$) which connect the points b and $\bar{b}$ with the point at infinity (we suppose that F lies in the half-plane Re z$\le$ 3, and $\gamma\backslash F$ in the half-plane Re z$>$ 3). Under the proper orientation of γ we have

$$e^{-nz} = \frac{1}{2\pi i}\int_{\gamma}\frac{e^{-nt}dt}{z-t} = \frac{1}{2\pi i}\int_{F}\frac{e^{-nt}dt}{z-t} + \Delta_n(z),\ z\in E,$$

moreover,

$$\overline{\lim_n}\ \|\Delta_n\|_E^{1/n} \le e^{-3}.$$

From here, (on account of (20)) it follows that the statement of theorem 2 is equivalent to the assertion that <u>the following limit exists</u>

$$\lim_n \rho_n\left(\int_F\frac{e^{-nt}dt}{z-t},E\right)^{1/n} = v, \qquad (21)$$

<u>where</u> v <u>is the solution of equation</u> (19) (we use the fact that $e^{-3}<v$; the selection of the number 3 is related only with this inequality).

In order to use theorem 1, given E and $\varphi(z)=\frac{1}{2}$Re z, we must construct F so that $(E,F,\varphi)\in S$ and find the corresponding constant $v=e^{-2w}$, $w=w(E,F,\varphi)$ (more precisely, we must prove that this constant is the solution of (19)).

Let us sketch the solution of this problem. For any a, Im a$<$ 0, put P(z,a) = 4z(z−a)(z−$\bar{a}$) and

$$\omega(a) = \int_0^{+\infty}\frac{dt}{\sqrt{P(t,a)}}\ (>0).$$

The following conditions determine a unique number a, Im a$<$0:

$$\frac{\pi i}{\omega(a)}\int_{\infty}^{a}\frac{dt}{t\sqrt{P(t,a)}} = -\frac{1}{2}; \qquad (22)$$

here and in the following, the branch of the root in a region of type $D=G\backslash\Gamma$ (Γ is a Jordan arc in $G=\mathbb{C}\backslash E$ which connects the points a and $\bar{a}$) is chosen so that $\sqrt{P(z,a)}>0$ on the upper side of the cut E. For this a, we put P(z,a)=P(z), $\omega(a)=\omega$. From (22) follows, in particular, the relation

$$\int_a^{\bar{a}}\frac{dt}{t\sqrt{P(t)}} = 0. \qquad (23)$$

From here, we obtain that the formula

$$g(z) = \frac{\pi i}{\omega} \int_{\infty}^{z} \frac{dt}{t\sqrt{P(t)}}, \quad z\in D$$

defines a (singlevalued) holomorphic function in any region D of the type indicated above. Now, we fix an (analytic and symmetrical with respect to $\mathbb{R}$) arc Γ from a to $\bar{a}$ using the condition

$$\mathrm{Re}\ (g(z)-g(a))dz = 0 \Leftrightarrow \mathrm{Im} \int_{a}^{z} \frac{dt}{t\sqrt{P(t)}}\, dz = 0, \quad z\in\Gamma;$$

at the same time the region $D=G\setminus\Gamma$ is also fixed. Using Cauchy's formula, we obtain

$$g(z) = \int_{\Gamma} \frac{d\lambda_{\Gamma}(t)}{t-z} - \int_{E} \frac{d\lambda_{E}(t)}{t-z}, \quad z\in D,$$

where

$$d\lambda_{\Gamma}(t) = \frac{1}{\omega} \int_{a}^{t} \frac{d\tau}{\tau\sqrt{P(\tau)}}\, dt, \quad t\in\Gamma,$$

$$d\lambda_{E}(t) = \frac{1}{\omega} \int_{t}^{+\infty} \frac{d\tau}{\tau\sqrt{P(\tau)}}\, dt, \quad t\in E,$$

(integration takes place in the positive direction with respect to the region D). It is not difficult to prove that the last formulas define unique measures $\lambda_{\Gamma}\in M(\Gamma)$ and $\lambda_{E}\in M(E)$; furthermore, $\lambda = \lambda_{\Gamma} - \lambda_{E} \in M(\Gamma,E)$ and

$$g(z) = \int \frac{d\lambda(t)}{t-z}, \quad z\in D.$$

Now, suppose that $F\subset G$ is the contour consisting of Γ and the segments $[a,b]$, $[\bar{a},\bar{b}]$, $b = 3 + i\ \mathrm{Im}\ a$ (with this we fix the curve F which figures in (21)). Using well known results which connect the logarithmic potentials with Cauchy type integrals, the following statement can be proved: <u>the charge constructed above is the equilibrium charge for the equipped condenser</u> $(E,F,\frac{1}{2}\mathrm{Re}z)$; <u>furthermore</u>, $\Gamma=s(\lambda_F)$ <u>and</u> $(E,F,\frac{1}{2}\mathrm{Re}z)\in S$.

The relation between the logarithmic potential V^{λ} of the equilibrium charge λ and function g constructed by us is given by the formula

$$V^{\lambda}(z) = \mathrm{Re}\ \tilde{V}(z),\ \tilde{V}(z) = \int_{\infty}^{z} g(t)dt = \int \log\frac{1}{t-z}\, d\lambda(t), \quad z\in D.$$

Let us now find the constant $w = w(E,F,\frac{1}{2}\mathrm{Re}z)$. We have

$$w = \mathrm{Re}\ (\tilde{V}(a) + \frac{a}{2}) = \mathrm{Re}\ (\frac{\pi i}{\omega}\int_{\infty}^{a} dt \int_{\infty}^{t} \frac{d\tau}{\tau\sqrt{P(\tau)}} + \frac{a}{2});$$

integrating by parts (considering (22)) we obtain

$$\frac{\pi i}{\omega}\int_{\infty}^{a} dt \int_{\infty}^{t} \frac{d\tau}{\tau\sqrt{P(\tau)}} = -\frac{a}{2} - \frac{\pi i}{\omega}\int_{\infty}^{a} \frac{dt}{\sqrt{P(t)}}.$$

Hence,

$$-w = \mathrm{Re}\ \frac{\pi i\omega'}{\omega} = \pi i(\frac{\omega'}{\omega} - \frac{1}{2}), \qquad \omega' = \int_{\infty}^{a} \frac{dt}{\sqrt{P(t)}},$$

and we obtain the following expression for the constant $v=e^{-2w}$ which interests us

$$v = -h^2, \qquad h = \exp(\frac{\pi i\omega'}{\omega}).$$

The equation for v can now be obtained directly from relation (22). Let (u) be the Weierstrass function corresponding to the periods 2ω and $2\omega'$ ($\mathrm{Im}\ \frac{\omega'}{\omega} > 0$); the polynomial $4x^3 - g_2x - g_3$ connected with has real coefficients g_2, g_3, one real root e_1 and a pair of conjugated roots e_2, e_3 (this polynomial coincides with $P(x-e_1)$, $\bar{a} = e_2 - e_1$, $a = e_3 - e_1$). Relation (23) can be rewritten as follows

$$\int_{\omega'}^{\omega'+\omega} \frac{du}{(u) - e_1} = 0.$$

Using formula (see [14], page 271)

$$(u+\omega) - e_1 = \frac{(e_2-e_1)(e_3-e_1)}{(u) - e_1},$$

we obtain

$$0 = \int_{\omega'}^{\omega'+\omega} (e_1 - (u+\omega))du = e_1\omega + \zeta(\omega'+2\omega) - \zeta(\omega'+\omega) = e_1\omega + \eta,$$

where $\zeta(u)$ is Weierstrass' zeta function, $\eta = \zeta(\omega)$. Hence, (23) is equivalent to the relation

$$e_1 = -\eta/\omega \tag{24}$$

The following formula for e_1 ([14], page 78),

$$e_1 = -\frac{\eta}{\omega} + (\frac{\pi}{\omega})^2(\frac{1}{4} + 2\sum_{n=1}^{\infty} \frac{h^{2n}}{(1+h^{2n})^2}), \quad h = \exp(\frac{\pi i\omega'}{\omega}), \tag{25}$$

shows that relation (24) is equivalent to the next equation for h

$$\sum_{n=1}^{\infty} \frac{h^{2n}}{(1+h^{2n})^2} = -\frac{1}{8}.$$

Making the substitution $v=-h^2$ we obtain

$$\mathcal{F}(v) = \sum_{n=1}^{\infty} \frac{(-1)^{n-1} v^n}{(1+(-v)^n)^2} = \frac{1}{8} . \tag{26}$$

It rests to observe that the series standing on the left hand of (19) is a power expansion of function $\mathcal{F}$; from this expansion it is obvious that equation (26) has a unique positive solution.

We derived the equation for $h=\sqrt{-v}$ from (23); using (22) the point a can be found:

$$a = -4\vartheta_0^{-4} , \quad \vartheta_0 = 1 + 2 \sum_{n=1}^{\infty} (-h)^{n^2} .$$

An equation for v can also be expressed in terns of a theta-series. Using in place of (25) the formula

$$e_1 = -\frac{\eta}{\omega} - \frac{1}{\omega} \left(\frac{\vartheta_1'(t)}{\vartheta_1(t)}\right) \Bigg|_{t=1/2} , \quad (t=u/2\omega),$$

(compare with [14], page 77) taking account of (24) we obtain $\vartheta_1''(h)=0$ and

$$\sum_{n=0}^{\infty} (2n+1)^2 (-v)^{n(n+1)/2} = 0. \tag{27}$$

In December of 1986, we received a letter of A.Magnus ""1/9" at Segovia and me" (at the Conference of Segovia in September 1986 E.A. Rakhmanov announced the results above), in which (in particular) our approach to the solution of this problem is presented and different equations for v are discussed. He discovered that equation (27) and also the equation

$$\sum_{n=1}^{\infty} \frac{nv^n}{1-(-v)^n} = \frac{1}{8} ,$$

which is equivalent to (19) and (26), can be found in the book of Halphen [15] (of 1886)). Halphen arrived to equation (27) (and calculated the value of v with six digits) in connection with the study of the variations of the theta-function. It is interesting, that the question of the rate of rational approximation of the exponential on the half-line happened to be connected with this problem and that Halphen's constant gives the solution.

In conclusion, we note that theorem 1 allows to investigate other problems connected with the approximation of the exponential function, in particular: the problem of the rate of rational approximation of e^{-z} on $E_\theta = \{z: |\arg z| \le \theta < \pi/2\}$; the approximation of $e^{-p(x)}$ on $[0,+\infty]$, where $p(x)$ is an arbitrary polynomial with positive leading coefficient; and so forth. The answers can be given in theoretical potential terms, the corresponding equilibrium problems do not have such a simple solution as in the case considered above.

Bibliography

1. Gončar, A.A., Rational approximations of analytic functions, Proceedings of the International Congress of Mathematicians, Berkeley' 86, 1987.

2. Landkov, N.S., Foundations of Modern Potential Theory, Nauka, Moskva, 1966; Springer-Verlag, Berlin-Heidelberg-N. York, 1972.

3. Gončar, A.A., E.A. Rakhmanov, On the equilibrium problem for vector potentials, Uzpiehi Mat. Nauk, 1985, v.40, 4, 155-156.

4. Gončar, A.A., On the rate of rational approximation of analytic functions, Trudy MIAN, 1984, v.166, 52-60.

5. Gončar, A.A., On the speed of rational approximation of some analytic functions, Mat. Sb., 1978, v.105, 1, 147-163; Math. of the USSR Sb., 1978, v.34, 2, 131-145.

6. Stahl, H., Orthogonal polynomials with complex valued weight function, I,II, Constr. Approx., 1986, v.2, 3, 225-240, 241-251.

7. Gončar, A.A., E.A. Rakhmanov, Equilibrium measure and the distribution of the zeros of extremal polynomials, Mat. Sb., 1984, v.125, 1, 117-127; Math. of the USSR Sb., 1986, v.53, 1, 119-130.

8. Varga, R.S., Topics in Polynomial and Rational Interpolation and Approximation, Univ. Montreal,1982.

9. Opitz, H.U., K. Scherer, On the rational approximation of e^{-x} on

[0,∞), Constr. Approx., 1985, v.1, 3, 195-216.

10. Trefethen, L.N., M. Gutknecht, The Caratheodory-Fejer method for real rational approximation, SIAM J. Numer. Anal., 1983, 20, 420-436.

11. Carpenter, A.J., A. Ruttan, R.S. Varga, Extended computations on the "1/9"-conjecture in rational approximation theory, Springer Verlag, Lecture Notes in Math. 1105, 1984, 383-411.

12. Magnus, A.P., CFGT determination of Varga's constant "1/9", Inst. Math., U.C.L., B-1348, 1986, (preprint).

13. Nuttall, J., Asymptotics of diagonal Hermite-Padé polynomials, J. Appr. Theory, 1984, v.42, 4, 299-386.

14. Akhiezer, N.I., Elements of the Theory of Elliptic Functions, Nauka, Moskva, 1970.

15. Halphen, G.H., Traité des Fonctions Elliptiques et de leurs Applications, I, Gauthier-Villars, Paris, 1886.

PARAMETRIC OPTIMIZATION: PATHFOLLOWING WITH JUMPS

Jürgen Guddat[1)], Hubertus Th. Jongen[2,3)], Dieter Nowack[1)]

1) Humboldt University
Dept. of Mathematics
PSF 1297
1086 Berlin
German Democratic Republic

2) University of Twente
Faculty of Applied Mathematics
P.O. Box 217
7500 AE Enschede
The Netherlands

__Abstract__. We consider finite dimensional optimization problems depending on one real parameter t. Recently, Jongen/Jonker/Twilt [9] studied the generic behaviour of such problems. Based on this investigation, we propose a partial concept for finding a suitably fine discretization $0 = t_o < \ldots < t_{i-1} < t_i < \ldots < t_N = 1$ of the interval [0,1], and corresponding local minima $x(t_i)$, $i = 1,\ldots,N$; here, information on the point $x(t_{i-1})$ is used in order to compute $x(t_i)$. Mainly, socalled continuation methods can be exploited. However, at some parameter values, the branch of local minima used might have an endpoint; at such points one has to jump to another branch of local minima in order to continue the execution of the desired process. In case that the feasible set in a neighborhood of such a mentioned endpoint remains nonempty for increasing parameter values, it will be shown how a jump can be realized.

3) Honorary Professor, University of Hamburg, Federal Republic of Germany

1. Introduction

We consider the following optimization problem depending on one real parameter:

$$\mathcal{P}(t) : \text{Minimize } f(\cdot,t) \text{ on } M(t) \qquad , t \in \mathbb{R} \quad , \tag{1.1}$$

where the feasible set M(t) is defined by

$$M(t) = \{x \in \mathbb{R}^n | h_i(x,t) = 0,\ i \in I,\ g_j(x,t) \geq 0,\ j \in J\}, \tag{1.2}$$

$I = \{1,...,m\}$, $m < n$, and $J = \{1,...,s\}$.

The main goal is to find a local minimum for $\mathcal{P}(t)$, for all $t \in [0,1]$ (if possible). The motivation for developping solution algorithms for $\mathcal{P}(t)$, $t \in [0,1]$, is manifold, e.g.:

(i) Solving naturally given parametric optimization problems (an example for the optimal economic dispatch of energy power stations is given in the survey paper by Guddat [5].

(ii) Globalization of locally convergent algorithms (see e.g. [2], [5]).

(iii) Multiobjective optimization based on parametric optimization (see e.g. [4], [14]).

(iv) Stochastic optimization based on parametric optimization (see e.g. [4].

(v) Multilevel optimization (cf. [13]).

Unless otherwise specified, we assume that $f, h_i, g_j \in C^3(\mathbb{R}^n \times \mathbb{R}, \mathbb{R})$, the space of three times continuously differentiable functions from $\mathbb{R}^n \times \mathbb{R}$ to $\mathbb{R}$. The general point from $\mathbb{R}^n \times \mathbb{R}$ is denoted by z, and $z = (x,t)$, where t stands for the parameter. A point z is called a <u>generalized critical point</u> (g.c. point) if $x \in M(t)$ and if the set $\{D_x f,\ D_x h_i,\ i \in I,\ D_x g_j,\ j \in J_o(z)\}_{|z}$ is linearly dependent (cf. [8], [9]). Here, $D_x f$ stands for the row vector of first partial derivatives, and $J_o(z)$ denotes the index set of active (=binding) inequality constraints, i.e. $J_o(z) = \{j \in J | g_j(z) = 0\}$. In particular, $z = (x,t)$ is a g.c. point whenever x is a local minimum for $\mathcal{P}(t)$. Let Σ denote the set of g.c. points. In [9] the local structure of Σ is completely described for $(f, h_i,...,g_j,..)$ belonging to a C^3_s-open and dense subset $\mathcal{F}$ from $C^3(\mathbb{R}^n \times \mathbb{R}, \mathbb{R})^{1+m+s}$; the topology C^3_s refers to the strong (or Whitney-) C^3-topology (cf. [7],[11]). For omitted details on the set $\mathcal{F}$ we refer to [9].

From now on we assume: $(f,h_i,\dots,g_j,\dots) \in \mathcal{F}$.

The points from Σ can be divided into five types.

A point $\bar{z} = (\bar{x},\bar{t})$ is of Type 1 if the following conditions hold (then, $\bar{x}$ is also called a nondegenerate critical point for $\mathcal{P}(\bar{t})$, cf. [10], but also [12]):

$$D_x f = \sum_{i\in I} \bar{\lambda}_i D_x h_i + \sum_{j\in J_o(\bar{z})} \bar{\mu}_j D_x g_j{}_{|\bar{z}} \tag{1.3}$$

$$\left.\begin{array}{l} \{D_x h_i,\ i \in I,\ D_x g_j,\ j \in J_o(\bar{z})\}_{|\bar{z}} \text{ is linearly independent} \\ \text{(linear independence constraint qualification)} \end{array}\right\} \tag{1.4.a}$$

$$\left.\begin{array}{l} \text{The numbers } \bar{\mu}_j,\ j \in J_o(\bar{z}) \text{ are unequal zero} \\ \text{(strict complementarity)} \end{array}\right\} \tag{1.4.b}$$

$$D_x^2 L(\bar{z})/T(\bar{z}) \text{ is nonsingular.} \tag{1.4.c}$$

Condition (1.4.c) needs some explanation: $D_x^2 L$ is the matrix of second partial derivatives - with respect to x - for the Lagrange function L, where $L = f - \sum_{i\in I} \bar{\lambda}_i h_i - \sum_{j\in J_o(\bar{z})} \bar{\mu}_j g_j$, the numbers $\bar{\lambda}_i$, $\bar{\mu}_j$ being taken from (1.3). Furthermore, $T(\bar{z})$ denotes the tangent space of $M(\bar{t})$ at $\bar{x}$, i.e. $T(\bar{z}) = \{\zeta \in \mathbb{R}^n | D_x h_i(\bar{z})\zeta = 0,\ i \in I,\ D_x g_j(\bar{z})\zeta = 0,\ j \in J_o(\bar{z})\}$. Now, $D_x^2 L(\bar{z})/T(\bar{z})$ stands for $V^T \cdot D_x^2 L(\bar{z}) \cdot V$, where V is a matrix whose columns (n-vectors) form a basis for $T(\bar{x})$.

If $\bar{z}$ is a point of Type 1, then the local behaviour of $f(\cdot,\bar{t})_{|M(\bar{t})}$ around $\bar{x}$ is completely determined by means of four characteristic numbers ("indices"): the number of negative/positive numbers $\bar{\mu}_j$, $j \in J_o(\bar{z})$, and the number of negative/positive eigenvalues of $D_x^2 L(\bar{z})/T(\bar{z})$ (i.e. the corresponding number for $V^T \cdot D_x^2 L(\bar{z})V$, cf. [9]).

The set Σ is pieced together from one-dimensional C^2-manifolds and it is the closure of the set of all points of Type 1; moreover, the points of Type 2-5 constitute a discrete subset of Σ. The points of Type 2-5 represent three basic degeneracies (compare also the interesting paper [12]). In fact, Type

2, Type 3 and Type 4,5 refer to the violation of (1.4.b), (1.4.c) and (1.4.a), respectively. All possible changes in the four characterizing indices, when passing points of Type 2-5 along Σ, are given in [9].

Let Σ_{loc} denote the subset of points $z = (x,t)$ of Type 1 for which x is a local minimum for $f(\cdot,t)_{|M(t)}$, and let $\bar{\Sigma}_{loc}$ stand for the closure of Σ_{loc}. In fact, a point $\bar{z}$ of Type 1 belongs to Σ_{loc} if and only if both $\bar{\mu}_j > 0$, $j \in J_o(\bar{z})$ and $D_x^2 L(\bar{z})$ positive definite on $T(\bar{z})$. Based on the investigations in [9] we have the following possibilities for the local structure of $\bar{\Sigma}_{loc}$, as depicted in Fig. 1. In Fig. 1 the point $\bar{z}$ under consideration is identified by an exposed point, whereas the full line stands for the curve of local minima.

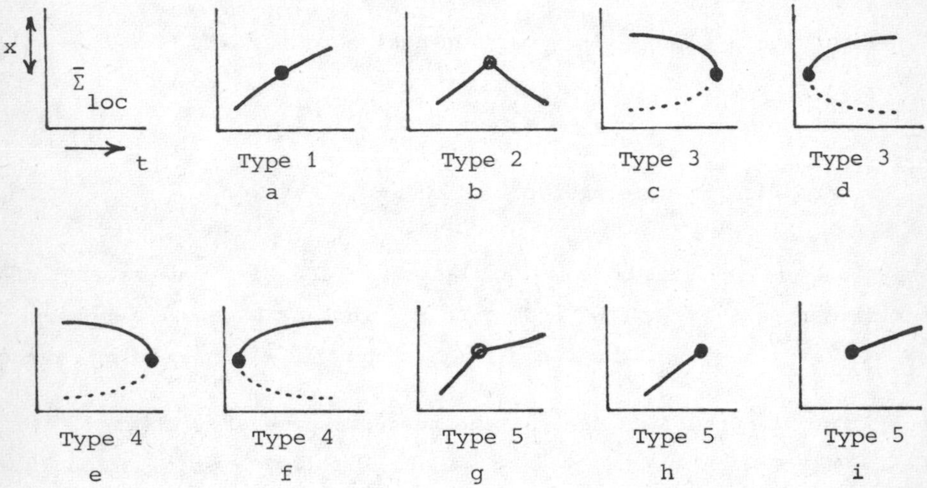

Fig. 1.

Now, we turn to the problem of finding local minima $x(t)$ for $\mathcal{P}(t)$, $t \in [0,1]$; or, more precisely, the problem of obtaining a fine discretization $0 = t_o < t_1 < \cdots < t_{i-1} < t_i < \cdots < t_N$ of the interval $[0,1]$ and for each t_i a local minimum $x(t_i)$ for $\mathcal{P}(t_i)$. The information on $x(t_{i-1})$ will be used to obtain $x(t_i)$.

To this aim we assume that $x(0)$ belongs to Σ_{loc}; moreover, we assume that there exists a <u>compact</u> set $K \subset \mathbb{R}^n$ which contains $M(t)$ for all $t \in [0,1]$.

In Section 2 we refer to pathfollowing methods for the situation of Fig. 1.a,b,g. In Section 3 we describe how to jump to another branch of $\bar{\Sigma}_{loc}$ at a point of Type 3, Fig. 1.c. Finally, in Section 4 we explain the situation of Fig. 1.e,h, and we indicate a possible jump. Since we will walk along branches of $\bar{\Sigma}_{loc}$ for increasing values of t, the situations of Fig. 1.d,f,i will not occur.

2. The pathfollowing part

Let us consider Fig. 1 again. As long as we are in the situation of Fig. 1.a,b,g, we walk on a branch of local minima which can be computed by means of pathfollowing methods. We restrict ourselves to a few clarifying remarks. For further reading we refer to the survey paper of Guddat [5] and the forthcoming book [3].

In the situation of Fig. 1.a,b, the linear independence constraint qualification is satisfied, whereas at the breakpoint in Fig. 1.g only the (weaker) Mangasarian-Fromovitz constraint qualification (shortly MFCQ) holds.

We recall that MFCQ is fulfilled at $\bar{x} \in M(\bar{t})$ if both the set $\{D_x h_i,\ i \in I\}_{|\bar{z}-(\bar{x},\bar{t})}$ is linearly independent and if there exists a vector $\zeta \in \mathbb{R}^n$ solving the system

$$D_x h_i(\bar{z})\zeta = 0,\ i \in I \qquad , \ Dg_j(\bar{z})\zeta > 0,\ j \in J_o(\bar{z}).$$

In case that MFCQ holds, it is well known that a local minimum $\bar{x}$ for $\mathcal{P}(\bar{t})$ satisfies the Karush-Kuhn-Tucker conditions, i.e. there exist λ_i, $i \in I$ and μ_j, $j \in J$ solving at $\bar{z} = (\bar{x},\bar{t})$ the following system (arguments being omitted):

$$\left.\begin{array}{ll} D_x f = \sum\limits_{i\in I} \lambda_i D_x h_i + \sum\limits_{j\in J} \mu_j D_x g_j & \\ h_i = 0,\ i \in I & \mu_j g_j = 0,\ j \in J \\ \mu_j \geq 0,\ j \in J & g_j \geq 0,\ j \in J \end{array}\right\} \qquad (2.1)$$

The system (2.1) defines a piecewise differentiable curve in (x,λ,μ,t)-space which can be computed by means of a pathfollowing method.
At the breakpoint in Fig. 1.b, for increasing t, either a new inequality constraint becomes active, or an active inequality constraint becomes inactive. At the breakpoint in Fig. 1.g, the number of active constraints becomes exactly n+1, but apart from the breakpoint the number of active

constraints equals n.

3. The jump at a point of Type 3

At a point of Type 3 the linear independence constraint qualification as well as the strict complementarity is satisfied. The only degeneracy lies in the fact that the restricted Hessian $D_x^2L(z)/T(z)$ becomes singular. So, we can still compute along the quadratic turning point (Fig. 1.c) by means of a pathfollowing method for solving system (2.1) in (x,λ,μ,t)-space. But, as we pass the turning point, a local minimum switches into a saddle point since (exactly) one eigenvalue of $D^2L(z)/T(z)$ changes from positive to negative sign; see also [11, Chapter 10] for a detailed discussion.

Let $\bar{z} = (\bar{x},\bar{t})$ denote the turning point. From the foregoing remarks we conclude that $\bar{x}$ cannot be a global minimum for $\mathcal{P}(\bar{t})$. Since the feasible set $M(\bar{t})$ was assumed to be compact, the existence of a global minimum $\tilde{x}$ for $\mathcal{P}(\bar{t})$ is assured. So, if we use a descent method starting at $\bar{x}$, we can arrive at a local minimum $\hat{x}$, and the point $(\hat{x},\bar{t})$ lies on another branch of $\bar{\Sigma}_{loc}$. Starting at $(\hat{x},\bar{t})$, a pathfollowing procedure can again be exploited. The actual jump consists in the transition from $(\bar{x},\bar{t})$ to $(\hat{x},\bar{t})$.

The problem now consists in finding - in an effective way - a tangential direction of descent.

Our proposal is based on the following observation; compare Fig. 2. Let t be near $\bar{t}$, $t < \bar{t}$, and let $x_m(t)$ and $x_s(t)$ be the local minimum and saddle point, respectively. Then, as t tends to $\bar{t}$, the vector $v(t) := (x_s(t) - x_m(t))/||x_s(t) - x_m(t)||$ tends to a tangential vector, say $\bar{v}$, which is a direction of descent. Hence, for t near $\bar{t}$, $t < \bar{t}$, the vector $x_s(t)-x_m(t)$ provides an approximatively tangential direction of descent.

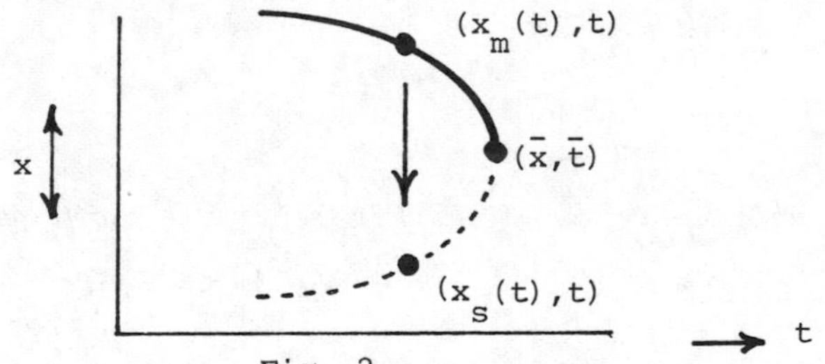

Fig. 2.

The above proposal has the following formulation. In view of the validity of the linear independence constraint qualification and the strict complementarity, we can restrict ourselves (in new local C^3-coordinates) to the unconstrained case. Using the characteristics of our point of Type 3 (cf. [9]), we then deal with a C^3-function $f(x,t)$ enjoying the following properties at $\bar{z} = (\bar{x},\bar{t})$:

(i) $D_x f = 0$.

(ii) $D_x^2 f$ is positive semi-definite with exactly one vanishing eigenvalue.
Let $D_x^2 f \cdot \zeta = 0$, $\zeta \neq 0$.

(iii) $D_x^3 f(\zeta,\zeta,\zeta) \neq 0$ and $D_t(D_x f \cdot \zeta) \neq 0$.

From a Taylor's expansion we see that either ζ or $-\zeta$ is a direction of <u>cubic descent</u>.

A further clarification can be given with the use of singularity theory. In fact, let us suppose that f is a C^∞-function. Then, the preceding conditions (i) – (iii) and unfolding theory (cf. [1]) provide the existence of a local C^∞-coordinate change $(y,u) = \Phi(x,t)$, having the structure

$$\Phi(x,t) = (\psi(x,t), \eta(t)) \quad , \quad \eta'(\bar{t}) > 0, \tag{3.1}$$

such that $\Phi(\bar{x},\bar{t}) = 0$ and

$$f\circ\Phi^{-1}(y,u) = y_1^3 + uy_1 + \sum_{i=2}^{n} y_i^2 + \delta(u), \tag{3.2}$$

where $\delta(u)$ represents the functional value at $y = 0$.

From (3.2) we see that $D_y f\circ\Phi^{-1}$ vanishes iff both $y_i = 0$, $i = 2,\dots,n$, and $3y_1^2 + u = 0$ (defining a parabola). For u varying from negative to positive values we have depicted the level lines of $f\circ\Phi^{-1}(\cdot,u)$ in Fig. 3. This clarifies the direction $\bar{v}$ in the preceding proposal.

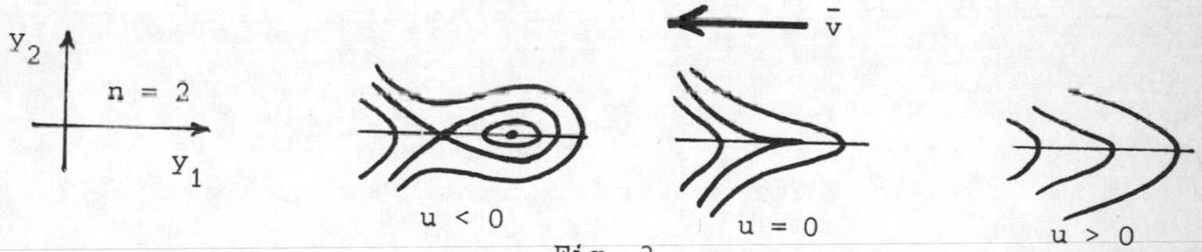

Fig. 3

4. The jump at a point of Type 4

Let $\bar{z} = (\bar{x},\bar{t})$ be a point of Type 4. Then, at $\bar{x} \in M(t)$, the Mangasarian-Fromovitz constraint qualification is not satisfied. Moreover, if we approach a point of Type 4 as t increases (Fig. 1 .e), then the corresponding Lagrange parameter vector (λ,μ) tends to infinity. This can be identified by a pathfollowing method.

Now, at $\bar{z}$ the set Σ consisting of generalized critical points has a quadratic turning point; moreover, when passing $\bar{z}$ along Σ, the local minimum switches into a local maximum (this can be derived using the index relations given in [9]). For a further insight into a point of Type 4 it suffices to consider the case in which only one (in)equality constraint is present (cf. [8], [9]). So, let us consider one inequality constraint $g(x,t) \geq 0$, and recall that we are interested in the situation of Fig. 1.e. For simplicity assume that g is a C^∞-function. Then, a local C^∞-coordinate transformation of the form (3.1), sending $(\bar{x},\bar{t})$ onto the origin, can be constructed such that g in these new coordinates takes the following form (cf. [11, Chapter 10])

$$g(x,t) = -\sum_{i=1}^{k} x_i^2 + \sum_{j=k+1}^{n} x_j^2 + \delta t, \tag{4.1}$$

where $\delta \in \{+1, -1\}$.

We have to distinguish two cases, namely $\delta > 0$ and $\delta < 0$.

Case I : $\delta > 0$. Recall that we approach the point $\bar{z}$ along a branch consisting of local minima. Then, from the index relations in [9, Fig. 4] we learn that the number of positive squares for δg, g as in (4.1), equals 1 or n (so k = 0 or k = n-1). The partial derivatives $\frac{\partial f}{\partial x_i}$ at the origin should satisfy the inequality

$$-\sum_{i=1}^{k} \left(\frac{\partial f}{\partial x_i}\right)^2 + \sum_{j=k+1}^{n} \left(\frac{\partial f}{\partial x_j}\right)^2 > 0$$

(compare the characteristic number α in [9]). This gives rise to two situations, depicted in Fig. 4; the feasible set M(t) is shaded.

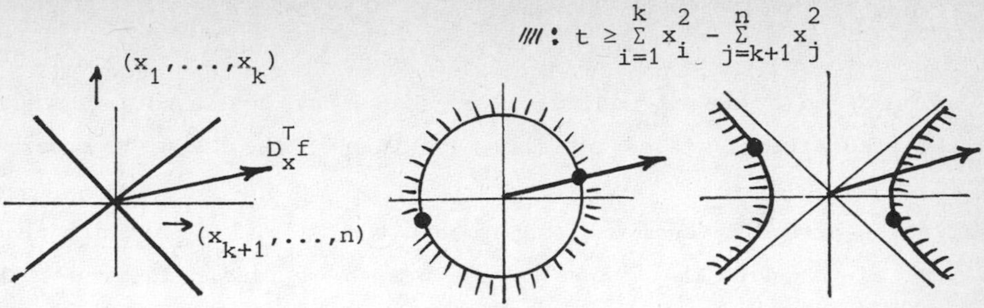

$t < 0,\ k = 0$ a $\qquad$ $t < 0,\ k = n - 1$ b

Fig. 4.

From Fig. 4 we see that only the situation of Fig. 4.b gives rise to the occurrement of a local minimum. Moreover, note that the corresponding local maximum has a functional value of f which is less than the value at the local minimum.

Case II : $\delta < 0$. Now, the number of positive squares for δg equals k and so, k must be equal to 1 or n. The partial derivatives $\frac{\partial f}{\partial x_i}$ at the origin should satisfy the inequality

$$\sum_{i=1}^{k} \left(\frac{\partial f}{\partial x_i}\right)^2 - \sum_{j=k+1}^{n} \left(\frac{\partial f}{\partial x_j}\right)^2 > 0$$ (compare again the number α in [9]).

This gives rise to two situations, depicted in Fig. 5.

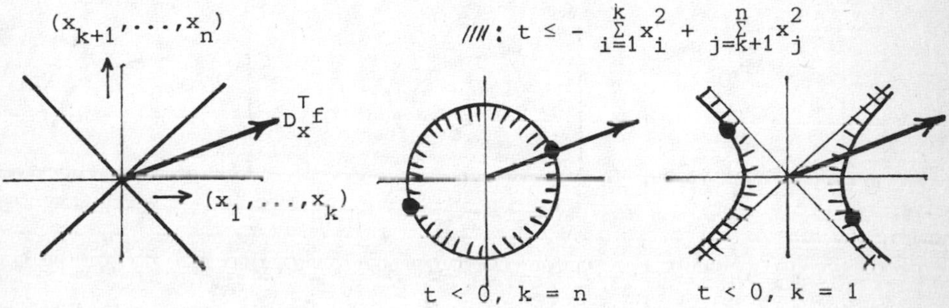

$t < 0,\ k = n$ a $\qquad$ $t < 0,\ k = 1$ b

Fig. 5.

From Fig. 5 we see that only the situation of Fig. 5.a gives rise to the occurrement of a local minimum. Now, however, the corresponding local maximum has a functional value of f which is greater than the value at the local minimum.

We emphasize that the situation of one equality constraint (rather than inequality constraint) is essentially contained in the above consideration.

Based on the preceding analysis, we propose the following partial procedure (cf. Fig. 6). Suppose that we walk on a branch of local minima and that we approach a point of Type 4 as t increases. If the functional value of f meanwhile decreases, then start a descent procedure at a point $\tilde{x}$ with $(\tilde{x},\tilde{t}) \in \Sigma$ on the other side of the turning point (Fig. 6.a). Then, it is guaranteed that a new local minimum for $f(\cdot,\tilde{t})_{|M(\tilde{t})}$ is found, say x^*, and $(x^*,\tilde{t})$ lies on another branch of $\bar{\Sigma}_{loc}$. Starting at $(x^*,\tilde{t})$, a pathfollowing procedure can again be exploited. In order to get $\tilde{x}$, one has to compute around the turning point of Σ (the set of generalized critical points). This can be done by using the system

$$\left.\begin{aligned} &\lambda D_x f = \sum_{i\in I}\lambda_i d_x h_i + \sum_{j\in J}\mu_j D_x g_j \\ &h_i = 0,\ i \in I \quad \mu_j g_j = 0,\ j \in J \\ &\lambda^2 + \sum_{i\in I}\lambda_i^2 + \sum_{i\in J}\mu_j^2 = 1 \end{aligned}\right\} \tag{4.2}$$

instead of (2.1). Formula (4.2) refers directly to the definition of a generalized critical point. We mention that the system (4.2) is also basic in the study of Poore and Tiahrt [12].

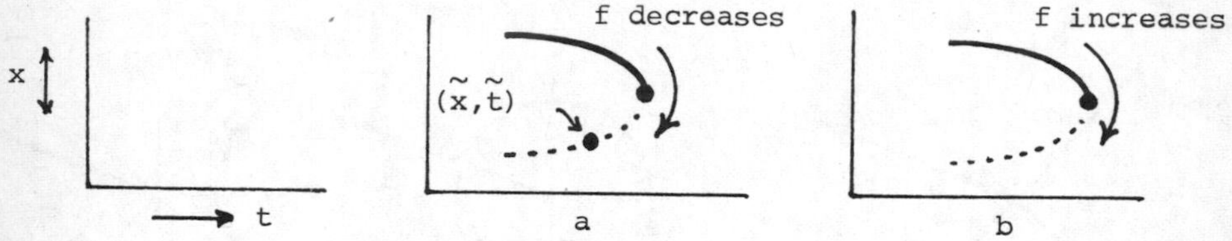

Fig. 6.

However, if we are in the situation of Fig. 6.b, i.e. f increases, then the corresponding component of the feasible set M(t) shrinks to a point, and becomes empty for increasing t. In that case we don't have a proposal which works, up to now. A similar problem appears if we reach a point of Type 5 as in the situation of Fig. 1.h. Also in that case the corresponding component of the feasible set shrinks to a point and then disappears. As an example, in dimension two, take $f(x) = x_1 + x_2$, $g_1(x) = x_1 \geq 0$, $g_2(x) = x_2 \geq 0$ and $g_3(x,t) = -x_1 - x_2 - t \geq 0$.

References

[1] Bröcker, Th.: Differentiable germs and catastrophes; translated by L. Lander. London Math. Soc. Lecture Notes 17, Cambridge Univ. Press (1975).

[2] Gfrerer, H., Guddat, J., Wacker, Hj., Zulehner, W.: Path-following methods for Kuhn-Tucker curves by an active index set strategy. In: Systems and Optimization (A. Bagchi, H.Th. Jongen, eds.), Lecture Notes in Control and Information Sciences, Vol. 66, Springer Verlag, Berlin-Heidelberg-New York (1985), 111-132.

[3] Gfrerer, H., Guddat, J., Jongen, H.Th., Wacker, Hj., Zulehner, W.: Parametric optimization and continuation methods (forthcoming).

[4] Guddat, J., Guerra Vasquez, F., Tammer, K., Wendler, K.: Multiobjective and stochastic optimization based on parametric optimization. Akademie Verlag, Berlin (1985).

[5] Guddat, J.: Parametric optimization: pivoting and predictor-corrector continuation, a survey. In [6].

[6] Guddat, J., Jongen, H.Th., Kummer, B., Nožička, F. (eds.): Parametric optimization and related topics. Akademie-Verlag Berlin (to appear).

[7] Hirsch, M.W.: Differential topology, Springer Verlag (1976).

[8] Jongen, H.Th., Jonker, P., Twilt, F.: One-parameter families of optimization problems: equality constraints. J. Optimization Theory and Appl., Vol. 48 (1986), 141-161.

[9] Jongen, H.Th., Jonker, P., Twilt, F.: Critical sets in parametric optimization. Mathematical Programming 34 (1986), 333-353.

[10] Jongen, H.Th., Jonker, P., Twilt, F.: Nonlinear optimization in $\mathbb{R}^n$, I. Morse theory, Chebyshev approximation. Peter Lang Verlag, Frankfurt a. M., Bern, New York (1983).

[11] Jongen, H.Th. Jonker, P., Twilt, F.: Nonlinear optimization in $\mathbb{R}^n$, II. Transversality, flows, parametric aspects. Peter Lang Verlag, Frankfurt a.M., Bern, New York (1986).

[12] Poore, A.B., Tiahrt, C.A.: Bifurcation problems in nonlinear parametric programming. Preprint, Colorado State University (1986).

[13] Tammer, K.: The application of parametric optimization and imbedding for the foundation and realization of a generalized primal decomposition approach. In [6].

[14] Wierzbicki, A.P.: On the completeness and constructiveness of parametric characterizations to vector optimization problems, OR-Spektrum 8 (1986), 73-87.

OPTIMIZATION PROBLEMS IN THE ROBUSTNESS ANALYSIS OF LINEAR STATE SPACE SYSTEMS

Diederich Hinrichsen
Institut für Dynamische Systeme

Matthias Motscha
Regionales Rechenzentrum

Universität Bremen
FRG

ABSTRACT

In this paper we report on some recent results concerning the distance of a stable matrix A from the set of unstable matrices. Related optimization and optimal control problems are discussed in detail and new algorithms are presented for their solution.

§1 INTRODUCTION

Methods of approximation and optimization play an important role in the theory of dynamical systems. Conversely, the theory of dynamical systems is an interesting field of application for both, approximation and optimization theory. This is evidenced by such areas as *optimal control* which has interacted strongly with optimization theory in Banach spaces, or *model reduction* where e.g. the results of *Adamjan, Arov and Krein* [1] on the approximation of Hankel operators have recently made a strong impact, leading to new algorithms for approximating high order systems by systems of lower dimension, see [15].
In this paper we will illustrate the importance of approximation and optimization problems in systems theory by another area which is not so well known but recently has attracted a lot of attention, the area of *robust control*.

Since no mathematical model is an exact representation of the real process whose dynamics it describes, controllers have to work in the presence of plant perturbations or model uncertainties. Roughly speaking, a controller is robust if it achieves the required performance criteria (e.g. stability) not only for the nominal plant model but also for a large set of perturbed models.

Most of the recent work on robust control problems is based on transform methods (frequency response techniques), see [5], [30], [19]. In contrast with this work, we will discuss a *state space approach* to robustness analysis.

Consider a time-invariant linear system

$$\dot{x} = Ax \qquad (1.1)$$

where $A \in K^{n\times n}$ $(K = \mathbb{R}$ or $\mathbb{C})$ is asymptotically stable, i.e. has spectrum $\sigma(A) \subset \mathbb{C}_- = \{s \in \mathbb{C};\ \mathrm{Re}\ s < 0\}$. For short, these matrices will be called *stable* in this paper. Suppose that the dimension n of the system is precisely known but that - as a first step - *all* the entries of A are subject to uncertainty. This means that the true system behaves like

$$\dot{x} = (A+D)x \tag{1.2}$$

where D is any $n\times n$ disturbance matrix. Surprisingly, the theory of differential equations provides little quantitative information as to which upper bound on $||D||$ will guarantee stability of the perturbed system. In recent years, however, several bounds of this sort have been derived in the control theoretic literature, see [20], [29],[22]. The question arises which of these bounds are tight. To answer this question one has to know the smallest norm of a destabilizing perturbation $D \in K^{n\times n}$ of A, i.e. the distance of A from the set of unstable systems

$$\mathcal{U}_n(K) = \{U \in K^{n\times n};\ \sigma(U) \cap \overline{\mathbb{C}}_+ \neq \emptyset\} \tag{1.3}$$

where $\overline{\mathbb{C}}_+ = \{s \in \mathbb{C};\ \mathrm{Re}\ s \geq 0\}$. This distance may be regarded as a plausible *measure of robustness* of stability of (1.1). Since $\mathcal{U}_n(K)$ is a complicated semi-algebraic set (described via the Routh-Hurwitz conditions, see [8,ch.XV]), the determination of

$$\mathrm{dist}(A,\mathcal{U}_n(K)) = \min\{||A-U||;\ U \in \mathcal{U}_n(K)\} \tag{1.4}$$

is a complicated global non-convex minimization problem. *Van Loan* [24] analyzed this problem from a computational point of view, with respect to the Frobenius norm on $K^{n\times n}$. Independently, *Hinrichsen and Pritchard* [10],[11] studied the same problem for the operator norm (spectral norm) on $K^{n\times n}$. They introduced the concept of stability radius and extended it to structured perturbations where only a part of the entries of the system matrix A is perturbed. Recently, this approach to robustness analysis was extended to a class of infinite-dimensional linear systems (described by semi-groups of operators on a Hilbert space) and to time-varying linear systems, see [21] and [9]. The theory is still very much in its beginning and there is a host of open problems.

This paper has two objectives. First, it gives a survey of recent robustness results in state space analysis [10],[11],[24]. On the other hand, it presents some new material, including a characterization of the distance from instability for an extensive class of norms, new algorithms for the computation of stability radii and a detailed case study. Section 2 to 4 deal with unstructured perturbations. In section 2 basic definitions and results are extended to a comprehensive set of norms on $K^{n\times n}$ which contains all the matrix norms of practical use in system theory. The relationship between stability radii and eigenvalues is investigated and it is shown that the distance of the spectrum $\sigma(A)$ from the imaginary axis which is traditionally taken as an indicator of robustness by control engineers may give a false impression about how robust the system is.

In section 3 a new minimization algorithm for computing the unstructured stability radius is described. Section 4 contains a detailed study of the real and the complex stability radius in the two-dimensional case. In section 5 the structured stability radius is introduced and related to a parametrized linear quadratic optimal control problem. Finally, an associated non-standard algebraic Riccati equation is considered (in section 6) and the results obtained are used for designing an algorithm for determining the structured stability radius.

§2 STABILITY RADIUS FOR UNSTRUCTURED PERTURBATIONS

Let $K = \mathbb{R}$ or $K = \mathbb{C}$. To provide a unifying framework for the various results recently obtained in the literature, we consider an arbitrary norm $\|\cdot\|_{K^n}$ on K^n and an arbitrary norm $\|\circ\|$ on $L(K^n) \simeq K^{n\times n}$. This norm may be different from the operator norm $\|\circ\|_{L(K^n)}$ induced by $\|\cdot\|_{K^n}$. Our aim is to determine, for any stable matrix $A \in K^{n\times n} \setminus \mathcal{U}_n(K)$ the (unstructured) *stability radius*

$$r_K(A) = \min\{\|A-U\| \; ; \; U \in \mathcal{U}_n(K)\} . \tag{2.1}$$

Clearly, $r_K(A)$ depends upon the specific norm $\|\cdot\|$. However, all these norms induce the same topology on $K^{n\times n}$. Henceforth $K^{n\times n}$ is supposed to be provided with this topology. Since the spectrum $\sigma(A)$ depends continuously upon the matrix A, $\mathcal{U}_n(K)$ is closed in $K^{n\times n}$ with boundary

$$\partial\mathcal{U}_n(K) = \{U \in K^{n\times n} \; ; \; \operatorname{Re}\sigma(U) \le 0 \text{ and } \sigma(U) \cap i\mathbb{R} \neq \emptyset\} . \tag{2.2}$$

Therefore the "min" in (2.1) is justified and there exists a destabilizing perturbation $D \in K^{n\times n}$ of minimal norm such that

$$\sigma(A+D) \cap i\mathbb{R} \neq \emptyset \quad \text{and} \quad \|D\| = r_K(A) . \tag{2.3}$$

Since $\mathcal{U}_n(K)$ is a closed (non convex) cone, the following properties of the stability radius are obvious:

$$\left.\begin{aligned} & r_K(A) = 0 \iff A \in \mathcal{U}_n(K) \\ & r_K(\alpha A) = \alpha\, r_K(A) \quad \text{for all } A \in K^{n\times n},\ \alpha \ge 0 \\ & A \to r_K(A) \text{ is continuous on } K^{n\times n} . \end{aligned}\right\} \tag{2.4}$$

To obtain explicit formulas for the stability radius, the norm $\|\cdot\|$ on $K^{n\times n}$ has to be related to $\|\cdot\|_{K^n}$.

Definition 2.1

Let $\|\cdot\|_{K^n}$ be a norm on K^n. A norm $\|\cdot\|$ on $K^{n\times n}$ is said to be *strongly compatible* with $\|\cdot\|_{K^n}$ if the following two conditions are satisfied

(C 1) $\quad \|Ax\|_{K^n} \le \|A\|\cdot\|x\|_{K^n}$ for all $A \in K^{n\times n}$, $x \in K^n$

(C 2) For any pair of vectors $x,y \in K^n$, $x \neq 0$ there exists $T \in K^{n\times n}$ satisfying

$$Tx = y \quad \text{and} \quad ||T||\cdot||x||_{K^n} = ||y||_{K^n}.$$

Lemma 2.2

For any norm $||\cdot||_{K^n}$ on K^n, the associated operator norm $|||\cdot|||_{L(K^n)}$ on $L(K^n) \simeq K^{n\times n}$ is strongly compatible with $||\cdot||_{K^n}$.

Proof:

Condition (C 1) holds by definition. To prove (C 2), let $x \in K^n, ||x||_{K^n} = 1$ and $y \in K^n$. By the Hahn-Banach Theorem there exists a linear form $f: K^n \to K$ such that

$$f(\lambda x) = \lambda ||x||_{K^n}, \ \lambda \in K \quad \text{and} \quad |f(z)| \leq ||z||_{K^n}, \ z \in K^n$$

Then $T \in K^{n\times n}$ defined by $Tz = f(z)y$, $z \in K^n$ satisfies (C 2).

□

Example 2.3

For $p \in [1,\infty]$ let $||\cdot||_p$ be the ℓ_p-norm on $\mathbb{C}^n$:

$$||x||_p = (\sum_{k=1}^{n} |x_k|^p)^{1/p}, \ p < \infty; \ ||x||_\infty = \lim_{p\to\infty} ||x||_p = \max_k |x_k|, \quad x \in \mathbb{C}^n \tag{2.5}$$

The associated operator norm on $\mathbb{C}^{n\times n}$ will also be denoted by $||\cdot||_p$. It is well known that for any $A = (a_{ij}) \in \mathbb{C}^{n\times n}$

$$||A||_1 = \max_k \sum_{i=1}^{n} |a_{ik}|, \ ||A||_\infty = \max_i \sum_{k=1}^{n} |a_{ik}|, \ ||A||_2 = [\lambda_1(A^*A)]^{1/2} \tag{2.6}$$

Here A^* is the Hermitian conjugate of A and $\lambda_1(A^*A) \geq \ldots \geq \lambda_n(A^*A)$ denote the eigenvalues of the Hermitian matrix A^*A. By definition, the singular values $s_1(A) \geq \ldots \geq s_n(A) \geq 0$ are the square roots of these eigenvalues. If A is nonsingular

$$s_1(A) = ||A||_2 \quad \text{and} \quad s_n(A) = ||A^{-1}||_2^{-1} \tag{2.7}$$

By Lemma 2.2, each matrix norm $|||\cdot|||_p$ is strongly compatible with the associated vector norm $||\cdot||_p$.

□

Besides the operator norms, the Hölder norms are of special interest in numerical linear algebra.

Example 2.4 (Hölder norms)

For any pair $p,r \in [1,\infty]$ the Hölder norm $||\cdot||_{p,r}$ on $\mathbb{C}^{n\times n}$ is defined by

$$||A||_{p,r} = [\sum_k(\sum_i |a_{ik}|^p)^{r/p}]^{1/r}, \quad A = (a_{ik}) \in \mathbb{C}^{n\times n} \tag{2.8}$$

if $p,r < \infty$, otherwise by taking the limit as $p \to \infty$ or $r \to \infty$, see (2.5). So

$$||A||_{p,r} = ||[||A_1||_p,\dots,||A_n||_p]||_r$$

where $A_1,\dots,A_n$ denote the column vectors of A. Now let $p,q \in [1,\infty]$, $1/p + 1/q = 1$ then (see [18])

$$||Ax||_p \le ||A||_{p,q}||x||_p. \tag{2.9}$$

Thus $||\cdot||_{p,q}$ is compatible with the vector norm $||\cdot||_p$. To prove (C 2), let $x,y \in \mathbb{C}^n$, $||x||_p = 1$. There exists $z \in \mathbb{C}^n$ such that $z^*x = ||z||_q = 1$ and $|z^*v| \le ||z||_q||v||_p$ for all $v \in \mathbb{C}^n$. Hence $T = yz^*$ satisfies $Tx = y$ and

$$||T||_{p,q} = ||[||z_1y||_p,\dots,||z_ny||_p]||_q = ||y||_p.$$

Therefore $||\cdot||_{p,q}$ is strongly compatible with $||\cdot||_p$. In particular, the Frobenius norm $||\cdot||_F = ||\cdot||_{2,2}$ is strongly compatible with the Euclidian norm $||\cdot||_2$. □

It is well-known that the smallest singular value of a matrix $A \in K^{n\times n}$ measures the distance of A from singularity with respect to the operator norm $||\cdot||_2$:

$$s_n(A) = \min\{||A-S||_2;\ S \in K^{n\times n},\ \det S = 0\}.$$

The following lemma extends this result to strongly compatible norms on $K^{n\times n}$.

<u>Lemma 2.5</u>

Let $||\cdot||_{K^n}$ be any norm on K^n and $||\cdot||$ any strongly compatible matrix norm on $K^{n\times n}$. Then the distance of $A \in Gl_n(K)$ from the set of singular matrices with respect to the norm $||\cdot||$ is given by

$$\begin{aligned} &\min\{||Q||;\ Q \in K^{n\times n},\ \det(A-Q) = 0\} \\ &= \min\{||Ax||_{K^n};\ x \in K^n,\ ||x||_{K^n} = 1\} = \frac{1}{||A^{-1}||_{L(K^n)}}. \end{aligned} \tag{2.10}$$

<u>Proof:</u>

$\det(A-Q) = 0$ if and only if there exists $x \in K^n$ such that $Ax = Qx$, $||x|| = 1$. Hence Q is a solution of the minimization problem (2.10) if and only if it is a solution of

$$\min_{x\in K^n,\,||x||_{K^n} = 1} \min\{||Q||;\ Q \in K^{n\times n},\ (A-Q)x = 0\}$$

Now (2.10) follows by means of (C 1) and (C2). □

Before we apply this lemma to our initial problem, note that there are two stability radii associated with every stable matrix $A \in \mathbb{R}^{n\times n}$, the *real stability radius* $r_{\mathbb{R}}(A)$ and the *complex stability radius* $r_{\mathbb{C}}(A)$. Since

$\mathcal{U}_n(\mathbb{C}) \supset \mathcal{U}_n(\mathbb{R}) \supset \{S \in \mathbb{R}^{n\times n};\ \det S = 0\}$ the following inequalities hold, by Lemma 2.5, for every stable matrix $A \in \mathbb{R}^{n\times n}$:

$$0 \le r_{\mathbb{C}}(A) \le r_{\mathbb{R}}(A) \le 1/\|A^{-1}\|_{L(\mathbb{R}^n)}. \tag{2.11}$$

These estimates are valid with respect to any norm on $\mathbb{C}^{n\times n}$ which is strongly compatible with $\|\cdot\|_{\mathbb{C}^n}$.

For the *complex* stability radius we obtain the following characterization.

Proposition 2.6

Let $\|\cdot\|_{\mathbb{C}^n}$ be any norm on $\mathbb{C}^n$ and $\|\cdot\|$ a strongly compatible matrix norm on $\mathbb{C}^{n\times n}$. Then the stability radius of a stable matrix $A \in \mathbb{C}^{n\times n}$ with respect to the norm $\|\cdot\|$ is

$$r_{\mathbb{C}}(A) = \min_{\omega\in\mathbb{R}} \min_{\substack{x\in\mathbb{C}^n \\ \|x\|=1}} \|i\omega x - Ax\|_{\mathbb{C}^n} = \min_{\omega\in\mathbb{R}} \frac{1}{\|(i\omega I_n - A)^{-1}\|_{L(\mathbb{C}^n)}}.$$

Proof:

$D \in \mathbb{C}^{n\times n}$ satisfies (2.3) if and only if $\sigma(A+D)\cap i\mathbb{R} \neq \emptyset$ and

$$\min_{\omega\in\mathbb{R}} \min\{\|Q\|;\ Q \in \mathbb{C}^{n\times n},\ \det(i\omega I - A - Q) = 0\} = \|D\|. \tag{2.12}$$

Hence the proposition follows by Lemma 2.5.

□

As a consequence of the previous proposition we conclude that for any given norm $\|\cdot\|_{\mathbb{C}^n}$ on $\mathbb{C}^n$ the complex stability radius $r_{\mathbb{C}}(A)$ is the same for all strongly compatible norms $\|\cdot\|$ on $\mathbb{C}^{n\times n}$. It follows easily from our case study in section 4 that an analogous result does *not* hold for the real stability radius.

For the Euclidean norm $\|\cdot\|_2$ on $\mathbb{C}^n$ one obtains the following characterization which reduces the minimization problem (2.1) to a one-dimensional minimization problem. This problem can be solved by a reliable and efficient algorithm (section 3).

Corollary 2.7

Suppose the norm $\|\cdot\|$ on $\mathbb{C}^{n\times n}$ is strongly compatible with $\|\cdot\|_2$. Then

$$r_{\mathbb{C}}(A) = \min_{\omega\in\mathbb{R}} s_n(i\omega I_n - A) \tag{2.13}$$

for any stable matrix $A \in \mathbb{C}^{n\times n}$.

A formula of similar simplicity is still missing for the real stability radius. The "simplest" characterization is available for the Frobenius norm and has been derived by *Van Loan* [24]. We state his result without proof.

Proposition 2.8 (*Van Loan*)

Suppose $A \in \mathbb{R}^{n\times n}$ is stable and $\mathbb{R}^{n\times n}$ is provided with the Frobenius norm. If

$r_{\mathbb{R}}(A) < s_n(A)$ then

$$r_{\mathbb{R}}(A) = \min_{\substack{x,y\in\mathbb{R}^n \\ ||x||_2^2=||y||_2^2=1 \\ \langle x,y\rangle=0,(x^TAy)(y^TAx)\le 0}} ||Ax||_2^2 + ||Ay||_2^2 - (x^TAy)^2 - (y^TAx)^2 \tag{2.14}$$

(2.14) is a difficult constrained minimization problem. To our knowledge, there does not yet exist a reliable algorithm for its solution. If $\mathbb{R}^{n\times n}$ is provided with the operator norm $||\cdot||_2$ the available formulas are even more complicated, see [10]. Therefore it is of interest to have easily verifiable criteria that guarantee $r_{\mathbb{R}}(A) = r_{\mathbb{C}}(A)$.

<u>Proposition 2.9</u>

Let $\mathbb{C}^{n\times n}$ be provided with the operator norm $||\cdot||_2$. If $A \in \mathbb{R}^{n\times n}$ is stable and normal then

$$r_{\mathbb{C}}(A) = r_{\mathbb{R}}(A) = \operatorname{dist}(\sigma(A), i\,\mathbb{R}). \tag{2.15}$$

<u>Proof:</u>

Let $\alpha = \operatorname{dist}(\sigma(A), iR)$. Then $\sigma(A-\alpha I)\cap i\mathbb{R} \neq \emptyset$ and so $r_{\mathbb{R}}(A) \le \alpha$. By (2.11) it suffices to show $r_C(A) = \alpha$. There exists a unitary matrix $Q \in \mathbb{C}^{n\times n}$ such that

$$Q^*AQ = \operatorname{diag}(\lambda_1,\ldots,\lambda_n) =: \Lambda.$$

For each $\omega \in \mathbb{R}$, $s_n(i\omega I-A) = s_n(i\omega I-\Lambda) = \min_j |i\omega-\lambda_j|$. Hence, Corollary 2.7 and Lemma 2.2 imply $r_{\mathbb{C}}(A) = \min_{\omega\in\mathbb{R}} s_n(i\omega I-A) = \alpha$.

□

Proposition 2.9 shows that for *normal* stable matrices $A \in \mathbb{R}^{n\times n}$ the distance of A from instability is measured by the distance of its spectrum from the imaginary axis. However, in case A is *not* normal $\operatorname{dist}(\sigma(A), i\mathbb{R})$ can be a very misleading indicator of robustness. This is illustrated by means of the following

<u>Example 2.10</u>

Let

$$A_k = -\begin{bmatrix} k & k^3 \\ 0 & k \end{bmatrix}, \quad D_k = \begin{bmatrix} 0 & 0 \\ -k^{-1} & 0 \end{bmatrix}.$$

Then $\det(A_k+D_k) = 0$. Hence $r_{\mathbb{R}}(A_k) \le k^{-1} \to 0$ whereas $\operatorname{dist}(\sigma(A_k), i\,\mathbb{R}) \to \infty$ if $k \to \infty$.

□

For arbitrary n, the distance of almost every similarity orbit in $\mathbb{C}^{n\times n}$ from $\mathcal{U}_n(C)$ is zero. More precisely,

<u>Proposition 2.11</u>

Let $||\cdot||_{\mathbb{C}^n}$ be a norm on $\mathbb{C}^n$ and $||\cdot||$ a strongly compatible norm on $\mathbb{C}^{n\times n}$. If $A \in \mathbb{C}^{n\times n}$ is stable with at least two different eigenvalues then

$$\sup_{T\in Gl_n(\mathbb{C})} r_{\mathbb{C}}(T^{-1}AT) = \operatorname{dist}(\sigma(A), i\mathbb{R}), \quad \inf_{T\in Gl_n(\mathbb{C})} r_{\mathbb{C}}(T^{-1}AT) = 0 \tag{2.16}$$

Proof:

By Prop. 2.6 we may assume $||\cdot|| = ||\cdot||_{L(\mathbb{C}^n)}$. Let $\alpha = \operatorname{dist}(\sigma(A), i\mathbb{R})$. Since $\sigma(A+\alpha I) \cap i\mathbb{R} \neq \emptyset$, we have $r_{\mathbb{C}}(A) \leq \alpha$. There exists a sequence of non-singular matrices $T_k \in Gl_n(\mathbb{C})$ such that $T_k^{-1}AT_k$ converges to a diagonal matrix Λ with $\sigma(\Lambda) = \sigma(A)$. By continuity of $r_{\mathbb{C}}(\cdot)$, the first equality of (2.16) follows. The second one is a consequence of Prop. 2.6 since

$$\inf_{T\in Gl_n(\mathbb{C})} r_{\mathbb{C}}(T^{-1}AT) \leq \frac{1}{\sup_{T\in Gl_n(\mathbb{C})} ||T^{-1}A^{-1}T||} = 0 \quad \text{if} \quad \operatorname{card}\sigma(A) \geq 2.$$

□

§ 3 A MINIMIZATION ALGORITHM FOR COMPUTING THE UNSTRUCTURED STABILITY RADIUS

Let $\mathbb{C}^n$ be provided with the usual Euclidean norm and $A \in \mathbb{C}^{n\times n}$ any stable $n\times n$ matrix. By Corollary 2.7, the stability radius of A with respect to any strongly compatible matrix norm is given by

$$r_{\mathbb{C}}(A) = \min_{\omega\in\mathbb{R}} s_n(i\omega I - A). \tag{3.1}$$

From a numerical point of view, this formula is attractive since there are efficient and reliable algorithms for computing the singular values of a matrix, see [4],[26]. To determine $r_{\mathbb{C}}(A)$, a one-dimensional global nonconvex minimization problem has to be solved. Applying a one dimensional minimizer to the function $\underline{s}(\omega) = s_n(i\omega I-A)$ will in general only yield a local minimum (cf. the algorithm described in [24, p.469/470]). To obtain a global minimizer we first have to study some properties of the function $\underline{s}(\cdot)$ or, equivalently,

$$\Lambda_{min}: \mathbb{R} \to \mathbb{R}_+, \quad \omega \mapsto \lambda_{min}(S(\omega))$$

where

$$S(\omega) = (i\omega I-A)^*(i\omega I-A) = \omega^2 I + \omega(iA-iA^*) + A^*A, \quad \omega \in \mathbb{R}.$$

$\lambda_{min}(S(\omega))$ denotes the minimal eigenvalue of $S(\omega)$. For arbitrary $\omega \in \mathbb{C}$ the matrix $S(\omega) := \omega^2 I + \omega(iA-iA^*) + A^*A$ depends holomorphically on ω. Since I, $iA-iA^*$ and A^*A are Hermitian, $S(\cdot)$ is Hermitian in the sense that $S(\omega)^* = S(\omega)$ for all $\omega \in \mathbb{R}$. Thus we may apply the well developed perturbation theory of Hermitian operator functions to $S(\cdot)$. The following theorem summarizes some properties which are direct consequences of results in [13,p.71,p.80-81,p.120]. The derivative of $S(\omega)$ is denoted by

$$T(\omega) = (iA-A^*) + 2\omega I.$$

Theorem 3.1

The eigenvalues of the family of Hermitian matrices

$$S: \mathbb{R} \to \mathbb{C}^{n\times n},\ \omega \mapsto \omega^2 I_n + \omega(iA-iA^*) + A^*A$$

can be written in the form $\lambda_1(\omega),\ldots,\lambda_r(\omega)$, $r = \text{const} \le n$ such that:

(i) $\lambda_k: \omega \to \lambda_k(\omega)$ is analytic on $\mathbb{R}$ for $k \in \underline{r} = \{1,\ldots,r\}$.

(ii) For each $i \in \underline{r}$, $E_i = \{\omega \in \mathbb{R};\ \lambda_j(\omega) = \lambda_i(\omega) \text{ for some } j \neq i\}$ is a discrete subset of $\mathbb{R}$.

(iii) If $(v_{i1},\ldots,v_{im_i})$ are orthonormal bases of the eigenspaces $V_i(\omega) = \ker(\lambda_i(\omega)I - S(\omega))$, $\omega \in \mathbb{R}\setminus E_i$, $i \in \underline{r}$ then for any $\omega \in \mathbb{R}\setminus(\bigcup_i E_i)$

(a) $\lambda_k'(\omega) = \frac{1}{m_k} \sum_{i=1}^{m_k} v_{ki}^* \, T(\omega)\, v_{ki}$

(b) $\lambda_k''(\omega) = 2 - \frac{2}{m_k} \sum_{\substack{j=1\\ j\neq k}}^{r} \sum_{i=1}^{m_k} \sum_{q=1}^{m_j} (\lambda_j(\omega)-\lambda_k(\omega))^{-1} |v_{ki}^* T(\omega) v_{jq}|^2$.

The following corollary has important consequences for the numerics of the minimization problem (3.1).

Corollary 3.2

(i) The function

$$\Lambda_{min}: \mathbb{R} \to \mathbb{R}_+,\ \omega \mapsto \lambda_{min}(S(\omega))$$

is continuous and piecewise analytic.

(ii) If Λ_{min} has a local minimum at ω_o then it is differentiable at ω_o and one of the branches $\lambda_i(\omega)$ has a local minimum at ω_o.

(iii) Every extremum ω_o of one of the branches λ_i (in particular, the local minima of Λ_{min}) belong to the interval $I_A = [-\lambda_{max}(\frac{iA-iA^*}{2}),\ -\lambda_{min}(\frac{iA-iA^*}{2})]$

(iv) If $\omega \in \mathbb{R} \setminus \bigcup_i E_i$ then $\Lambda_{min}''(\omega) \le 2$.

Proof:

(i),(ii) and (iv) are direct consequences of the previous proposition. To prove (iii), let $k \in \underline{r}$ arbitrary and $\omega \in \mathbb{R} \setminus \bigcup_i E_i$. By Theorem 3.1,(iiia)

$$\lambda_k'(\omega) = \frac{1}{m_k} \sum_{i=1}^{m_k} v_{ki}^* \, T(\omega) v_{ki} .$$

Since for any $z \in \mathbb{C}^n$, $||z||_2 = 1$

$$\lambda_{min}(iA-iA^*) + 2\omega \le z^*T(\omega)z \le \lambda_{max}(iA-iA^*) + 2\omega$$

we conclude that

$$\lambda_{min}(iA-iA^*) + 2\omega \le \lambda_k'(\omega) \le \lambda_{max}(iA-iA^*) + 2\omega .$$

Hence $\lambda_k'(\omega_o) = 0$ implies $\omega_o \in I_A$. □

Let us briefly discuss some aspects of these results. (iii) shows that the minimization problem (3.1) can be restricted to the finite interval

$$I_A = [-\lambda_{max}(\frac{iA-iA^*}{2}), -\lambda_{min}(\frac{iA-iA^*}{2})].$$

This choice of the initial interval is in general much better than the choice of $[-2||A||, 2||A||]$ which was suggested in [24, Lemma 2.1]. In particular, if A is Hermitian, then $I_A = \{0\}$ and so the minimization problem (3.1) is solved:

$$r_{\mathbb{C}}(A) = s_n(A) \quad \text{if} \quad A = A^*.$$

In view of (ii) it is of some interest for our minimization problem to know an upper bound to the number of all extrema of the branches $\lambda_i(\cdot)$ in the basic interval I_A. It can be shown (see [17]) that this number is always smaller than $2n(n-1) + n(n+1)/2$. For every n, one can construct stable matrices A for which Λ_{min} has at least n local minima.

The bound on $\Lambda''_{min}(\omega)$ specified in (iv) is basic for our algorithm. It allows to derive a lower bound for $\underline{s}(\cdot)$ on $[\alpha,\beta]$ from the values $\underline{s}(\alpha)$, $\underline{s}(\beta)$ in the extremal points α,β.

<u>Proposition 3.3</u>

If $I = [\alpha,\beta]$ then a lower bound of $\underline{s}|I$ is given by

$$\nu(I) = \begin{cases} \min\{\underline{s}(\alpha),\underline{s}(\beta)\} & \text{if } |\underline{s}^2(\alpha)-\underline{s}^2(\beta)| \geq (\beta-\alpha)^2 \\ \max\{0, \left[\underline{s}^2(\alpha) - \dfrac{[\underline{s}^2(\beta)-\underline{s}^2(\alpha)-(\beta-\alpha)^2]^2}{4(\beta-\alpha)^2}\right]^{1/2}\} & \text{otherwise}. \end{cases}$$

<u>Proof:</u> follows from elementary calculations. □

In particular, the above proposition yields an easily verifiable test whether, in the following minimization algorithm, a subinterval I can be neglected or not.
In each iteration of this algorithm new subintervals are generated via bisection and the minimum μ of $\underline{s}(\cdot)$ on the extrema of these subintervals is determined. Those subintervals I in which a global minimum could exist, i.e. $\nu(I) < \mu$, are called *relevant* and are collected in the set V. The algorithm proceeds by the following steps.

<u>Algorithm 3.4</u>

1. Input: matrix A, tolerance $\varepsilon > 0$
2. Compute $\alpha = -\lambda_{max}(\frac{iA-iA^*}{2})$, $\beta = -\lambda_{min}(\frac{iA-iA^*}{2})$

   and set $V = \{[\alpha,\beta]\}$, $\mu = \min\{\underline{s}(\alpha),\underline{s}(\beta)\}$
3. Set $W = \{[\alpha,\frac{\alpha+\beta}{2}],[\frac{\alpha+\beta}{2},\beta];\ [\alpha,\beta] \in V\}$
4. Determine $\mu = \underline{s}(\omega^*) = \min\{\underline{s}(\alpha),\underline{s}(\beta);[\alpha,\beta] \in W\}$
5. Set $V = \{[\alpha,\beta] \in W;\ \nu([\alpha,\beta]) < \mu\}$

6. If $V \neq \emptyset$ and $\mu - \min_{I \in V} \nu(I) \geq \varepsilon$ go to 3.

7. Stop: $\mu = \underline{s}(\omega^*)$ is an estimate of $r_{\mathbb{C}}(A)$.

It can be shown [17] that the order of convergence of ω^* to a global minimum of $\underline{s}(\cdot)$ (of second importance) is linear whereas the order of convergence of μ to $r_{\mathbb{C}}(A)$ (of prime importance) is quadratic. Since there exist reliable and numerically stable procedures for the computation of singular values (cf. [26],[4]) the algorithm can be implemented with high reliability and accuracy. An error analysis can be found in [17].

The following example illustrates the algorithm and the previous result.

Example 3.5

Let

$$A = \begin{bmatrix} 0 & 0.9 & 0 & 0 & 0 & 0 \\ -0.9 & -0.9 & 0 & 0 & 0 & 0 \\ 0 & 0 & 0 & -0.6 & 0 & 0 \\ 0 & 0 & 3.0 & -3.0 & 0 & 0 \\ 0 & 0 & 0 & 0 & 0 & 1.8 \\ 0 & 0 & 0 & 0 & -8.1 & -0.9 \end{bmatrix}$$

Then

$$\lambda_{min}(iA-iA^*) = -9.9, \quad \lambda_{max}(iA-iA^*) = 9.9$$

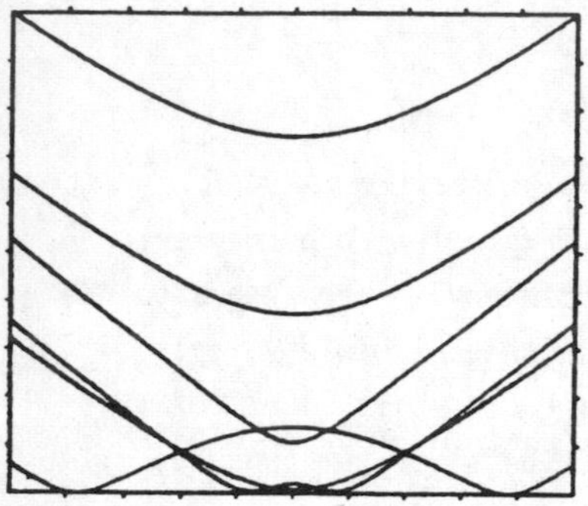

Fig. 1 Graph of $\omega \to s_k(i\omega I-A)$, $k=1,\ldots,6$ in $[-4.95,4.95]\times[0.34,10.84]$

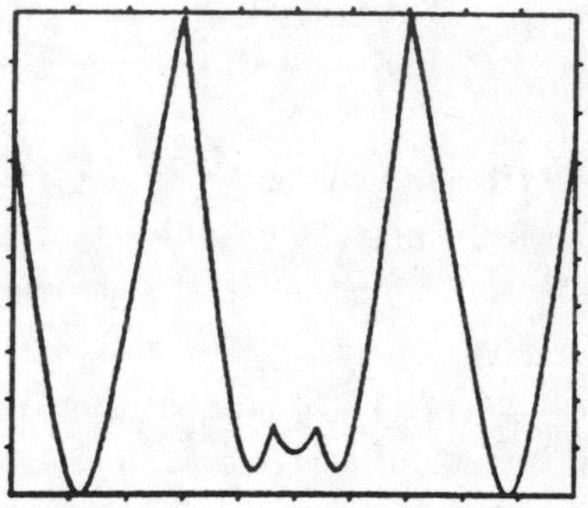

Fig. 2 Graph of $\omega \to \underline{s}(\omega)$ in $[-4.95,4.95]\times[0.34,1.22]$

Figure 3 shows the union of the relevant subintervals at each of the first nine iterations of the algorithm .

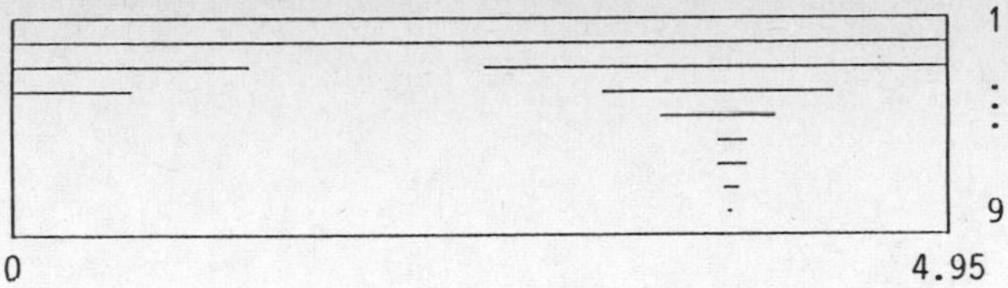

Fig. 3 Convergence of the algorithm

Since A is real the function $\underline{s}(\cdot)$ is even and so the algorithm may start with the initial interval $[0,4.95]$. After 9 iterations the computed value of $r_{\mathbb{C}}(A)$ coincides with the exact value 0.3447061 up to the sixth digit.

□

§ 4 CASE STUDY: STABILITY RADII FOR 2-DIMENSIONAL SYSTEMS

The 2-dimensional case has many special features which greatly simplify the analysis. It certainly does not offer a representative picture of the complications which may arise in higher dimensions. On the other hand it allows to derive explicit formulas and yields a first testing ground for conjectures concerning the real or complex stability radius and their relationship.

In this section we provide K^2 with the Euclidean norm $\|\cdot\|_2$ and consider only the corresponding operator norm on $K^{2\times 2}$. In contrast with the general case, the real stability radius is more easily determined than the complex one, if $n = 2$.

Proposition 4.1

Let $A \in \mathbb{R}^{2\times 2}$ be stable. Then

$$r_{\mathbb{R}}(A) = \min\{|\mathrm{tr}A|/2,\ s_2(A)\}. \tag{4.1}$$

Proof:
$D \in \mathbb{R}^{2\times 2}$ is a destabilizing perturbation of minimal norm if and only if either $\det(A+D) = 0$ or $i\omega \in \sigma(A+D)$ for some $\omega \neq 0$, i.e. $\mathrm{tr}(A+D) = 0$. In the first case $r_{\mathbb{R}}(A) = s_2(A)$ whereas in the second case $r_{\mathbb{R}}(A) = |\mathrm{tr}A|/2$. □

A formula of similar simplicity apparently does not exist for the complex stability radius.

We will now derive formulae for the real and complex stability radii from the singular value decomposition $A = USV^T$ of A where $U = [u_1,u_2]$, $V = [v_1,v_2]$ are orthogonal and $S = \mathrm{diag}(s_1(A),s_2(A))$. We denote by $\kappa(A) = s_1(A)/s_2(A)$ the condition number of A and by $\theta = \theta(u_1,v_1)$ the angle between u_1 and v_1.

Proposition 4.2

If $A \in \mathbb{R}^{2\times 2}$ is stable then

$$r_{\mathbb{R}}(A) = \begin{cases} s_2(A) & \text{if } \kappa(A) \geq \frac{2}{|\cos\theta|} - 1 \\ (s_1(A)+s_2(A))\frac{|\cos\theta|}{2} & \text{otherwise} \end{cases} \tag{4.2}$$

and

$$r_{\mathbb{C}}(A) = \begin{cases} s_2(A) & \text{if } \kappa(A) \geq \frac{2}{\cos^2\theta} - 1 \\ \{s_1(A)s_2(A)-[(s_1(A)+s_2(A))\frac{\cos\theta}{2}]^2\}^{1/2}|\cot\theta| & \text{otherwise} \end{cases} \tag{4.3}$$

Proof:

Let $K = \mathbb{R}$ or $\mathbb{C}$. Then

$$r_K(A) = r_K(SV^TU) = r_K(\begin{bmatrix} s_1(A)\cos\theta & -s_1(A)\sin\theta \\ s_2(A)\sin\theta & s_2(A)\cos\theta \end{bmatrix}) \tag{4.4}$$

For $K = \mathbb{R}$, formula (4.2) follows from (4.4) by application of Proposition 4.1. Formula (4.3) follows by Corollary 2.7 via a lengthy but elementary calculation. □

The following corollary specifies conditions under which equality holds in (2.11).

Corollary 4.3

If $A \in \mathbb{R}^{2\times 2}$ is stable then

(i) $r_{\mathbb{R}}(A) = s_2(A) \Leftrightarrow \kappa(A) \geq \frac{2}{|\cos\theta|} - 1$

(ii) $r_{\mathbb{C}}(A) = s_2(A) \Leftrightarrow \kappa(A) \geq \frac{2}{\cos^2\theta} - 1$

(iii) $r_{\mathbb{R}}(A) = r_{\mathbb{C}}(A) \Leftrightarrow \kappa(A) = 1$ or $r_{\mathbb{C}}(A) = s_2(A)$.

If $s_2(A)$ is normalized at 1 then $r_{\mathbb{R}}(A), r_{\mathbb{C}}(A)$ are functions of $\sigma(A)$.

Corollary 4.4

If A is stable with eigenvalues $-\alpha \pm i\beta$, $\alpha,\beta > 0$ and $s_2(A) = 1$ then

$$r_{\mathbb{R}}(A) = \begin{cases} 1 & \text{if } \alpha \geq 1 \\ \alpha & \text{if } 0 < \alpha < 1 \end{cases} \tag{4.5}$$

$$r_{\mathbb{C}}(A) = \begin{cases} 1 & \text{if } \beta^2 \leq \alpha^2 - 1 \\ \frac{2\alpha\beta}{[(\alpha^2+\beta^2+1)^2-4\alpha^2]^{1/2}} & \text{if } \beta^2 > \alpha^2 - 1 \end{cases} \tag{4.6}$$

Figures 4 and 5 illustrate the dependences of $r_{\mathbb{C}}$ and $r_{\mathbb{R}}$ on the real and imaginary parts of the eigenvalues of the matrix A satisfying $s_2(A) = 1$ and $\sigma(A) \cap \mathbb{R} = \emptyset$. (Note that $\alpha^2+\beta^2 \geq 1$ as a consequence of the first property.)

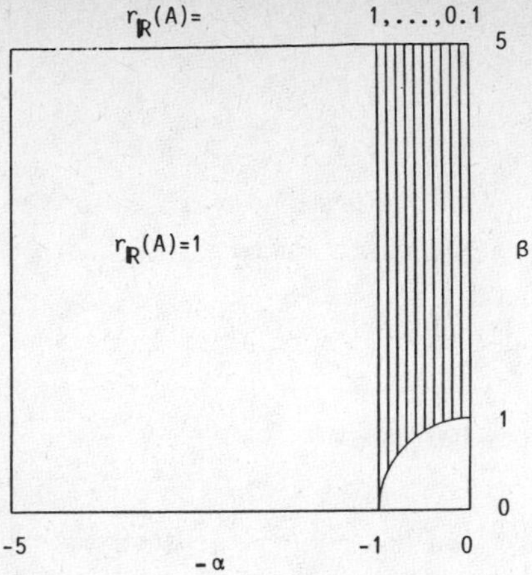

Fig. 4 level lines $r_{\mathbb{R}} = \ell$
$\ell = 0.1,0.2,\dots,1.0$

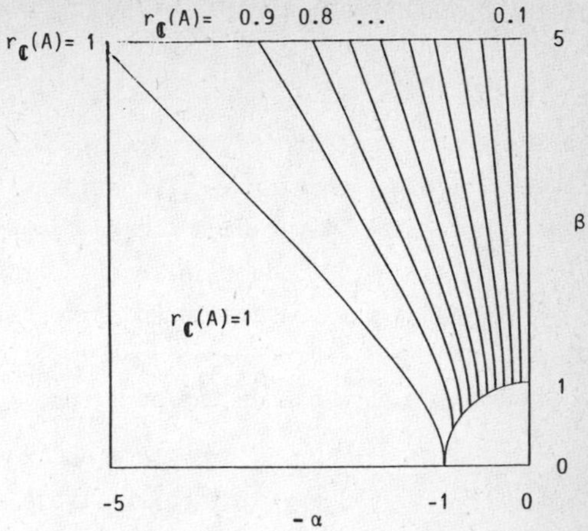

Fig. 5 level lines $r_{\mathbb{C}} = \ell$
$\ell = 0.1,0.2,\dots,1.0$

An interesting question is to determine the range of variation of the quotient $r_{\mathbb{C}}(A)/r_{\mathbb{R}}(A) \le 1$.

Proposition 4.5

(i) For every $q \in (0,1]$ there exists a stable matrix $A \in \mathbb{R}^{2\times 2}$ such that $r_{\mathbb{C}}(A)/r_{\mathbb{R}}(A) = q$.

(ii) Let (A_k) be a sequence of stable matrices $A_k \in \mathbb{R}^{2\times 2}$ with eigenvalues $-\alpha_k \pm i\beta_k$, $k \in \mathbb{N}$. Then

$$\lim_{k\to\infty} \frac{r_{\mathbb{C}}(A_k)}{r_{\mathbb{R}}(A_k)} = 0 \iff \lim_{k\to\infty} \frac{\beta_k}{\alpha_k} = \infty \text{ and } \lim_{k\to\infty} \frac{\beta_k}{s_2(A_k)} = \infty.$$

Instead of a proof we illustrate the dependence of the quotient $q(A) = r_{\mathbb{C}}(A)/r_{\mathbb{R}}(A)$ on the eigenvalues for stable matrices $A \in \mathbb{R}^{2\times 2}$ with $s_2(A) = 1$.

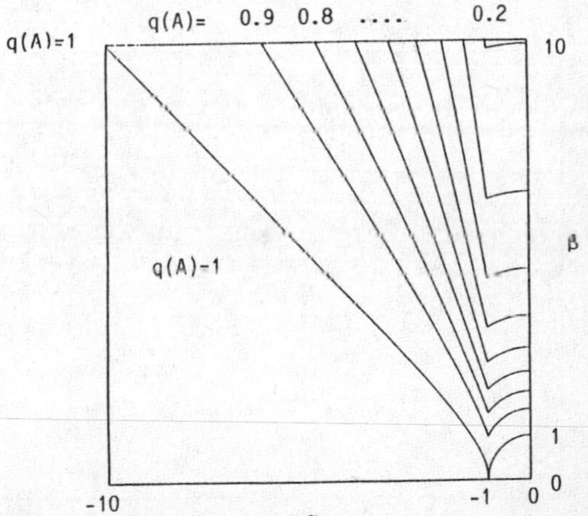

Fig. 6 level lines of $q(A)$

§ 5 STRUCTURED STABILITY RADIUS AND OPTIMAL CONTROL

In many applications the uncertainty about the system matrix A is *structured* in the sense that some of its entries are exactly known (e.g. fixed zeros or ones) whereas the other entries are (functions of) unknown parameters. In these cases the unstructured stability radius $r_K(A)$ is no longer an adequate measure of robustness. Following [11] we introduce in this section the concept of *structured stability radius* for a special class of structured perturbations and discuss its relation to a linear quadratic optimal control problem.

Throughout, we assume that the perturbed system equation has the form

$$\dot{x} = (A + BDC)x \tag{5.1}$$

where $A \in K^{n\times n}$ is asymptotically stable, $D \in K^{m\times p}$ an unknown disturbance matrix and $B \in K^{n\times m}$, $C \in K^{p\times n}$ are known matrices describing the structure of the perturbations.

Example 5.1

The linear oscillator $\ddot{\xi} + a_1\dot{\xi} + a_2\xi = 0$ is represented by the state space model

$$\dot{x} = Ax = \begin{bmatrix} 0 & 1 \\ -a_2 & -a_1 \end{bmatrix} x .$$

Here the first row of A does not contain uncertain parameters. If both physical parameters a_1, a_2 are of equal uncertainty this can be modelled by choosing $B = \begin{bmatrix} 0 \\ -1 \end{bmatrix}$ $C = I_2$, whereas if only a_2 is uncertain one has to choose B as before and $C = [1\ 0]$.

□

The *structured stability radius* of the stable matrix A (with respect to perturbations of the given structure (B,C)) is defined by

$$r_K = r_K(A;B,C) = \inf\{\|D\|;\ \sigma(A+BDC) \cap \overline{\mathbb{C}}_+ \neq \emptyset\}. \tag{5.2}$$

Note that choosing $m = p = n$ and $B = C = I_n$ in (5.2), one obtains the unstructured stability radius defined in (2.1).

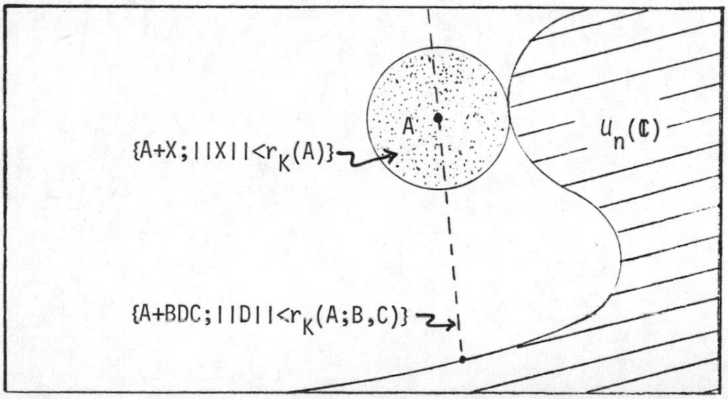

Fig. 7 Structured and unstructured stability radius

To simplify the exposition we will only consider the Euclidian norms on $\mathbb{C}^p$, $\mathbb{C}^m$ and the corresponding operator norm on $L(\mathbb{C}^p,\mathbb{C}^m)$. Moreover we will restrict our analysis to the complex case $K = \mathbb{C}$.

It follows directly from the above definition that the structured stability radius is invariant with respect to similarity transformations

$$r_{\mathbb{C}}(A;B,C) = r_{\mathbb{C}}(T^{-1}AT;\ T^{-1}B,\ CT),\quad T \in Gl_n(\mathbb{C}) \tag{5.3}$$

Note that not only the system matrix A but also the "scaling matrices" B and C are transformed in (5.3) whereas in Proposition 2.11 we considered the unstructured stability radius $r_{\mathbb{C}}(T^{-1}AT) = r_{\mathbb{C}}(T^{-1}AT;\ I_n, I_n)$ where the matrices $B = C = I_n$ have been fixed.

The following proposition extends the characterization given in Corollary 2.7 to the structured stability radius.

Proposition 5.2

Let $G(s) = C(sI_n-A)^{-1}B$. Then

$$r_{\mathbb{C}}(A;B,C) = [\max_{\omega\in\mathbb{R}} ||G(i\omega)||]^{-1} \tag{5.4}$$

(In particular, $r_{\mathbb{C}} = \infty$ if $G = 0$).

Proof: First suppose that for some $D \in \mathbb{C}^{m\times p}$, $x \in \mathbb{C}^n$, $x \neq 0$, $\omega \in \mathbb{R}$

$$(A + BDC)x = i\,\omega\,x\ .$$

Then $(i\omega I_n-A)x = BDCx$ and so $y = G(i\omega)Dy \neq 0$.
This implies $||G(i\omega)||\,||D|| \geq 1$ and hence the inequality $\geq$ in (5.4).
Conversely, suppose that the maximum of $\omega \mapsto ||G(i\omega)||$ occurs at ω_o. Let $s_1 = ||G(i\omega_o)||$ be the largest singular value of $G(i\omega_o)$ and (u_1,v_1) a pair of singular vectors corresponding to s_1. If we define $D = s_1^{-1}\ v_1u_1^*$ then $G(i\omega_o)Du_1 = u_1$. Thus $x = (i\omega_o I_n-A)^{-1}BDu_1$ satisfies $Cx = u_1 \neq 0$, hence $x \neq 0$ and $(i\omega_o I_n-A)x = BDCx$, i.e. $i\omega_o \in \sigma(A+BDC)$. Since $||D|| = [\max_{\omega\in\mathbb{R}}||G(i\omega)||]^{-1}$, this proves $\leq$ in (5.4).

□

The bounded linear operator $G\colon L^2(\mathbb{R},\mathbb{C}^m) \to L^2(\mathbb{R},\mathbb{C}^p)$ defined by multiplication with $G(i\cdot)$

$$(Gu)(\omega) = G(i\omega)u(\omega),\quad \omega \in \mathbb{R}$$

is similar - via the Fourier transform - to the convolution operator

$$\begin{aligned} &L\colon L^2(0,\infty;\mathbb{C}^m) \to L^2(0,\infty;\mathbb{C}^p)\\ &(Lu)(t) = \int_0^t Ce^{A(t-\tau)}Bu(\tau)d\tau,\quad t \geq 0 \end{aligned} \tag{5.5}$$

L is called the *perturbation operator* associated with the perturbed system (5.1).
It is not difficult to show that the norm of the multiplication operator $G\colon u(\cdot) \to G(i\cdot)u(\cdot)$ from $L^2(\mathbb{R};\mathbb{C}^m)$ into $L^2(\mathbb{R};\mathbb{C}^p)$ is $\max_{\omega\in\mathbb{R}}||G(i\omega)||$. Note that

this is actually the H^∞-norm of G. Thus (5.4) implies the following characterization of $r_{\mathbb{C}}$ in terms of L

$$r_{\mathbb{C}}(A;B,C) = \frac{1}{||L||} . \tag{5.6}$$

This formula enables us to establish an interesting relation between the structured stability radius $r_{\mathbb{C}}(A;B,C)$ and the following auxiliary linear quadratic *optimal control problem* parametrized by $\rho \in \mathbb{R}$:

$$(OCP_\rho) \qquad \text{Minimize} \quad J_\rho(x_0,u) = \int_0^\infty (||u(\tau)||^2 - \rho||y(\tau)||^2)d\tau$$

$$\text{where} \quad u \in L^2(0,\infty;\mathbb{C}^m)$$
$$\dot{x} = Ax + Bu, \quad x(0) = x_0$$
$$y = Cx$$

Note that this is not the usual linear quadratic regulator problem (cf.[14],[28]), if $\rho > 0$. In fact, if $\rho > 0$ then the deviation of the "output" $y(t)$ from zero is not penalized but awarded. So one will expect that optimal controls will keep $y(t)$ away from zero as much as possible (with a given control energy). Anticipating that the optimal control has feedback form, this means that the optimal feedback will tend to *destabilize* the system with increasing ρ. The conjecture will be fully confirmed in section 6.

The next proposition characterizes $r_{\mathbb{C}}$ by the minimal costs associated with the initial state $x_0 = 0$.

Proposition 5.3

$$\inf_{u \in L^2(0,\infty;\mathbb{C}^m)} J_\rho(0,u) = 0 \Leftrightarrow \rho \le r_{\mathbb{C}}^2(A;B,C) \tag{5.7}$$

Proof:

Since $y(t) = Cx(t) = C \int_0^t e^{A(t-\tau)}Bu(\tau)d\tau = (Lu)(t)$ (if the system is initialized at $x_0 = 0$) it follows that

$$J_\rho(0,u) = ||u||^2_{L^2(0,\infty;\mathbb{C}^m)} - \rho||Lu||^2_{L^2(0,\infty;\mathbb{C}^p)} .$$

Hence the equivalence (5.7) is a direct consequence of (5.6). □

If $\rho > r_{\mathbb{C}}^2(A;B,c)$ there exists, by the above proposition, $v(\cdot) \in L^2(0,\infty;\mathbb{C}^m)$ such that $J_\rho(0,v) < 0$ and so $\inf_{u \in L^2} J_\rho(0,u) = -\infty$ since $J_\rho(0,\alpha v) = \alpha^2 J_\rho(0,v)$ for $\alpha \in \mathbb{R}$.

The following proposition shows that, conversely, the minimal costs are finite for arbitrary initial states x_0, if $\rho < r_{\mathbb{C}}(A;B,C)$.

Proposition 5.4

Suppose $r_{\mathbb{C}}(A;B,C) < \infty$. Then for all $\rho \in (-\infty, r_{\mathbb{C}}^2)$ there exists a unique solution

$\hat{u}(\cdot)$ of the optimal control problem (OCP_ρ). Moreover there exists a constant $c_\rho > 0$ such that

$$\min_{u \in L^2(0,\infty;\mathbb{C}^m)} J_\rho(x_0,u) \geq -c_\rho ||x_0||^2 \tag{5.9}$$

Proof:

Let $x_0 \in \mathbb{C}^n$ and $y_0(t) = Ce^{At}x_0$, $t \geq 0$. Since for all $a,b \in L^2(0,\infty;\mathbb{C}^p)$ and $\alpha > 0$

$$2 \operatorname{Re} \langle a,b \rangle_{L^2} \leq \alpha ||a||^2_{L^2} + \alpha^{-1} ||b||^2_{L^2}$$

we have

$$J_\rho(x_0,u) = ||u||^2_{L^2} - \rho ||y_0 + Lu||^2_{L^2}$$

$$= ||u||^2_{L^2} - \rho ||Lu||^2_{L^2} - \rho ||y_0||^2_{L^2} - 2\rho \operatorname{Re} \langle Lu, y_0 \rangle_{L^2}$$

$$\geq ||u||^2_{L^2} - \rho(1+\alpha) ||Lu||^2_{L^2} - \rho(1+\alpha^{-1}) ||y_0||^2_{L^2} .$$

Hence for sufficiently small α and suitable $c_\rho > 0$

$$J_\rho(x_0,u) \geq -\rho(1+\alpha^{-1}) ||y_0||^2_{L^2} \geq -c_\rho ||x_0||^2, \quad u \in L^2.$$

On the other hand

$$J_\rho(x_0,u) \geq (1 - \rho(1+\alpha) ||L||^2) ||u||^2_{L^2} - c_\rho ||x_0||^2, \quad u \in L^2 . \tag{5.10}$$

Let (u_k) be a minimizing sequence for the optimal control problem (OCP_ρ). Then (u_k) is bounded in $L^2(0,\infty;\mathbb{C}^m)$ by (5.10). Hence we may extract a subsequence (u_{k_j}) which converges weakly to some $\hat{u} \in L^2(0,\infty,\mathbb{C}^m)$. It is easy to see that $J_\rho(x_0,\cdot)$ is strictly convex and coercive and so (cf. [6,p.36]) $\hat{u}(\cdot)$ is the unique optimal control. □

§ 6 THE STRUCTURED STABILITY RADIUS AND A PARAMETRIZED ALGEBRAIC RICCATI EQUATION

In this section we establish a relation between the structured stability radius $r_{\mathbb{C}}(A;B,C)$ and the parametrized algebraic Riccati equation

$$(ARE_\rho) \qquad A^*P + PA - \rho C^*C - PBB^*P = 0$$

We then use this relationship to design an algorithm for the computation of $r_{\mathbb{C}}(A;B,C)$. As in section 5 we suppose $\sigma(A) \subset \mathbb{C}_-$.

(ARE_ρ) is a quadratic matrix equation which does not possess Hermitian solutions P for all values of ρ. In fact $r^2_{\mathbb{C}}(A;B,C)$ can be characterized as the supremum of all the values of ρ for which (ARE_ρ) admits a Hermitian solution. This is a consequence of the following theorem which summarizes results from [11].

Theorem 6.1

(i) If $\rho < r_{\mathbb{C}}^2(A;B,C)$ then there exists a unique Hermitian solution P_ρ of (ARE_ρ) satisfying

$$\sigma(A-BB^*P_\rho) \subset \mathbb{C}_-.$$

The optimal solution of (OCP_ρ) is given by

$$\hat{u}_\rho(t) = -B^*P_\rho\hat{x}(t) \tag{6.1}$$

where $\hat{x}(\cdot)$ solves the differential equation $\dot{x} = (A-BB^*P_\rho)x$ with $x(0) = x_o$ and the minimal costs are

$$J(x_o,\hat{u}_\rho) = \langle x_o, P_\rho x_o\rangle.$$

(ii) If $\rho = r_{\mathbb{C}}^2(A;B,C)$ then there exists a unique Hermitian solution P_ρ such that $\sigma(A-BB^*P_\rho) \subset \bar{\mathbb{C}}_-$.

(iii) If $\rho > r_{\mathbb{C}}^2(A;B,C)$ then (ARE_ρ) has no Hermitian solution.

Proof:

(iii) is a consequence of Proposition 5.2. In fact; if $P = P^*$ is a solution of (ARE_ρ) then

$$(A-i\omega I)^*P + P(A-i\omega I) - \rho C^*C - PBB^*P = 0, \quad \omega \in \mathbb{R}$$

and so

$$0 \le (B^*P(A-i\omega I)^{-1}B-I)^*(B^*P(A-i\omega I)^{-1}B-I) = I - \rho G^*(i\omega)G(i\omega), \quad \omega \in \mathbb{R}.$$

Thus $\rho \le r_{\mathbb{C}}^2(A;B,C)$ by (5.4).

We will not derive the statements in (i), (ii) but only comment on the methods employed in their proof. If $\rho < 0$, the optimal control problem (OCP_ρ) (cf. section 5) is identical with the linear quadratic regulator problem [12] and (i) is a standard result of modern control theory. It is usually proved via a dynamic programming or Hamilton-Jacobi approach, cf. [12],[14]. Starting from a finite time version of (OCP_ρ) with final time $t_1 \to \infty$ and applying Proposition 5.4, it is possible - although by no means trivial - to extend the classical result to *positive* $\rho < r_{\mathbb{C}}^2(A;B,C)$. This has been done (for time-varying linear systems) in [9]. Alternatively, (i) and (ii) can be proved by extending a theorem of *Brockett* [2] to non-controllable systems (A,B,C), see [11]. □

The above theorem shows that for $\rho < r_{\mathbb{C}}^2$ the optimal control has feedback form (6.1). Applying this feedback control one obtains the *closed loop system*

$$\dot{x} = A_\rho x, \qquad A_\rho = A-BB^*P_\rho. \tag{6.3}$$

Since, for $\rho > 0$, the cost functional decreases if the output norm $||y(\cdot)||_{L^2}$ increases, one would expect that the optimal feedback law (6.1) tends to deteriorate the stability of A. This is confirmed by the following theorem which determines the distance of the closed loop system (6.3) from instability.

Theorem 6.2

Let $r_{\mathbb{C}} = r_{\mathbb{C}}(A;B,C)$. Then

(i) The maps $\rho \to P_\rho$, $\rho \to A_\rho$ are differentiable on $(-\infty, r_{\mathbb{C}}^2)$ and continous on $(-\infty, r_{\mathbb{C}}^2]$.

(ii) $\rho_1 \le \rho_2 \le r_{\mathbb{C}}^2 \Rightarrow P_{\rho_1} \ge P_{\rho_2}$.

(iii) If $\rho \le r_{\mathbb{C}}^2$ then

$$r_{\mathbb{C}}^2(A_\rho;B,C) = r_{\mathbb{C}}^2(A;B,C) - \rho . \tag{6.4}$$

We omit the proof which can be found in [11].

As a consequence of (6.4) one obtains

$$\lim_{\rho \to -\infty} r_{\mathbb{C}}(A_\rho;B,C) = \infty, \quad \lim_{\rho \to r_{\mathbb{C}}^2} r_{\mathbb{C}}(A_\rho;B,C) = 0 . \tag{6.5}$$

Of special interest is the behaviour of the eigenvalues of the closed loop system as $\rho \to -\infty$ and $\rho \to r_{\mathbb{C}}^2$. This behaviour is illustrated for a four-dimensional example in Fig. 8.

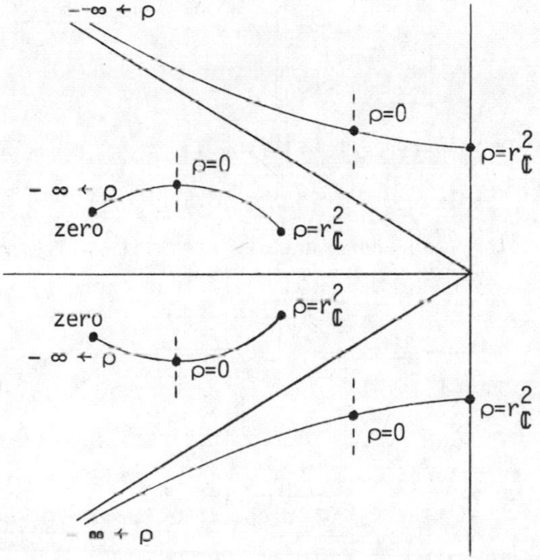

Fig. 8 Locus of $\sigma(A_\rho)$, $-\infty < \rho \le r_{\mathbb{C}}^2$

Let us first consider the case $\rho \to -\infty$. The optimal control $\hat{u}_\rho$ minimizes the cost functional

$$\frac{1}{\rho} J_\rho(x_0,u) = \int_0^\infty [\frac{1}{-\rho} ||u(t)||^2 + ||y(t)||^2]dt$$

in which the control costs $(-\rho)^{-1}||u||^2_{L^2}$ are less and less important as $\rho \to -\infty$. The behaviour of the closed loop spectrum $\sigma(A_\rho)$ for $\rho \to -\infty$ has been investigated as the so-called *cheap control problem*, see [7],[28]. In the scalar case $m = p = 1$ the behaviour of the spectrum $\sigma(A_\rho)$ is completely known. If the rational function $g(s) = C(sI-A)^{-1}B$ has k zeros then k of the eigenvalues of A_ρ tend to zeros of $g(s)$ whereas the remaining $n-k$ eigenvalues tend to ∞ along prescribed asymptotes as $\rho \to -\infty$ (cf.[7],[28]). Robustness analysis, conversely, deals with the behaviour of $\sigma(A_\rho)$ as $\rho \to r_{\mathbb{C}}^2$. Theorems 6.1 and 6.2 show that the locus of $\sigma(A_\rho)$ terminates when the first branch hits the imaginary axis, i.e. when $\rho = r_{\mathbb{C}}^2$, see Fig. 8.

We conclude the paper with an algorithm for the computation of $r_{\mathbb{C}}(A;B,C)$. The algorithm is based on Theorem 6.1 and makes use of a well studied relationship between algebraic Riccati equations and Hamiltonian matrices, see [16],[23],[27].
The following proposition is a consequence of Thm. 5.1 in [27].

Proposition 6.3
The parametrized algebraic Riccati equation (ARE_ρ) possesses a Hermitian solution P_ρ satisfying $\sigma(A-BB^*P_\rho) \subset \mathbb{C}_-$ if and only if $\sigma(H_\rho) \cap i\mathbb{R} = \emptyset$ where H_ρ is the Hamiltonian matrix associated with (ARE_ρ):

$$H_\rho = \begin{bmatrix} A & -BB^* \\ \rho C^*C & -A^* \end{bmatrix} \tag{6.6}$$

As a consequence of Theorem 6.1 and Proposition 6.3

$$\rho < r_{\mathbb{C}}^2(A;B,C) \Leftrightarrow \sigma(H_\rho) \cap i\mathbb{R} = \emptyset \tag{6.7}$$

Thus $r_{\mathbb{C}}(A;B,C)$ can be determined approximately by an iterative procedure which increases ρ in a suitable way when $\sigma(H_\rho) \cap i\mathbb{R} = \emptyset$ and decreases it when $\sigma(H_\rho) \cap i\mathbb{R} \neq \emptyset$.

Algorithm 6.4 (for computing $r_{\mathbb{C}}(A;B,C)$)

1. Input: $A,B,C,\rho,\rho_0^- = 0$, $\rho_0^+ = \infty,\delta,\varepsilon$ (ρ_0 is the initial value of ρ, e.g. $\rho_0 = s_n(A)$ in the unstructured case; ρ_k^- and ρ_k^+ are the lower resp. upper approximates of $r_{\mathbb{C}}^2$ at the k-th step, $k \geq 1$; $\delta > 0$ and $\varepsilon > 0$ are tolerances). Set $k = 1$
2. Compute $\sigma(H_{\rho_{k-1}})$
3. If $\text{dist}(\sigma(H_{\rho_{k-1}}),i\mathbb{R}) < \varepsilon$ then $\rho_k^+ = \rho_{k-1}$, $\rho_k^- = \rho_{k-1}^-$ and $\rho_k = \frac{1}{2}(\rho_{k-1} + \rho_{k-1}^-)$
4. If $\text{dist}(\sigma(H_{\rho_{k-1}}),i\mathbb{R}) \geq \varepsilon$ then $\rho_k^- = \rho_{k-1}$, $\rho_k^+ = \rho_{k-1}^+$ and
$$\rho_k = \begin{cases} 2\,\rho_{k-1} & \text{if } \rho_{k-1}^+ = \infty \\ \frac{1}{2}(\rho_{k-1} + \rho_{k-1}^+) & \text{otherwise} \end{cases}$$
5. If $|\rho_k - \rho_{k-1}| \geq \delta$, set $k = k+1$ and go to 2

6. Stop. $\sqrt{\rho_k}$ is an estimate of $r_{\mathbb{C}}(A;B,C)$.

To compute $\sigma(H_\rho)$ in step 2 one can use either an algorithm for computing eigenvalues of arbitrary matrices, e.g. a QR-procedure, or an algorithm which exploits the Hamiltonian structure of H_ρ, e.g. *Van Loan's symplectic method*, see [25]. The following computations were performed on a Siemens 7880 using an EISPACK QR routine in step 2, with double precision arithmetic.

Example 6.5

Let

$$A = \begin{bmatrix} -0.201 & 0.755 & 0.351 & -0.075 & 0.033 \\ -0.149 & -0.696 & -0.160 & 0.110 & -0.048 \\ 0.081 & 0.004 & -0.189 & -0.003 & 0.001 \\ -0.173 & 0.802 & 0.251 & -0.804 & 0.056 \\ 0.092 & -0.467 & -0.127 & 0.075 & -1.162 \end{bmatrix}$$

Using algorithm 6.4 with $\varepsilon = \delta = 10^{-8}$, $\rho_0 = \mathrm{dist}(\sigma(A), i\mathbb{R}) = 0.15811926$ one obtains $r_{\mathbb{C}}(A) = 0.11158202$. This value coincides to 7 digits with the value 0.11158200 obtained by application of algorithm 3.4.

The unstructured stability radius is considerably smaller than the structured stability radii $r_{\mathbb{C}}(A;B,C)$ where the structure defining matrices B and C are standard unit vectors. The following table contains the computed values of $r_{\mathbb{C}}(A;B,C)$ where $C = [1,0,0,0,0]$ is fixed and B is any one of the 5 standard unit vectors

B^T	$r_{\mathbb{C}}(A;B,C)$
(1,0,0,0,0)	0.31038547
(0,1,0,0,0)	0.26467890
(0,0,1,0,0)	0.32408486
(0,0,0,1,0)	5.00045978
(0,0,0,0,1)	19.12825531

□

The spectrum of a real Hamiltonian matrix is symmetric with respect to both the real and the imaginary axes. The following example illustrates how the spectrum $\sigma(H_\rho)$ typically changes with increasing ρ.

Example 6.6

$$A = \begin{bmatrix} 0 & 1 \\ -5 & -2 \end{bmatrix}, \quad B = C = I_2, \quad r_{\mathbb{C}}(A) = 2/3$$

Figure 9 shows the real parts of the eigenvalues of $\sigma(H_\rho)$ as functions of $\rho \in [0, 1.4]$ while Figure 10 presents the locus of the eigenvalues of H_ρ in the complex plane as ρ varies from 0 to 1.4.

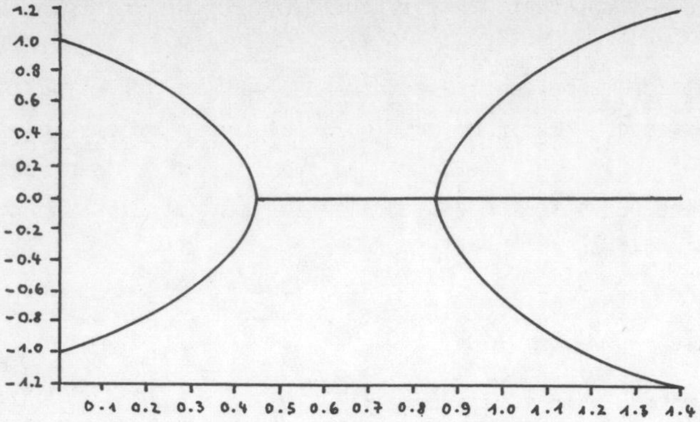

Fig. 9 The graph of $\rho \to \operatorname{Re} \sigma(H_\rho)$, $0 \le \rho \le 1.4$

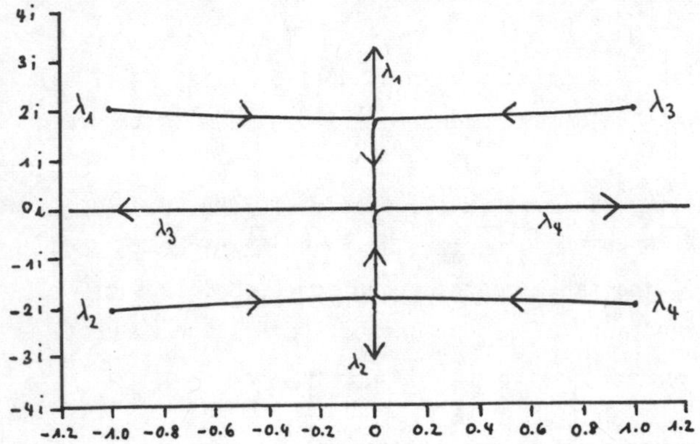

Fig. 10 Spectrum of the Hamiltonian H_ρ, $0 \le \rho \le 1.4$

Note that, as ρ increases, some eigenvalues of H_ρ remain only temporarily on the imaginary axis splitting away from the axis where two eigenvalues collide. But, in accordance with Propos. 6.3, there always is at least one pair of eigenvalues on $i\mathbb{R}$ for all $\rho \ge r^2_{\mathbb{C}}(A;B,C)$. □

ACKNOWLEDGEMENT

We would like to thank *A. Linnemann* for communicating to us the idea of algorithm 6.4 and *B. Kelb* for implementing the algorithm and computing Examples 6.5 and 6.6.

LITERATURE

[1] V.M. Adamjan, D.Z. Arov and M.G. Kreĭn, Analytic properties of Schmidt pairs for a Hankel operator and the generalized Schur-Takagi problem, Math. USSR Sbornik 15, 31-73,(1971)

[2] R.W. Brockett, Finite Dimensional Linear Systems, J. Wiley, New York (1970)

[3] W.A. Coppel, Matrix quadratic equations, Bull. Austral. Math. Soc. 10, 377-401, (1974)

[4] J.J. Dongarra, J.R. Bunch, C.B. Moler and G.W. Stewart, LINPACK User's Guide, SIAM Publications Philadelphia, PA. (1978)

[5] J.C. Doyle and G. Stein, Concepts for a classical/modern synthesis, IEEE Trans. Aut. Control 26, 4-16, (1981)

[6] I. Ekeland and R. Temam, Convex Analysis and Variational Problems, North-Holland, Amsterdam (1976)

[7] B.A. Francis, The optimal linear quadratic regulator with cheap control, IEEE Trans. Aut. Control AC-24, 616-621, (1979)

[8] E.R. Gantmacher, The Theory of Matrices, vol. 2, New York, (1959)

[9] D. Hinrichsen, A. Ilchmann, A.J. Pritchard, Robustness of stability of time-varying linear systems, Report Nr. 161, Institut für Dynamische Systeme, U. Bremen, (1987)

[10] D. Hinrichsen, A.J. Pritchard, Stability radii of linear systems, Systems & Control Letters 7, 1-10 (1986a)

[11] D. Hinrichsen, A.J. Pritchard, Stability radius for structured perturbations and the algebraic Riccati equation. Systems & Control Letters 8, 105-113 (1986b)

[12] R.E. Kalman, P.L. Falb, M.A. Arbib, Topics in Mathematical System Theory, McGraw-Hill, (1969)

[13] T. Kato, Perturbation theory for linear operators, Grundlehren der mathematischen Wissenschaften 132, Springer, Berlin, Heidelberg, New York, 2nd edition, (1976)

[14] H.W. Knobloch, H. Kwakernaak, Lineare Kontrolltheorie, Springer, Berlin, Heidelberg, New York, Tokyo, (1985)

[15] S.Y. Kung, D.W. Lin, Recent progress in linear system model-reduction via Hankel matrix approximation, in: Circuit Theory and Design (The Hague, 1981) North Holland (1981)

[16] P. Lancaster, L. Rodman, Existence and uniqueness theorems for the algebraic Riccati equation, Int. J. Control 32, 285-309, (1980)

[17] M. Motscha, An algorithm to compute the complex stability radius, Report 168, Institut für Dynamische Systeme, U. Bremen, (1987)

[18] A. Ostrowski, Über Normen von Matrizen, Math. Z. 63, 2-18, (1955)

[19] D.H. Owens, A. Chotai, Robust controller design for linear dynamic systems using approximate models, IEE Proc. 130, 45-56, (1983)

[20] R.V. Patel, M. Toda, Quantitative measures of robustness for multivariable system, Proc. Joint Autom. Control Conf., TD8-A, (1980)

[21] A.J. Pritchard, S. Townley, Robustness of infinite dimensional systems, Control Theory Centre, Report No. 138, U. Warwick, (1986)

[22] L. Qui, E.J. Davison, New perturbation bounds for the robust stability of linear state space models, Proceedings of 25th Conference on Decision and Control, Athens, Greece, 751-755,(1986)

[23] M. Shayman, Geometry of the algebraic Riccati equation, Part I, SIAM J. Control and Optimization 21, 375-394,(1983)

[24] C. Van Loan, How near is a matrix to an unstable matrix, Contemporary Mathematics, 47, 465-478,(1985)

[25] C. Van Loan, A symplectic method for approximating all the eigenvalues of a Hamiltonian matrix, LAA 61, 233-251,(1984)

[26] H.H. Wilkinson, C. Reinsch, Linear algebra, Handbook for automatic computation II, Springer, New York, (1971)

[27] H.K. Wimmer, The algebraic Riccati equation without complete controllability, SIAM J. Alg. Disc. Meth. 3, 1-12,(1982)

[28] W.M. Wonham, Linear Multivariable Control: a Geometric Approach, 2nd edition, Springer-Verlag, Heidelberg (1979)

[29] R.K. Yedavallie, Perturbation bounds for robust stability in linear state space models, Int. J. Control 42, 1507-1517,(1985)

[30] G. Zames, B. Francis, Feedback minimax sensitivity and optimal robustness, IEEE Trans. Aut. Control 28, 585-601,(1983)

A PRINCIPLE OF CONTAMINATION IN BEST POLYNOMIAL APPROXIMATION

by

E.B. Saff*
Institute for Constructive Mathematics
Department of Mathematics
University of South Florida
Tampa, Florida 33620
USA

Abstract. We discuss the qualitative behavior of the polynomials $p_n^*(z)$ of best uniform approximation to a function f that is continuous on a compact set E of the z-plane, analytic in the interior of E, but not analytic at some point of the boundary of E. Particularly, we survey results on the asymptotic behavior of the zeros of the $p_n^*(z)$ and the extreme points for the error $f(z) - p_n^*(z)$. The theorems and examples presented support a "principle of contamination," which roughly states that the existence of one or more singularities of f on the boundary of E adversely affects the behavior over the *whole* boundary of E of a subsequence of the best approximants $p_n^*(z)$

§1. Introduction.

Let E be a compact set in the complex plane $\mathbb{C}$ whose complement $\overline{\mathbb{C}} \setminus E$ with respect to the extended plane $\overline{\mathbb{C}} := \mathbb{C} \cup \{\infty\}$ is connected. By the classical theorem of Mergelyan (cf. [18]), every function f that is continuous on E and analytic in the (2-dimensional) interior $\mathring{E}$ of E can be uniformly approximated on E, as closely as desired, by algebraic polynomials. To measure the rate of this approximation we consider the sequence of polynomials of best approximation.

---

*The research of the author was supported, in part, by the National Science Foundation under grant DMS-862-0098.

Let $\mathcal{P}_n$ denote the collection of all algebraic polynomials of degree $\leq n$, and $\|\cdot\|_E$ denote the sup norm over E. Then for each $n = 0,1,2,\ldots$, there exists a $p_n^* = p_n^*(f) \in \mathcal{P}_n$ satisfying

$$(1.1) \qquad \|f - p_n^*\|_E = \inf_{p\in\mathcal{P}_n} \|f - p\|_E ,$$

and p_n^* is uniquely determined by (1.1) provided $\operatorname{card}(E) \geq n + 1$. We set

$$(1.2) \qquad e_n = e_n(f,E) := \|f - p_n^*\|_E ,$$

and note that, with the previously mentioned assumptions on E and f, Mergelyan's theorem asserts that

$$(1.3) \qquad e_n \downarrow 0.$$

As is well-known, the rate of decay in (1.3) is intimately related to the smoothness properties of f. For example, if $A(E)$ denotes the collection of all functions f analytic on the compact set E, then under mild geometric assumptions on E the theorem of Bernstein-Walsh (cf. [18]) asserts that

$$(1.4) \qquad f \in A(E) \iff \limsup_{n\to\infty} [e_n]^{1/n} < 1.$$

That is, the error in best polynomial approximation decays exponentially (geometrically) if and only if f has an analytic continuation to some open set containing E. Furthermore, there exist, in this complex setting, more refined theorems of the Bernstein-Jackson type that relate the modulus of continuity of f over E to the rate of decay of e_n (cf. [1], [2]). The following two examples of this slower than geometric convergence are probably familiar to the reader:

If $f_1(x) = |x|$ on $E := [-1,1]$, then

$$(1.5) \qquad \frac{c_2}{n} \leq e_n(f_1) \leq \frac{c_1}{n} , \qquad n = 1,2,\ldots .$$

Here and below c_1, c_2 denote positive constants independent of n.

If $f_2(z) = \sqrt{z}$ on the disk $E := \{z : |z - 1| \leq 1\}$, then

$$(1.6) \qquad \frac{c_2}{\sqrt{n}} \leq e_n(f_2) \leq \frac{c_1}{\sqrt{n}}, \qquad n = 1,2,\ldots .$$

It is the purpose of this paper to study the *qualitative* behavior of the polynomials of best approximation to f in the case when f has one or more singularities on the boundary of E (such as the functions f_1, f_2 defined above). We shall present theorems that support the following general property:

Principle of Contamination. *Let f be continuous on E and analytic in $\mathring{E}$, where E is a compact set with connected and regular complement. If f has one or more singularities on the boundary of E (i.e., $f \notin A(E)$), then these singularities adversely affect the behavior over the* **whole** *boundary of E of a subsequence of the best polynomial approximants p_n^* to f on E.*

By the assumption that $\overline{\mathbb{C}} \setminus E$ is *regular* we mean that this set possesses a classical Green's function with pole at infinity. In particular, regularity holds if $\overline{\mathbb{C}} \setminus E$ is simply connected; that is, if E is a continuum (not a single point).

Of course the principle of contamination is not a mathematical theorem. Rather it is a rough summary of rigorous theorems to be discussed below. It is hoped that the statement of this principle will lead to further supporting theorems as well as to comparisons with other methods of approximation for which "the contamination" is nonexistent or less severe.

The outline of this paper is as follows. In Section 2 we discuss theorems and examples concerning the asymptotic behavior of the zeros of the best approximants $p_n^*(f)$. Such results are intimately related to the possibility of using these approximants to obtain analytic continuations of f. In Section 3 we consider the behavior of the extreme points for the error $f - p_n^*(f)$. The latter results are significant for purposes of comparing e_n with the rate of convergence on a subset of the boundary of E.

Before embarking on the theorems that support the principle of contamination, we wish to emphasize three important limitations of the principle.

(i) The principle refers only to best *polynomial* approximation. As we shall see below, best *rational* approximants may not exhibit the same ill effects.

(ii) The principle applies specifically to *best* polynomial approximants. For example, a sequence of "near-best" polynomials $q_n \in \mathcal{P}_n$, $n = 0,1,\dots,$ satisfying

$$\|f - q_n\|_E \leq 2\,\|f - p_n^*(f)\|_E, \qquad n = 0,1,2,\dots,$$

may behave qualitatively better than the $p_n^*(f)$ at those points of the boundary of E where f is analytic.

(iii) The principle refers only to some *subsequences* of the polynomials $\{p_n^*\}_1^\infty$. It is possible that there are other subsequences with less contaminated behavior.

Although we shall deal mainly with best *uniform* approximants, there do exist theorems that support the principle of contamination for sequences of best L^p polynomial approximants to f.

§2. Zeros of Best Polynomial Approximants

We assume here and throughout that E is a compact subset of $\mathbb{C}$ with connected and regular complement. For $f \in C(E) \cap A(\mathring{E})$; that is, f is continuous on E and analytic in the (possibly empty) interior $\mathring{E}$ of E, we ask the following question. What can be said about the locations (in the complex plane) of the zeros of the polynomials $p_n^*(f)$ of best uniform approximation to f on E? This question was studied by J. L. Walsh [19], [20], for the case when f is analytic *on* E ($f \in A(E)$) but not entire. Walsh's results are analogues of the classical theorem of Jentzsch [11] which states that the partial sums s_n of a power series (about $z = 0$) with finite radius of convergence $r > 0$ have the property that every point on the circle $|z| = r$ is a limit point of the set of zeros of the polynomials s_n, $n = 0,1,2,\dots$.

More recently Blatt and Saff [5] investigated the zeros of $p_n^*(f)$ for the more delicate case when f has one or more singularities on the boundary ∂E of E. They proved the following.

THEOREM 2.1 ([5]). *Suppose* $f \in C(E) \cap A(\mathring{E})$, *but* f *is not analytic on* E. *Assume further that* f *does not vanish identically on any component of the interior* $\mathring{E}$. *Then every boundary point of* E *is a limit point of the set of zeros of the sequence of best uniform approximants* $\{p_n^*(f)\}_1^\infty$ *to* f *on* E.

In particular, if f is continuous on the interval $I := [a,b]$ of the real axis and if f is not the restriction to I of a function analytic in a neighborhood of I, then every point of I attracts zeros of the sequence of best approximants $\{p_n^*(f)\}_1^\infty$ to f on I. In Figure 2.1 we illustrate this fact for $f_1(x) = |x|$ on $[-1,1]$ by plotting the zeros of $p_{26}^*(f_1)$.

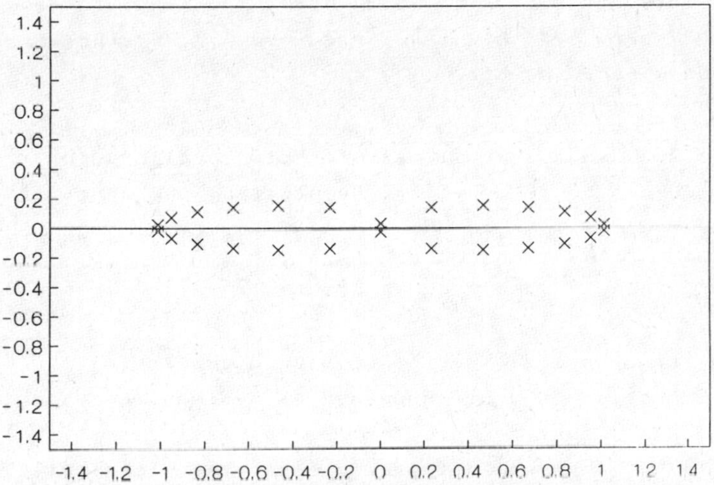

Figure 2.1 Zeros of $p_{26}^*(f_1)$, where $f_1(x) = |x|$ on $E = [-1,1]$

Notice that in Theorem 2.1 we assume $f \notin A(E)$ and so, from (1.4), the errors $e_n = e_n(f,E)$ of (1.2) satisfy

$$\limsup_{n\to\infty} [e_n]^{1/n} = 1. \tag{2.1}$$

Thus there is a subsequence $\Lambda = \Lambda(f,E)$ of integers for which

(2.2) $$\lim_{n\to\infty} [e_{n-1} - e_n]^{1/n} = 1, \quad n \in \Lambda.$$

In the proof of Theorem 2.1 it is shown, more generally, that, for any subsequence Λ satisfying (2.2), the Jentzsch-type behavior holds for the zeros of $\{p_n^*(f)\}_{n\in\Lambda}$. We further remark that Theorem 2.1 applies not only to the zeros of the $p_n^*(f)$ but also to their α-values; that is, to the roots of $p_n^*(f,z) = \alpha$, where α is any complex constant. Indeed, the function $f_\alpha(z) := f(z) - \alpha$ satisfies $f_\alpha \in C(E) \cap A(\mathring{E})$, $f_\alpha \notin A(E)$, and the polynomials of best uniform approximaton to f_α are just $p_n^*(f) - \alpha$. Thus we have

COROLLARY 2.2. *Suppose* $f \in C(E) \cap A(\mathring{E})$, $f \notin A(E)$, *and* f *is not identically constant on any component of* $\mathring{E}$. *Let* z_0 *be any boundary point of* E, $U(z_0)$ *a neighborhood (open disk) about* z_0, *and* Λ *a subsequence of integers for which* (2.2) *holds. Then, for every constant* α *and every sufficiently large* $n \in \Lambda$, *the equation* $p_n^*(f,z) = \alpha$ *has a root in* $U(z_0)$.

In Figures 2.2 and 2.3 we illustrate this corollary for the case of $f_1(x) = |x|$ on $E = [-1,1]$ by plotting the roots of $p_{26}^*(f_1,z) = -5$ and $p_{26}^*(f_1,z) = -i$, respectively. The reader should note the strong resemblance in Figures 2.1, 2.2, and 2.3.

Recalling the classical theorem of Picard concerning the behavior of an analytic function in a neighborhood of an isolated essential singularity, we can summarize Corollary 2.2 by saying that the sequence $\{p_n^*(f)\}_{n\in\Lambda}$ has an "asymptotic essential singularity" at each point of ∂E. In the context of *normal families* of analytic functions, we see that no point of ∂E is a normal point for the sequence $\{p_n^*(f)\}_1^\infty$. As shown in [5], this fact holds even when f is constant on some component of $\mathring{E}$; that is, we have

THEOREM 2.3. *Suppose* $f \in C(E) \cap A(\mathring{E})$, *but* $f \notin A(E)$. *Then the sequence* $\{p_n^*(f)\}_1^\infty$ *does not converge uniformly in any neighborhood of a boundary point of* E.

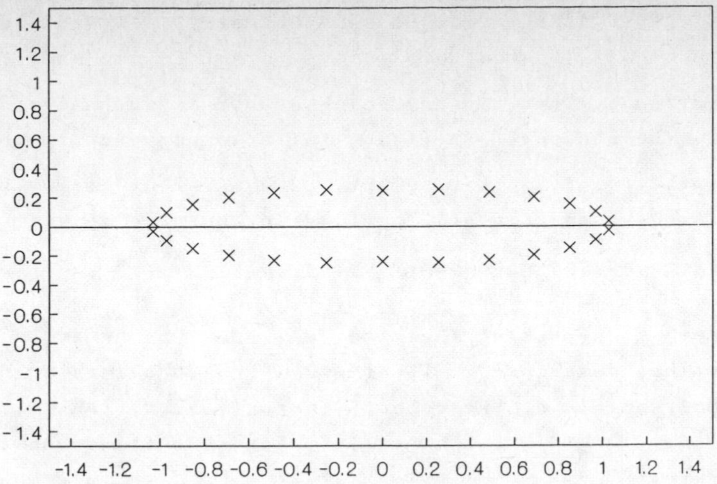

Figure 2.2 Roots of $p^*_{26}(f_1,z) = -5$, where $f_1(x) = |x|$ on $E = [-1,1]$

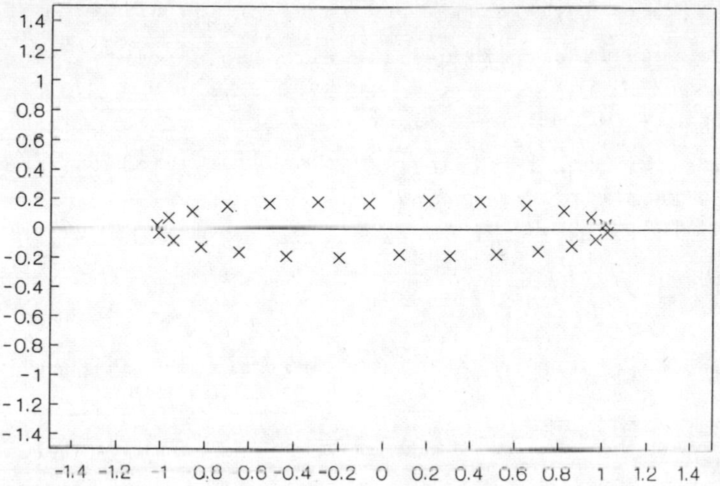

Figure 2.3 Roots of $p^*_{26}(f_1,z) = -i$, where $f_1(x) = |x|$ on $E = [-1,1]$.

The above results show that polynomials of best uniform approximation have an undesirable property: Consider a function

$f \in C(E) \cap A(\mathring{E})$ that is analytic at some boundary points of E, but is not analytic at all the boundary points (e.g. $f_1(z)$ and $f_2(z)$ in the Introduction). Then, in any neighborhood of an analytic boundary point, the sequence $\{p_n^*(f)\}_1^\infty$ fails to converge to the analytic continuation of f. We remark, however, that it may be possible for some subsequence of $\{p_n^*(f)\}$ to converge in a neighborhood of an analytic boundary point.

It is somewhat surprising that the Jentzsch-type behavior of zeros and α-values described in Theorem 2.1 and Corollary 2.2 need not hold for polynomials of "near-best" approximation; that is, for polynomials $\{q_n\}_1^\infty$, $q_n \in \mathcal{P}_n$, that satisfy for some fixed constant $K > 1$,

$$\|f - q_n\|_E \leq K\|f - p_n^*(f)\|_E = Ke_n, \qquad n = 1,2,\ldots . \tag{2.3}$$

Examples of this type were constructed by Grothmann and Saff [10] and Saff and Totik [15] where the only boundary points of E that attract zeros of the q_n are the singular points of f. To be more specific we describe the results of [15] dealing with the absolute value function.

Let $f(x) = |x|$ on $E = [-1/2,1/2]$ and let $g(z)$ be the analytic extension of f defined by

$$g(z) := \begin{cases} z, & \text{for } \operatorname{Re} z \geq 0 , \\ -z, & \text{for } \operatorname{Re} z < 0 . \end{cases} \tag{2.4}$$

In [15] a sequence $\{q_n\}_1^\infty$, $q_n \in \mathcal{P}_n$, is constructed such that

$$\|g - q_n\|_{E_1} \leq \frac{C}{n}, \qquad n = 1,2,\ldots, \tag{2.5}$$

where

$$E_1 := \{z \in \mathbb{C} : |\operatorname{Re} z| \leq 1, \quad |\operatorname{Im} z| \leq |\operatorname{Re} z|^2\}$$

(i.e., E_1 is a parabolic region with $z = 0$ a double point of ∂E_1). Since $E \subset E_1$ and f is the restriction to E of g, we have

$$\|f - q_n\|_E \le \frac{C}{n} \le Ke_n(f,E), \qquad n = 1,2,\ldots,$$

where the last inequality follows easily from (1.5). Also, from (2.5), we see that the q_n converge uniformly to g on the region E_1. Thus, since $g(z) \neq 0$ for $z(\neq 0) \in E_1$, it follows that *only one point* of E (namely, $z = 0$) can be a limit point of zeros of the sequence $\{q_n\}_1^\infty$.

The above example illustrates that "near-best" polynomial approximants may behave qualitatively better than the best polynomial approximants. Indeed, in [15], Saff and Totik have shown that for the class of *piecewise-analytic* functions f on [-1,1] it is possible to construct polynomials q_n satisfying (2.3) that converge uniformly and geometrically in some open disk about each point of [-1,1] where f is analytic (see Theorem 4.1).

In comparison to best polynomial approximants, best *rational* approximants can also have qualitatively much better behavior. To illustrate this statement we return to the approximation of the absolute value function $f(x) = |x|$ on $E = [-1,1]$. In [3], Blatt, Iserles and Saff proved the following.

PROPOSITION 2.4. *Let* $R_n^* = P_n/Q_n$ *denote the unique real rational function of degree at most* n *of best uniform approximation to* $|x|$ *on* [-1,1]. *Then for each* $n = 1,2,\ldots$, *all the zeros and all the poles of* R_n^* *lie on the imaginary axis. Moreover,*

$$\lim_{n\to\infty} R_n^*(z) = \begin{cases} z, & \textit{for} \quad \operatorname{Re} z > 0, \\ -z, & \textit{for} \quad \operatorname{Re} z < 0. \end{cases}$$

Thus, unlike the behavior of the zeros of the best polynomial approximants to $|x|$ on [-1,1], only one point of [-1,1] attracts the zeros of the sequence of best rational approximants. Furthermore, the best rational approximants converge to the analytic continuation of $|x|$ in the right-hand and left-hand planes.

Thus far we have discussed only Jentzsch-type theorems concerning the limit points of zeros of best polynomial approximants. Much more information is provided by Szegö-type theorems that concern the *limiting distributions* of these zeros. In order to state such results we need to introduce some terminology from potential theory (cf. [17]).

Let $\mathcal{M}(E)$ denote the collection of all unit measures that are supported on the compact set E where, as above, we assume that $\overline{\mathbb{C}} \setminus E$ is connected and regular. Then it is known that there exists a unique measure $\mu_E \in \mathcal{M}(E)$ that minimizes the energy integral

$$I[\mu] := \iint \log |z-t|^{-1} d\mu(t) d\mu(z)$$

over all $\mu \in \mathcal{M}(E)$. The measure μ_E is called the *equilibrium distribution* for E and (since $\overline{\mathbb{C}} \setminus E$ is regular) we have $\operatorname{supp}(\mu_E) = \partial E$. For example, if E is the closed disk $|z| \leq r$, then $d\mu_E = ds/2\pi r$, where ds is arclength measured on the circumference $|z| = r$. If $E = [-1,1]$, then $d\mu_E$ is the arcsine distribution, $d\mu_E = (1/\pi) dx/\sqrt{1 - x^2}$.

Next, to each nonconstant best polynomial approximant $p_n^*(f)$ we associate a discrete unit measure ν_n^*, called the *zero distribution* of $p_n^*(f)$, by

$$(2.6) \qquad \nu_n^*(B) := \frac{\text{number of zeros of } p_n^*(f) \text{ in } B}{\deg p_n^*(f)},$$

for Borel sets $B \subset \mathbb{C}$, where we count the zeros according to their multiplicity.

We can now state the theorem of Blatt, Saff, and Simkani concerning the limiting distribution of the zeros of best polynomial approximants.

THEOREM 2.5 ([6]). *Suppose* $f \in C(E) \cap A(\mathring{E})$, *but* $f \notin A(E)$. *Assume further that* f *does not vanish identically on any component of* $\mathring{E}$. *Then* ν_n^* *converges in the weak-star topology to* μ_E *as* $n \to \infty$ *through a sequence* $\Lambda = \Lambda(f,E)$ *of positive integers.*

Consequently, for any Borel set $B \subset \mathbb{C}$,

$$\mu_E(\mathring{B}) \le \liminf_{n\to\infty} \nu_n^*(B) \le \limsup_{n\to\infty} \nu_n^*(B) \le \mu_E(\bar{B}), \quad n \in \Lambda. \tag{2.7}$$

In the above theorem, Λ is any sequence of integers for which (2.2) holds. We remark that since $\operatorname{supp}(\mu_E) = \partial E$, the assertion of Theorem 2.1 that each point of ∂E is a limit point of the sets of zeros of $\{p_n^*(f)\}_1^\infty$ is an immediate consequence of (2.7).

In Theorem 2.5 the convergence of the zero distributions to the equilibrium distribution of E holds only for suitable *subsequences*. In [10], Grothmann and Saff have shown that for any admissible compact set E, there exists a function f satisfying the hypotheses of Theorem 2.5 and a subsequence $\left\{\nu_{n_k}^*\right\}$ of the measures (2.6) such that

$$\lim_{k\to\infty} \nu_{n_k}^*(B) = 0 \quad \text{for all bounded sets } B \subset \mathbb{C}.$$

Although the results of this section have been concerned only with best *uniform* polynomial approximants, theorems analogous to Theorem 2.5 have been proved by Simkani [16] and Blatt, Saff and Simkani [6] that apply to best L^p polynomial approximants (as measured by a line integral over a rectifiable Jordan curve) and to best uniform rational approximants having a bounded number of free poles.

§3. Behavior of Extreme Points.

In this section we present further evidence for the principle of contamination by examining the behavior of extreme points in best polynomial and best rational approximation. We first recall the classical result that, for a real-valued f continuous on the interval $[-1,1]$, there exist $n+2$ points

$$-1 \le x_1^{(n)} < x_2^{(n)} < \cdots < x_{n+2}^{(n)} \le 1$$

such that

$$|(f - p_n^*)(x_k^{(n)})| = \|f - p_n^*\|_{[-1,1]}, \quad k = 1,\ldots,n+2 \tag{3.1}$$

$$(3.2) \qquad (f - p_n^*)(x_k^{(n)}) = -(f - p_n^*)(x_{k+1}^{(n)}) \, , \quad k = 1,\dots,n+1,$$

where $p_n^* = p_n^*(f,x)$ is the best uniform approximation to f on $[-1,1]$. A set of points $\{x_k^{(n)}\}_{k=1}^{n+2}$ satisfying (3.1) and (3.2) is called an *alternation set* for the error $f - p_n^*$. We remark that such an alternation set need not be unique. For the special case when $f(x) = x^{n+1}$, the error $x^{n+1} - p_n^*(x)$ is just the Chebyshev polynomial of the first kind and the corresponding alternation set is

$$(3.3) \qquad \left\{\cos\left(\frac{k\pi}{n+1}\right)\right\}_{k=0}^{n+1} .$$

It turns out that for an arbitrary real-valued $f \in C[-1,1]$, there is always a *subsequence* of the errors $f - p_n^*$ for which the corresponding alternation sets have the same limiting distribution as the points (3.3); that is, the arcsine distribution. This fact was first proved by Kadec [12], who also provides estimates for the rate of this convergence. Further improvements in Kadec's result were obtained by Fuchs [9] and Blatt and Lorentz [4]. Unfortunately, the method of proof of Kadec's theorem (as well as its generalizations) relies heavily on the fact that consecutive points in an alternation set and the zeros of $p_n^*(f) - p_{n-1}^*(f)$ interlace; a fact that has no analogue for the approximation of complex-valued functions. To circumvent this difficulty, Kroó and Saff [13] used, instead, a potential theoretic argument to establish a Jentzsch-type result concerning the *denseness* of extreme points. To state their result we need to introduce some notation.

Let $E \subset \mathbb{C}$ be a compact set with connected and regular complement. Given $f \in C(E) \cap A(\mathring{E})$, we set

$$(3.4) \qquad A_n(f) := \{z \in E : |f(z) - p_n^*(f,z)| = \|f - p_n^*(f)\|_E\} ,$$

and refer to $A_n(f)$ as the set of extreme points for $f - p_n^*$. We remark that each set $A_n(f)$ consists of at least $n + 2$ points. To measure the denseness of a set A in a set B $(A, B \subset \mathbb{C})$ we define

$$(3.5)\qquad \rho(A,B) := \sup_{z\in B}\ \inf_{\xi\in A} |z - \xi|.$$

The main result of Kroó and Saff asserts that there is a subsequence of integers n for which the sets $A_n(f)$ become dense in the boundary ∂E. More precisely, we have

THEOREM 3.1([13]). *If* $f \in C(E) \cap A(\mathring{E})$, *then*

$$(3.6)\qquad \liminf_{n\to\infty} \rho(A_n(f),\partial E) = 0.$$

Furthermore, there exists an entire function g *such that*

$$(3.7)\qquad \limsup_{n\to\infty} \rho(A_n(g),\partial E) > 0.$$

In the proof of this theorem it is shown, moreover, that if Λ is any increasing sequence of integers for which the errors e_n of (1.2) satisfy

$$(3.8)\qquad \lim_{n\to\infty}\left[\frac{e_{n-1} - e_n}{e_{n-1} + e_n}\right]^{1/n} = 1\ ,\qquad n \in \Lambda,$$

then

$$(3.9)\qquad \lim_{n\to\infty} \rho(A_n(f),\partial E) = 0\ ,\qquad n \in \Lambda.$$

The second part of Theorem 3.1 asserts that denseness need not hold for all subsequences - a fact that was first proved by G.G. Lorentz [14] for the Kadec case of a real function on a real interval.

Theorem 3.1 is the analogue of Theorem 2.1 concerning the denseness of zeros of best polynomial approximants. However, there is an important difference - namely, in the latter theorem we require that f have a singularity on ∂E ($f \notin A(E)$), while Theorem 3.1 applies as well to functions f that are analytic on E.

Theorem 3.1 shows that the polynomials $p_n^*(f)$ of best uniform approximation have another undesirable feature. Suppose, for example, that E is a Jordan arc or a Jordan region and consider a function

$f \in C(E) \cap A(\overset{\circ}{E})$ that is analytic at some but not all points of the boundary ∂E (such as the functions $f_1(z)$ and $f_2(z)$ in the Introduction). Since $f \notin A(E)$, the rate of decay of the errors e_n is slow (not geometric). However, it would seem reasonable to expect that on those subarcs of ∂E where f is analytic (say, a subarc Γ), the rate of convergence of $\{p_n^*(f)\}_1^\infty$ to f is faster. Unfortunately this is not the case since, by Theorem 3.1, for infinitely many n this subarc must contain points of $A_n(f)$; that is,

$$\|f - p_n^*(f)\|_\Gamma = \|f - p_n^*(f)\|_E = e_n$$

for infinitely many n.

What can be said about the denseness of extreme points for the case of best *rational* approximation? While this question has not yet been answered in full generality, there is an important case for which results are known - namely, for best real rational approximation to a real-valued function f on $[-1,1]$. We now describe these results in the context of the Walsh array.

Let $\mathcal{R}_{m,n}$ be the collection of all real rational functions with numerator in $\mathcal{P}_m$ and denominator in $\mathcal{P}_n$. For real-valued $f \in C[-1,1]$, we let $r_{m,n}^* = r_{m,n}^*(f)$ denote the best uniform approximation to f on $[-1,1]$ out of $\mathcal{R}_{m,n}$. These best approximants are typically displayed in the following doubly-infinite table called the *Walsh array*.

$$\begin{array}{|cccccc}
\hline
r_{00}^* & r_{10}^* & r_{20}^* & \cdot & \cdot & \cdot \\
r_{01}^* & r_{11}^* & r_{21}^* & \cdot & \cdot & \cdot \\
r_{02}^* & r_{12}^* & r_{22}^* & \cdot & \cdot & \cdot \\
\cdot & \cdot & \cdot & & & \\
\cdot & \cdot & \cdot & & & \\
\cdot & \cdot & \cdot & & &
\end{array}$$

Notice that the first row of this array consists of the polynomials of best uniform approximation $(r^*_{m,0} = p^*_m)$. We know, therefore, that extreme points are dense for the approximants in the first row. But what about other sequences formed from the array such as the diagonal, other rows, or a "ray sequence" that consists of rationals $\{r^*_{m,n}\}$ for which the ratio m/n has a finite limit as $m,n \to \infty$?

To state the main result [8] of Borwein, Grothmann, Kroó and Saff concerning the denseness of such extreme points, we first recall that, for each pair of nonnegative integers (m,n), the best approximant $r^*_{m,n} = p^*_{m,n}/q^*_{m,n}$ is characterized by the following equioscillation property: There are $m + n + 2 - d$ points

$$(3.10) \qquad -1 \le x_1^{(m,n)} < \dots < x_{m+n+2-d}^{(m,n)} \le 1 ,$$

where

$$d := d(m,n) := \min\{m - \deg p^*_{m,n} , n - \deg q^*_{m,n}\}$$

such that

$$(3.11) \qquad |(f-r^*_{m,n})(x_k^{(m,n)})| = \|f-r^*_{m,n}\|_{[-1,1]}, \qquad k = 1,\dots,m+n+2-d,$$

$$(3.12) \qquad (f-r^*_{m,n})(x_k^{(m,n)}) = -(f-r^*_{m,n})(x_{k+1}^{(m,n)}), \qquad k = 1,\dots,m+n+1-d.$$

Next, we measure the denseness of the extreme points $\{x_k^{(m,n)}\}_1^{m+n+2-d}$ in $[-1,1]$ by defining

$$(3.13) \qquad \rho_{m,n}(f) := \sup_{x\in[-1,1]} \min_k |x - x_k^{(m,n)}| .$$

The main result of [8] is the following.

THEOREM 3.2. *Let the sequence of nonnegative integers* $n = n(m)$ *satisfy*

$$(3.14) \qquad n(m) \le n(m+1) \le n(m) + 1 , \qquad n(m) \le m ,$$

for $m = 0,1,\dots$. *If* $f \in C[-1,1]$, $f \notin \mathcal{R}_{m,n(m)}$, $m = 0,1,\dots$, *then*

$$(3.15) \qquad \liminf_{m\to\infty} \left[\frac{m - n(m)}{\log m}\right] \rho_{m,n(m)}(f) < \infty .$$

Notice that Theorem 3.2 applies in the case of ray sequences of the form $n(m) = [cm]$ for any constant $c \leq 1$, where $[\cdot]$ denotes the greatest integer function. If $c < 1$, we deduce from (3.15) that

$$\liminf_{m\to\infty} \rho_{m,[cm]}(f) = 0$$

which means that the extreme points are dense in $[-1,1]$. In other words, if we proceed down the Walsh array at an angle of less than $\pi/4$ with the first row, then the Jentzsch-type (denseness) result holds for the extreme points.

Of course, Theorem 3.2 gives no information about the important case of diagonal sequences; that is, when we proceed down the table at an angle of $\pi/4$. In [8], it is shown that denseness need no longer hold for such sequences. In fact, for a diagonal sequence, all alternation points can occur in an arbitrarily small subinterval of $[-1,1]$. More precisely we have

THEOREM 3.3. *For every* $2 > \epsilon > 0$, *there is a function* $f \in C[-1,1]$ *such that for each* $n = 1,2,\ldots$, *the error* $f - r^{*}_{n-1,n}(f)$ *has no extreme points in* $(-1 + \epsilon, 1]$.

Returning to the behavior of extreme points for best polynomial approximation, a Szegö-type *limiting distribution* result has recently been obtained by Blatt, Saff and Totik [7]. To describe this result we must first associate a unit measure with each extremal set $A_n(f)$ in (3.4). For the Kadec case of real approximation on an interval, one can use the discrete unit measure that is supported in an $(n + 2)$ - point alternation set $\{x^{*}_{k}\}_{k=1}^{n+2}$ (cf. (3.1) and (3.2)). For complex approximation, the notion of alternation set is replaced by that of an *extremal signature*, which is a subset of $A_n(f)$ consisting of at most $2n + 3$ points. However, unlike the case of real approximation, the cardinality of an extremal signature can vary from $n + 2$ to $2n + 3$ points. Thus it is not immediately clear how to select $n + 2$ points from $A_n(f)$ in order to define a discrete measure. This difficulty was resolved in [7] by selecting as the subset of $A_n(f)$ an $(n + 2)$-point *Fekete subset* which we denote by $\mathcal{F}_{n+2}$. To be precise, $\mathcal{F}_{n+2}$ is an $(n + 2)$-point subset S of $A_n(f)$ for which the Vandermonde expression

$$V(S) := \prod_{\substack{z, t \in S \\ z \neq t}} |z - t|$$

is as large as possible.

Next, as in (2.6), we associate a unit measure λ_n with $\mathcal{F}_{n+2}$ by defining

$$(3.16) \qquad \lambda_n(B) := \frac{\text{number of points of } \mathcal{F}_{n+2} \text{ in } B}{n + 2},$$

for any Borel set $B \subset \mathbb{C}$.

With the above notation we can now state

THEOREM 3.4([7]). *Suppose* $f \in C(E) \cap A(\mathring{E})$ *and let* $\mathcal{F}_{n+2}$ *be an* $(n + 2)$*-point Fekete subset of the set of extreme points* $A_n(f)$ *defined in* (3.4). *Then the measures* λ_n *of* (3.16) *converge in the weak-star topology to the equilibrium distribution* μ_E *as* $n \to \infty$ *through a sequence* $\Lambda = \Lambda(f, E)$ *of positive integers.*

§4. Concluding Remarks

The results of the preceding sections have hopefully convinced the reader that best polynomial approximants have significant drawbacks. Moreover, in a sense, "near-best may be better than best." In [15] Saff and Totik construct such near-best polynomial approximants for the case of piecewise-analytic functions f on $[-1,1]$ (such as $f(x) = |x|$ on $[-1,1]$). The following is a sample of their results.

THEOREM 4.1 *Suppose* $f \in C^k[-1,1]$ *is piecewise analytic on* $[-1,1]$ *and* $\beta > 1$ *is given. Then there exist constants* $c, C > 0$ *and polynomials* $q_n \in \mathcal{P}_n$, $n = 1,2,\ldots$, *such that for every* $x \in [-1,1]$

$$(4.1) \qquad |f(x) - q_n(x)| \le \frac{C}{n^{k+1}} \exp(-cn[d(x)]^{\beta}) ,$$

where d(x) *denotes the distance from* x *to the nearest singularity of* f *in* [-1,1].

Notice from (4.1) that the q_n converge geometrically in an open interval and hence in an open disk about each point of [-1,1] where f is analytic.

ACKNOWLEDGEMENT. The author is grateful to Dr. Jon Snader and Mr. Hongzhu Qiao for generating the graphs in Section 2.

REFERENCES

[1] J.M. Anderson, A. Hinkkanen and F.D. Lesley, *On theorems of Jackson and Bernštein type in the complex plane*, Constr. Approx. (to appear).

[2] V.V. Andrievskii, *Approximation characterization of classes of functions on continua of the complex plane*, Math. U.S.S.R. Sbornik, 53 (1986), 69-87.

[3] H.-P. Blatt, A. Iserles and E.B. Saff, *Remarks on the behavior of zeros of best approximating polynomials and rational functions*, IMA Series No. 10, pp. 437-445, Oxford Univ. Press (1987).

[4] H.-P. Blatt and G.G. Lorentz, *On a theorem of Kadec* (to appear).

[5] H.-P. Blatt and E.B. Saff, *Behavior of zeros of polynomials of near best approximation*, J. Approx. Theory 46 No. 4 (1986), 323-344.

[6] H.-P. Blatt, E.B. Saff and M. Simkani, *Jentzsch-Szegő type theorems for the zeros of best approximants*, J. London Math. Soc. (to appear).

[7] H.-P. Blatt, E.B. Saff and V. Totik, *The distribution of extreme points in best complex polynomial approximation* (ICM Technical Report #87-0015) (to appear).

[8] P.B. Borwein, R. Grothmann, A. Kroó and E.B. Saff, *The density of alternation points in rational approximation* (ICM Technical Report #87-010) (to appear).

[9] W.H. Fuchs, *On Chebyshev approximation on sets with several components*, in "Proc. NATO Adv. Study Inst., Univ, Durham, 1979," pp. 399-408, Academic Press, New York, 1980.

[10] R. Grothmann and E.B. Saff, *On the behavior of zeros and poles of best uniform polynomial and rational approximants* (ICM Technical Report #87-009). To appear in: Proceedings of Antwerp Conference 1987 (A. Cuyt, ed.).

[11] R. Jentzsch, "Untersuchungen zur Theorie analytischer Funktionen," Inangural-dissertation, Berlin, 1914.

[12] M.I. Kadec, *On the distribution of points of maximum deviation in the approximation of continuous*, Amer. Math. Soc. Transl. 26 (1963), 231-234.

[13] A. Kroó and E.B. Saff, *The density of extreme points in complex polynomial approximation*, Proc. Amer. Math. Soc. (to appear).

[14] G.G. Lorentz, *Distribution of alternation points in uniform polynomial approximation*, Proc. Amer. Math. Soc., 92 (1984), 401-403.

[15] E.B. Saff and V. Totik, *Polynomial approximation of piecewise analytic functions* (ICM Technical Report #87-017) (to appear).

[16] M. Simkani, Asymptotic Distribution of Zeros of Approximating Polynomials, Ph.D. Dissertation, University of South Florida, Tampa (1987).

[17] M. Tsuji, Potential Theory in Modern Function Theory, 2nd ed., Chelsea Publ. Co., New York, (1958).

[18] J.L. Walsh, Interpolation and Approximation by Rational Functions in the Complex Domain, Amer. Math. Soc. Colloquium Publications, Vol. 20, (1935, 5th ed., 1969).

[19] J.L. Walsh, *Overconvergence, degree of convergence and zeros of sequences of analytic functions*, Duke Math. J. 13 (1946), 195-234.

[20] J.L. Walsh, *The analogue for maximally convergent polynomials of Jentzsch's theorem*, Duke Math. J. 26 (1959), 605-616.

NEARBY SETS AND CENTERS

Marco Baronti
Dipartimento di Matematica
University of Parma
I-43100 Parma - Italia

Pier Luigi Papini
Dipartimento di Matematica
University of Bologna
I-40126 Bologna - Italia

ABSTRACT. Let us consider a Banach space X and a class $\mathscr{C}$ of subsets in X. Let c be a (Chebyshev) center of $C\in\mathscr{C}$; it is known that under some assumptions on $\mathscr{C}$ and/or X, if $\{C_n\}$ is a sequence from $\mathscr{C}$ converging to C according to the Hausdorff metric $h(C_n,C)$, then c is the limit of a sequence of centers of C_n. Here we shall discuss some estimates of the distance between c_n and c, in terms of $h(C_n,C)$.

1. INTRODUCTION

Let X be a Banach space. Throughout the paper, we shall denote by A and B closed, bounded and nonempty subsets of X. Also, we set

$$h(A,B) = \max\{ \sup_{x\in A}\inf_{y\in B} \|x-y\|, \ \sup_{y\in B}\inf_{x\in A} \|x-y\| \} ;$$

$$d(A,B) = \inf\{ \|x-y\| ; \ x\in A ; \ y\in B \} ;$$

$$r(x,A) = \sup_{y\in A} \|x-y\| ; \qquad r_A = \inf_{x\in X} r(x,A) ; \qquad r'_A = \inf_{a\in A} r(a,A) .$$

The numbers r_A and r'_A are called the radius and -respectively- the self radius of A. A center [resp.: a self center] of A is a point $\bar{x}\in X$ [resp.: $\bar{a}\in A$] such that $r(\bar{x},A)=r_A$ [resp.: $r(\bar{a},A)=r'_A$] .

By $\delta(A)$ we shall denote the diameter of A.

As known, existence of centers is assured e.g. when X is a reflexive space, while no set may have at most one center if and only if X is "uniformly convex in every direction". For other general results on this matter we refer to [2].

Take a space X and a class of sets in it for which existence and unicity of center holds; then consider the map assigning to a set in this class its center. The continuity of such map for variable sets (with respect to Hausdorff metric) was studied in several papers: Section 6 of [2] can be used as a general reference; previous results had been considered by Belobrov (see [5]), and by Ward in Sections 2 and 4 of [9] ; in Section 4 of [9] also the behavior of centers with respect to nearby norms in finite dimensional spaces was examined.

Seemingly, it is more difficult to estimate constants relating e.g. $\|c_A-c_B\|$ to $h(A,B)$ for two arbitrary sets with centers. A discussion of this item is the subject of the present paper.

2. GENERAL FACTS AND SIMPLE REMARKS

The following inequalities are immediate.

$$d(A,B) \leqslant h(A,B) \leqslant d(A,B) + \max(\delta(A),\delta(B)) \qquad (2.1).$$

Note that $\delta(A) \leqslant 2r_A \leqslant 2\delta(A)$ always.

$$|r(x,A)-r(y,A)| \leqslant \|x-y\| \leqslant r(x,A)+r(y,A) \quad \text{for any set } A; \text{ any } x, y \text{ in } X \qquad (2.2)$$

$$\text{if } c_A \ [c_B] \text{ is a center of } A\ [B], \text{ then } \|c_A-c_B\| \leqslant r_A+r_B+d(A,B) \qquad (2.3).$$

Trivial examples show how the above estimates are sharp (in the sense that equality in them is possible). When A is strictly convex, equality in the right part of (2.2) holds only in very particular cases.

Given a set A, let $C(A)$ the set of its centers and $C'(A)$ the set of its self centers. Of course, $r'(A) \leqslant 2r(A)$ for any A. The sets $C(A)$ and $C'(A)$ can be also very large: it is possible to have $\delta(C(A)) = 2r_A$; $\delta(C'(A)) = \delta(A)$.

It is easy to prove the following

LEMMA 2.1. *For any* $x \in X$ *we have*

$$|r(x,A) - r(x,B)| \leqslant h(A,B) \qquad (2.4)$$

and also

$$|r_A-r_B| \leqslant h(A,B) \qquad (2.5).$$

Proof. Take $a \in A$ and let $\varepsilon>0$; choose $b_\varepsilon \in B$ such that $\|a-b_\varepsilon\| < h(A,B)+\varepsilon$. Then for any x we obtain: $\|a-x\| \leq \|a-b_\varepsilon\| + \|b_\varepsilon - x\| < h(A,B) + \varepsilon + r(x,B)$, therefore $r(x,A) \leq$ $\leq h(A,B)+r(x,B)$. A similar inequality holds with A and B exchanged, and this implies (2.4).

By using (2.4), we obtain $r_A = \inf_{x\varepsilon X} r(x,A) \leq h(A,B) + \inf_{x\varepsilon X} r(x,B) = h(A,B) + r(B)$, thus also (2.5). ■

If we consider self centers instead of centers, the analogue of (2.5) is not true, as the following example shows.

EXAMPLE 2.2. Let $X=C[0,1]$. Fix some $\varepsilon<\frac{1}{2}$ and set $A=\{f\varepsilon X;\ 0\leq f(x)\leq 1-\varepsilon$ for any x; $f(0)=0\}$; $B=\{g\varepsilon X;\ \frac{1}{2}-\varepsilon \leq g(x)\leq \frac{1}{2}$ for any x and $g(0)=\frac{1}{2}-\varepsilon\}$. Then we have $r'_A = 1-\varepsilon$; $r'_B=\varepsilon$; $h(A,B)=\frac{1}{2}-\varepsilon < 1-2\varepsilon = r'_A - r'_B$.

For self radii we can prove a weaker estimate, which is nevertheless sharp, as shown by the above example (when we let $\varepsilon\to 0$).

PROPOSITION 2.3. <u>For any</u> A <u>and</u> B <u>we have</u>

$$|r'_A - r'_B| \leq 2\,h(A,B) \tag{2.6}$$

<u>Proof</u>. Let $\varepsilon>0$; take $b_\varepsilon \varepsilon B$ such that $r(b_\varepsilon,B) < r'_B+\varepsilon$. Then take $a_\varepsilon \varepsilon A$ such that $\|a_\varepsilon - b_\varepsilon\| < h(A,B)+\varepsilon$. By using (2.2) and (2.4) we obtain $r(a_\varepsilon,A) \leq r(b_\varepsilon,A) + \|a_\varepsilon - b_\varepsilon\| <$ $< h(A,B) + r(b_\varepsilon,B) + h(A,B) + \varepsilon < 2\,h(A,B) + r'_B + 2\varepsilon$, which implies $r'_A - r'_B \leq 2\,h(A,B)$. By symmetry, we obtain (2.6). ■

For self centers, by using Lemma 2.1 we can prove the following

PROPOSITION 2.4. <u>If</u> c'_A <u>and</u> c'_B <u>are self centers of</u> A, B <u>respectively, then the following inequalities are true</u> ($h=h(A,B)$):

$$\|c'_A - c'_B\|^2 \leq (h+r'_A)\cdot(h+r'_B) \tag{2.7}$$

$$\|c'_A - c'_B\| \leq h + (r'_A + r'_B)/2 \tag{2.8}$$

<u>Proof</u>. By using (2.4) we obtain $\|c'_A - c'_B\| \leq r(c'_B,A) \leq r(c'_B,B) + h = r'_B + h$; similarly, $\|c'_B - c'_A\| \leq r'_A + h$. Now it is enough to multiply these two inequalities, or to add them then divide by 2 to obtain the thesis. ■

REMARK 2.5. The above estimates are sharp: in fact, consider e.g. in R^2, endowed with the max norm, the sets $A=\{(x,y);\ 0\leq x\leq 1;\ |y|\leq 1\}$ and $B=\{(x,y);\ -1\leq x\leq 0;\ |y|\leq 1\}$.

Then $(1,0)\varepsilon C'(A)$ and $(-1,0)\varepsilon C'(B)$.

REMARK 2.6. The analogue of Proposition 2.4 for (absolute) centers is probably false, as suggested by Example 2.7 below. We are only able to indicate that the following inequalities are true:

$$\|c_A-c_B\| \leqslant r(c_A, B\cup\{c_B\}) \leqslant r(c_A,A)+h(A,B\cup\{c_B\}) = r_A+h(A, B\cup\{c_B\})$$

and the analogue with A and B interchanged.

EXAMPLE 2.7. Take a strictly convex, smooth but not euclidean three-dimensional normed space. There exists a triangle T such that $c_T \notin T$, thus $d(\{c_T\}, T)>0$. Consider a hyperplane M separating c_T from T and such that $d(\{c_T\},M)=\overline{d}=d(T,M)>0$. But M is expressible as the union of an increasing sequence of balls $B(x_n,r_n)$; so we can choose $\overline{n}$ such that $T\subset B(x_{\overline{n}},r_{\overline{n}})$. If we set $A=\{x_{\overline{n}}\}$, then we have $\|c_A-c_T\| \geqslant r_{\overline{n}}+\overline{d} \geqslant$ $\geqslant h(A,T)+\overline{d}+r_A$.

REMARK 2.8. Each of the two numbers $h(A,B)$ and $\|c_A-c_B\|$ (or $\|c'_A-c'_B\|$), can be larger, equal or smaller than the other, also in condition of uniqueness. The study concerning existence of some constant k such that

$$\|c'_A-c'_B\| \leqslant k\cdot h(A,B) \tag{2.9}$$

for any pair of sets A, B, was initiated by Borwein and Keener, and deepened by Amir (see [2]). If we restrict the class of sets to be considered, at least in uniformly convex spaces some Lipshitz constant may exist: for example, taking the class of sets for which $B(c_A,r'_A)\cap B(c'_B,r'_B)=\emptyset$ (which implies $r'_B\leqslant h$), then (2.9) holds with $(1+\sqrt{5})/2\leqslant$ $\leqslant k\leqslant 2$ (recall (2.5)), while no analogue is true for centers: see § 6.4 of [2] and the references therein for these results. Moreover, $k=(1+\sqrt{5})/2$ is suitable exactly when X is a Hilbert space (see [1]), while the value 2 cannot be improved in general. Estimates similar to those indicated now could be obtained for relative centers with respect to a set $V\subset X$: $C_V(A)=\left\{\overline{v}\varepsilon V;\ \sup\{\|\overline{v}-a\|;\ a\varepsilon A\}=\inf_{v\varepsilon V}\sup\{\|v-a\|;\ a\varepsilon A\}\right\}$. Also, stability of relative radii for some fixed V (for example, a hyperplane) and variable A could be studied.

REMARK 2.9. If we substitute convergence in the sense of Hausdorff with some weaker kind of convergence, then stability results in this area are rather uncommon: think e.g. at the sequence of sets $C_n=[\theta,e_n]$ in ℓ^2, which "converges" to $\{\theta\}$.

3. A THEOREM IN HILBERT SPACES

The following result was proved in [4] (Corollary 2.5).

PROPOSITION 3.1. *If* X *is a Hilbert space, then* $y=c_A$ *for a set* A *if and only if there exists* $\bar{\varepsilon}>0$ *such that for* $0<\varepsilon<\bar{\varepsilon}$ *we have* $y\in\overline{co}\{a\in A;\ \|y-a\|>r(y,A)-\varepsilon\}$.

By using the above proposition we can give an extension of Theorem 1 from [8], whose proof used a lemma requiring a compactness assumption: in this way we answer the question raised in Remark 1 of [8].

THEOREM 3.2. *If* X *is a Hilbert space, then the following inequality holds for any* A *and* B $(h=h(A,B))$:

$$\|c_A-c_B\|^2\leqslant h(r_A+r_B+h) \tag{3.1}$$

Proof. Let c_A, c_B be the centers of A, B respectively. Denote by M the linear variety through c_A and orthogonal to the line joining c_A and c_B. By Proposition 3.1, for every $n\in N$ we can choose P_n in A such that $\|c_A-P_n\|>r_A-\frac{1}{n}$, in the halfspace determined by M and not containing c_B; therefore $\|P_n-c_B\|^2\geqslant(r_A-\frac{1}{n})^2+$ $+\|c_A-c_B\|^2$. On the other hand (see (2.4)) $\|P_n-c_B\|\leqslant r(c_B,A)\leqslant r_B+h$, thus we obtain: $\|c_A-c_B\|^2\leqslant\|P_n-c_B\|^2-(r_A-\frac{1}{n})^2\leqslant(r_B+h)^2-(r_A-\frac{1}{n})^2$. By letting $n\to\infty$, we obtain

$$\|c_A-c_B\|^2\leqslant(h+r_B)^2-r_A^2 \tag{3.2}$$

Since the role of A and B can be interchanged, we have also

$$\|c_A-c_B\|^2\leqslant(h+r_A)^2-r_B^2 \tag{3.3}$$

By adding (3.2) and (3.3) we obtain the thesis. ■

REMARK 3.3. As noticed in [8], (3.1) remains true for centers relative to vector subspaces of X. Also, (3.1) is precise when A and B reduce to singletons, while it does not hold in spaces where A=B can have more than one center. The more general estimate (2.7) has one term added with respect to (3.2) and continuity on h is lost.

Note (see [9], Example II.2.2) that $\|c_A-c_B\|$ is not Lipshitz continuous with respect to h (when $h\to0$): the Hölder constant 1/2 is as good as could be hoped for.

4. THE CASE OF UNIFORMLY CONVEX SPACES

Set, for $\varepsilon\in[0,2]$

$$\delta(\varepsilon)=\inf\{1-\frac{\|x+y\|}{2}\ ;\ \|x\|\leq 1\ ;\ \|y\|\leq 1\ ;\ \|x-y\|\geq\varepsilon\} \tag{4.1}$$

Let X be uniformly convex, i.e., $\delta(\varepsilon)$ is positive for any $\varepsilon>0$. In that case, $\delta(\varepsilon)$ is a continuous, strictly increasing function; so we can define its inverse in $(0,1]$:

$$\varepsilon(\delta)=\sup\{\|x-y\|\ ;\ \|x\|\leq 1\ ;\ \|y\|\leq 1;\ \frac{\|x+y\|}{2}\geq 1-\delta\} \tag{4.2}$$

Of course $\varepsilon(\delta)\to 0$ when $\delta\to 0$.

The assumption of uniform convexity was used in [3] (Lemma 2.1) to prove uniform continuity of centers with respect to Hausdorff metric for closed, bounded sets. Some estimates were used to evaluate $\|c_A-c_B\|$ in terms of $h(A,B)$. Following more or less the same lines as in that proof, we can formulate the following proposition, which gives very rough estimates (similar relations had been considered in [6], Lemma 3).

PROPOSITION 4.1. *Let* $h=h(A,B)$, *with* A *and* B *in a uniformly convex space. Also, set* $c=\|c_A-c_B\|$ (c_A, c_B *being the centers of* A *and* B); $\gamma=\max(r_A+h\ ,\ r_B+h)$. *Then, if* $0\neq c\leq 2\gamma$, *we have*

$$\gamma\,\delta(\frac{c}{\gamma})\leq h \tag{4.2}$$

and also

$$c\leq\gamma\,\varepsilon(\frac{h}{\gamma}) \tag{4.4}$$

Proof. Take any point $a\in A$; we have $\|a-c_A\|\leq r_A$. Also, $\|a-c_B\|\leq h+r_B$ (use the triangular inequality with points in B at a distance near h from a). We have (see (2.5)) $r_A\leq r_B+h\leq\gamma$; now let $a'=\frac{a-c_A}{\gamma}$, $a''=\frac{a-c_B}{\gamma}$; we have $\|a'\|\leq 1$; $\|a''\|\leq 1$; $\|a'-a''\|=$ $=\frac{\|c_B-c_A\|}{\gamma}=\frac{c}{\gamma}$; therefore, by using (4.1): $\left\|\frac{a'-a''}{2}\right\|=\frac{1}{\gamma}\left\|a-\frac{c_A-c_B}{2}\right\|\leq 1-\delta(\frac{c}{\gamma})$.

Therefore we obtain: $r_A\leq\sup_{a\in A}\left\|a-\frac{c_A+c_B}{2}\right\|\leq\gamma\left[1-\delta(\frac{c}{\gamma})\right]$, so $\gamma\cdot\delta(\frac{c}{\gamma})\leq\gamma-r_A$.

Similarly, we obtain $\gamma\,\delta(\frac{c}{\gamma})\leq\gamma-r_B$. In any case, we have $\gamma\cdot\delta(\frac{c}{\gamma})\leq\gamma-\max(r_A,r_B)=h$.

To prove (4.4), note that $\|x\|\leq\gamma$, $\|y\|\leq\gamma$ and $\|\frac{x+y}{2}\|\geq\gamma-\delta$ imply (see (4.2)) $\|x-y\|\leq$

$\leq \gamma\cdot\varepsilon\,(\frac{\delta}{\gamma})$. Now take $a\varepsilon A$ such that $r_A \leq \| a - \frac{c_A+c_B}{2} \|$: by using the above relation with $x=a-c_A$; $y=a-c_B$, we obtain $\|x-y\| = c \leq \gamma\cdot\varepsilon(\frac{\gamma-r_A}{\gamma})$; similarly, $c \leq \gamma\cdot\varepsilon\,(\frac{\gamma-r_B}{\gamma})$: thus $c \leq \gamma\cdot\varepsilon\left(\frac{\min\{\gamma-r_A,\gamma-r_B\}}{\gamma}\right) = \gamma\cdot\varepsilon\,(\frac{h}{\gamma})$. ■

REMARK 4.2. Recall that in any space we have

$$\varepsilon(\delta) \geq 2\sqrt{2\delta-\delta^2} \qquad \text{and} \qquad \delta(\varepsilon) \leq 1-\sqrt{1-\varepsilon^2/4} \tag{4.5}$$

while equality holds in (4.5) for Hilbert spaces.

Simple examples, e.g. in Hilbert spaces, show that the estimates given by Proposition 4.1 are in general very poor. For example, if $\|c_A-c_B\| \leq 2h$, then equality in (4.4) cannot hold. In fact, $2\gamma\sqrt{2\frac{h}{\gamma}-\frac{h^2}{\gamma^2}} < \gamma\,\varepsilon(\frac{h}{\gamma}) = c \leq 2h$ implies $\sqrt{8h\gamma-4h^2} \leq 2h$, thus $h \geq \gamma \geq \max\{r_A+h\,,\,r_B+h\}$ and so $r_A=r_B=0$; $h=\gamma$. If $\gamma\neq 0$, then $A=\{a\}$ and $B=\{b\}$ imply $h=\|a-b\| = h\,\varepsilon(1) = 2h$, thus an absurdity.

REMARK 4.3. We do not know if in uniformly convex spaces (2.7) and (2.8) are still precise, or if the inequalities obtained in Section 3 are true.

Finally, we recall another result (valid for spaces like L_p, $1<p<\infty$), proved recently in [7] (Theorem 4.1):

PROPOSITION 4.4. <u>Assume that a positive constant</u> c <u>exists such that</u>

$$\delta_X(\varepsilon) \geq c\,\varepsilon^q \quad (0<\varepsilon\leq 2) \quad \underline{\text{for some}}\ q\geq 2 \tag{4.6}$$

<u>Given a set</u> A, <u>if</u> $x=c_A$ <u>or</u> $x=c'_A$, <u>then we have</u>

$$r(x,A) \leq r(a,A) - k\,\|x-a\|^q \quad \underline{\text{for every}}\ a\varepsilon A \tag{4.7}$$

<u>where</u> k <u>is a constant depending only on</u> q <u>and</u> c.

REFERENCES

[1] J. ALONSO and C. BENITEZ, Some characteristic and non-characteristic properties of inner product spaces; Extracta Math. 1 (1986), 96-98.

[2] D. AMIR, Best simultaneous approximation (Chebyshev centers); Parametric Optimization and Approximation, Conf. Oberwolfach 1983, ISNM 72, Birkhäuser Verlag (Basel, 1985), 19-35.

[3] D. AMIR and F. DEUTSCH, Approximation by certain subspaces in the Banach space of continuous vector-valued functions; J. Approx. Theory 27 (1979), 254-270.

[4] D. AMIR and J. MACH, Chebyshev centers in normed spaces; J. Approx. Theory 40 (1984), 364-374.

[5] P.K. BELOBROV, On the Čebyšev point of a system of sets (in russian); Izv. Vysš. Učebn. Zaved. Matematika 55 (1966), n. 6, 18-24 (Amer. Math. Soc. Transl. 78 (1968), 119-126).

[6] T.-C. LIM, On asymptotic centers and fixed points of nonexpansive mappings; Canad. J. Math. 32 (1980), 421-430.

[7] B. PRUS and R. SMARZEWSKI, Strongly unique best approximations and centers in uniformly convex spaces; J. Math. Anal. Appl. 121 (1987), 10-21.

[8] P. SZEPTYCKI and F.S. VAN VLECK, Centers and nearest points of sets; Proc. Amer. Math. Soc. 85 (1982), 27-31.

[9] J.D. WARD, Existence and uniqueness of Chebyshev centers in certain Banach spaces; Doctoral Thesis, Purdue Univ., 1973.

APPROXIMATION BY LIPSCHITZ FUNCTIONS AND ITS APPLICATION TO BOUNDARY VALUE OF CAUCHY-TYPE INTEGRALS

Jorge Bustamante González
Department Theory of functions .Havana University.

Some problems in function theory deal with boundary values of analytic functions, specially they appear in the solution of singular integral equations and Riemann boundary problems. In this paper, necesary and sufficient conditions are given, in terms of some kind of approximation by Lipschitz functions,in order that the Cauchy-type integral has continuous boundary values.

The symbols L, L_n,and L^* will denote rectifiables Jordan curves on the complex plane. V is the set $[\,|z|<1\,]$ and V^* is the set $[\,|z|\leq 1\,]$. For a curve L, D(L) is the open bounded region limited by L and E(L) the unbounded one. C(L) is the class of continuous functions on L, Q(L) the class of analytic functions in E(L) which are continuous on $L\cup D(L)$. R(L) denotes the class of analytic functions in E(L) which are continuous in $L\cup E(L)$ and such that $f(\infty)=0$. S(L) is the direct sum of R(L) and Q(L) restricted to L. It's well known that each $f\in S(L)$ admits a unique decomposition of the form $f=f^{+}+f^{-}$, where $f^{+}\in Q(L)$ and $f^{-}\in R(L)$. Taking this into account it is possible to define in S(L) the norm

$$\|f\| = \|f^{+}\|_{\infty} + \|f^{-}\|_{\infty} \quad , \tag{1}$$

For some special curves there are characterizations of S(L) (see [1.2]). The class S(L) is connected with the Cauchy-type integrals in the following way. If $f = f^{+}+ f^{-}$, then $f^{+} = f + \hat{f}$ where

$$\hat{f}(t)=\frac{1}{2\pi i}\int_{L}\frac{f(x)-f(t)}{x-t}dx, \quad t\in L,$$

and the integral is understood in the sense of Cauchy principal value. For $f\in C(L)$ we define

$$F_f(z)=\frac{1}{2\pi i}\int_{L}\frac{f(x)}{x-z}dx, \quad z\in\mathbb{C}\setminus L.$$

Let H be the class of Lipschitz functions on L, that is, $f\in H$ iff there is a constan K_f such that , for $x,y\in L$

$$|\,f(x) - f(y)\,| \leq K_f\,|x - y\,|.$$

It is known that $H(L) \subseteq S(L)$ (see [3]).

The following lemma is relatively easy to prove.

<u>Lemma</u> Let (X,r), (Y,d) be metrics spaces, A a compact subset of X, and f_p ($p = 0,1,2\ldots$) a one-to-one continuous function on A such that $f_0(A) \subseteq f_p(A) \subseteq Y$ ($p = 1,2,\ldots$). If $f_n \longrightarrow f_0$ on A , then $f_p^{-1} \longrightarrow f_o^{-1}$ on $f_o(A)$ ($g_n \longrightarrow g$ stands for uniform convergence in the corresponding metric).

<u>Theorem 1</u> Let L be such that $0 \in D(L)$. Suppose that p_o is the conformal mapping that transforms E onto D, $p_o(\infty) = 0$, $p_o'(\infty) = 1$. There are functions $g_n \in R(L)$ such that $g_n|_L \in H(L)$ and $g_n \longrightarrow p_o$.

<u>Proof</u> Let L_n be a family of closed Jordan rectifiable curves such that

i) $0 \in D_n$, $E \cup L \subseteq D_n$, $D_n \subseteq D_{n+1}$, $n = 1,2,\ldots$;

ii) There is a one-to-one map $d_n(z)$ from L onto L_n such that $a_n \longrightarrow 0$ where

$$a_n = \sup_{z \in L} | z - d_n(z) |.$$

Let h_n be the conformal mapping that transforms E_n onto V such that $h_n(\infty) = 0$, $h_n'(\infty) = 1$; h the conformal mapping that transforms E onto V such that $h(\infty) = 0$ and $h'(\infty) = 1$; and ϕ the conformal mapping that transforms V onto D such that $\phi(0) = 0$ and $\phi'(0) = 1$.

It's clear that $p_o(w) = \phi(h(w))$ for $w \in E \cup L$, and ϕ in continuous in V^* and analytic in V , so there is a sequence of polynomials P_n such that $P_n \longrightarrow \phi$ in V . Make $\phi_n(w) = P_n(h_n)$ for $w \in E \cup L$.

Let us divide the proof in three parts:

a) $h_n \longrightarrow h$ on $E \cup L$.

Let $L_n^* = \{ \frac{1}{w},\ w \in L_n \}$ for $n = 0,1,\ldots$. L_n is a closed rectifiable Jordan curve . Put $F_n = D(L_n^*)$ and $g_n(w) = \left[h_n^{-1}(w)\right]^{-1}$,g_n transforms V conformally onto F_n . By Rado's Theorem [4] $g_n \longrightarrow g_o$ on $g_o(V^*)$.

Fix r ($0 < r \leq 1$),put $Q_r = \left\{z \ ;\ r \leq |z| \leq 1\right\}$,and $q_n(z)=(g_n(z))^{-1}$,$z\in Q_r$.

Obviously $q_n \longrightarrow q_o$, then by lemma 1 $(q_n)^{-1} \longrightarrow (q_o)^{-1}$. But

$$q_n(h_n z)) = (g_n(h_n(z)))^{-1} = h_n^{-1}(h_n(z)) = z,$$

then $q_n^{-1} = h_n \longrightarrow h_o$ on Q_r . Using Weierstrass' Theorem ,we obtain $h_n \longrightarrow h_o$ in $D \cup L$.

b) Fix $\varepsilon > 0$, since the family of polynomials (P_n),$n \in \mathbb{N}$, is equicontinuous then there is a $\delta(\varepsilon) > 0$ such that for $|x - y| < \delta$, $| x | \leq 1$ and $| y | \leq 1$, $| P_n(x) - P_n(y)| < \frac{\varepsilon}{2}$ for every $n \in \mathbb{N}$ .Let N_ε be such that for $n > N_\varepsilon$, $\| h_n - h \| < \delta$, and $\| \phi - P_n\| < \frac{\varepsilon}{2}$, then

$$|\phi(h_o(z))- P_n(h_n(z))|\leq | \phi(h_o(z)) - P_n(h_o(z)) | +$$

$$+|\, P_n(h_o(z)) - P_n(h_n(z))\, | < \varepsilon \, .$$

c) Since h_n is analytic on E_n, it has a bounded derivative on L, hence for $n = 1,2,\ldots,$ $h_n \in H(L)$.

From Riesz' Theorem and properties of the direct sum of normed spaces it is easy to verify the following :

Lemma 2 For every bounded linear functional Ψ on $S(L)$ there is a pair of regular mesures μ and λ such that

$$\Psi(f) = \int_L f^+ \, d\mu + \int_L f^- \, d\lambda$$

for every $f = f^+ + f^- \in S(L)$, where $f^+ \in D(L)$ and $f^- \in E(L)$.

Theorem 2 Let L be a closed rectifiable Jordan curve and $f \in C(L)$. Then $f \in S(L)$ iff there is a sequence $f_n \in H(L)$ such that

$$\| f_n + \hat{f}_n - f - \hat{f} \|_\infty + \| \hat{f}_n - \hat{f} \|_\infty \longrightarrow 0. \qquad (2)$$

Proof It may be supposed that $0 \in D(L)$.

a) First, let us prove the if part. Suppose Ψ is a linear bounded functional on $S(L)$ such that $\Psi(f) = 0$ for every $f \in H(L)$. It is sufficient to prove that $\Psi(f) = 0$ for $f \in S(L)$. Let μ and λ be regular mesure connected to Ψ according to lemma 2. Since every polynomial is in $H(L)$ and every function in $Q(L)$ can be uniformly approximated by polynomials, then $\int f^+ d\mu = 0$ for every $f \in S(L)$ (see lemma 2 for notation). Hence

$$\Psi(f) = \int f^- d\lambda = \Psi(f^-) \quad , \ f \in S(L) \, ,$$

and $f = f^+ + f^-$, where $f^+ \in Q(L)$ and $f^- \in R(L)$.

Let ϑ be the conformal mapping that transforms $E(L)$ onto $D(L)$, $\vartheta(\infty) = 0$ $\vartheta'(\infty) = 1$. By theorem 1 there is a sequence $\vartheta_n \in R(L)$ such that $\vartheta_n|_L \in H(L)$ and $\vartheta_n \longrightarrow \vartheta$. Then, since

$$|\Psi(\vartheta)| = |\int \vartheta d\lambda| \le \| \lambda \| \, \| \vartheta - \vartheta_n \|_\infty \, ,$$

it follows that $\Psi(\vartheta) = 0$.

For every positive integer q, $(\vartheta_n)^q \in H(L)$; $(\vartheta_n)^q \in H(L)$, $(\vartheta)^q \in R(L)$ and $(\vartheta_n)^q \longrightarrow (\vartheta)^q$, then $\Psi((\vartheta)^q) = 0$ for $q = 1,2,\ldots.$

Let $S_o = \{ f \in Q(L) : f(0) = 0 \}$ and

$$\Lambda(f) = \int_L f(\vartheta(z)) \, d\lambda(z) \qquad , \ f \in S_o \, .$$

Λ is a bounded linear functional on S_o (with the sup norm),

$$\sup \Big\{ \, | f(\vartheta(z)) | \, , \ z \in L \Big\} = \sup \Big\{ |f(z)| \, , \ z \in L \Big\} ,$$

and

$$\Lambda(z^q) = \int_L (\vartheta(z))^q \, d\lambda(z) = 0 \quad , \ q = 1,2,\ldots \, .$$

For $f \in S_o$ there is a sequence of polynomials P_n such that $P_n(0) = 0$ and $\| P_n - f \|_\infty \longrightarrow 0$ on $D \cup L$. This yields that $\Lambda(f) = 0$, $f \in S_o(L)$. Thus , $\Psi(f) = 0$ for $f \in S(L)$.

b) Let $f \in C(L)$ and suppose there is a sequence $f_n \in H(L)$ such that (2) takes place. It's known that $\hat{f}_n \in R(L)$ and $f_n + \hat{f}_n \in Q(L)$, so $f_n = f_n + \hat{f}_n - \hat{f}_n \in S(L)$. From (2) , it follows that $\| f_n - f \|_\infty \longrightarrow 0$, so $\hat{f} \in C(L)$

It's sufficient to prove that $f + \hat{f} \in R(L)$ (for $- f \in Q(L)$, the proof is similar).

Let $z \in D(L)$, $\alpha = d(z,L)$ is the distance from z to L . Let N_α be such that for $n > N_\alpha$

$$\| f_n + \hat{f}_n - f - \hat{f} \|_\infty + \| \hat{f}_n - \hat{f} \|_\infty < \alpha ,$$

then

$$| F_f(z) | \le | F_f(z) - F_{f_n}(z) | + | F_{f_n}(z) | \le$$

$$\frac{1}{2\pi}\int_L |dz| + \| f_n + \hat{f}_n \|_\infty \le$$

$$\frac{1}{2\pi}\int_L | dz | + \| f_n + \hat{f}_n - f - \hat{f} \|_\infty + \| f + \hat{f} \|_\infty \le$$

$$\frac{1}{2\pi}\int_L |dz| + \alpha + \| f + \hat{f} \|.$$

So $F_f(z)$ $(z \in D)$ is bounded .

Let ϑ be the conformal mapping that transforms V onto D . $F_f o\, \vartheta$ is a bounded analytic function on V . According to Privalov's Theorem [5] the boundary values of $F_f o \vartheta$ are equal (almost everywhere) to the continuous function $f o\vartheta(e^{i\theta}) + \hat{f} o\vartheta(e^{i\theta})$. $F_f o\vartheta \in H_1$ (H_1 denotes the usual Hardy space) then, by Fijtengolz's Theorem [4, 392],

$$g(re^{i\theta}) = F_f o\vartheta(e^{i\theta}) = \frac{1}{2\pi}\int_0^{2\pi} F_f o\vartheta(e^{i\theta})P(r,t-\theta)dt \quad , \quad 0 \le r < 1 ,$$

where $P(r,t-\theta)$ is Poisson's kernel. The functions $Re(F_f o\vartheta)$ and $Im(F_f O\vartheta)$ are continuous on $\left[\; |z|<1 \; \right]$, so the harmonic functions $Re(g(re^{i\theta}))$ and $Im(g(re^{i\theta}))$ are continuous on V^* . Then $F_f o\vartheta$ is continuous on $D \cup L$. It's clear that $F_f(w)$ is analytic on D. This completes the proof of the theorem.

REFERENCES

[1] V.V.Salaev ,A.O.Tokov, Neobjodimîe i dostatochnîe usloviya neprerîvnostî integrala Koshi v zamknutoy oblasti . Dokl. Aka. Nauk. Azerbai. S.S.R. Tom. XXXIX . No. 12, 1983, 7-10.

[2] B.Gonzalez D., Condiciones necesarias y suficientes para la continuidad hasta la frontera de la integral del tipo Cauchy para

curvas cerradas no suaves. Ciencias Matematica. Vol. V, No. 12, 1984. Cuba ,3-12.

[3] V.V. Salaev, Priamîe i obratnîe otzenki dlya osobogo integrala Koshi no zamknumoi kpivoi . Mat. Zam. Tom. 19, No 3,(1976),365-380.

[4] G.M.Goluzin, Geometicheskaya teoriya funktziî kompleksnogo peremennogo,1966.

[5] I.I.Privalov, Granichnîe svoystva analiticheskij funktziî , Gosmejizgat,1950.

ASYMPTOTICS FOR THE RATIO OF THE LEADING COEFFICIENTS OF ORTHOGONAL POLYNOMIALS ASSOCIATED WITH A JUMP MODIFICATION

M.A. Cachafeiro
Departamento de Matemática Aplicada
E.T.S. Ingenieros Industriales
36280 Vigo, España.

F. Marcellán
Departamento de Matemática Aplicada
E.T.S. Ingenieros Industriales
28006 Madrid, España.

Abstract: In this paper we study the behaviour of the quotient $\dfrac{e_n(\alpha)}{e_n(\beta)}$, where $(e_n(\alpha))^{-1/2}$ and $(e_n(\beta))^{-1/2}$ are the leading coefficients of the orthonormal polynomials $P_n(z,d\alpha)$ and $\hat{P}_n(z,d\beta)$ on the unit circle U with respect to the measures α and $\beta=\alpha+\varepsilon\delta(t)$ ($t\in U$, $\varepsilon>0$, $\delta(t)$ the unit measure supported at $z=t$). By proving that it is possible to build a sequence of orthonormal polynomials over U with prescribed values at an arbitrary point of U, some examples are given in which $\lim\limits_{n\to\infty}\dfrac{e_n(\alpha)}{e_n(\beta)}\neq 1$.

1. First at all we shall prove that it is possible to build a sequence of orthonormal polynomials over U with prescribed values at an arbitrary point of U.

Proposition 1:

Let $\{b_n\}_{n\in N}$ be a sequence of complex numbers such that $b_0=1$, $b_n\neq 0$ $\forall n$ and $\left|\dfrac{b_n}{b_{n-1}}-t\right|<1$ $\forall n\geq 1$ for some t in U. Then there is a sequence of monic orthogonal polynomials over U, $\{\tilde{P}_n(z)\}$, such that $\tilde{P}_n(t)=b_n$ $\forall n\geq 0$.

Proof:

$$\text{Put } a_{n-1}=\frac{\bar{b}_{n-1}}{\bar{t}^{n-1}b_{n-1}}\left(\bar{t}-\frac{\bar{b}_n}{\bar{b}_{n-1}}\right) \text{ for } n\geq 1. \tag{1}$$

Then $|a_{n-1}|<1$ $(n\geq 1)$, so that there is a sequence of monic orthogonal polynomials over U, $\{\tilde{P}_n(z)\}$, such that $-\tilde{P}_n(0)=a_{n-1}$ $(n\geq 1)$ (see [4]).

As is well known, the polynomials $\tilde{P}_n(z)$ satisfy the recurrence formula: $\tilde{P}_n(z) = z\tilde{P}_{n-1}(z)-\overline{a_{n-1}}\tilde{P}^*_{n-1}(z)$ $(n\geq 1)$ and $\tilde{P}_0(z) = 1$.

Taking z=t we obtain $a_{n-1} = \dfrac{\overline{\tilde{P}_{n-1}(t)}}{\bar{t}^{n-1}\tilde{P}_{n-1}(t)}\left(\bar{t}-\dfrac{\overline{\tilde{P}_n(t)}}{\overline{\tilde{P}_{n-1}(t)}}\right)$ $(n\geq 1)$ (2)

In particular $a_0=\bar{t}-\overline{\tilde{P}_1(t)}$ and from (1) that $\tilde{P}_1(t)=b_1$. By induction $\tilde{P}_n(t)=b_n$ $\forall n$.

Proposition 2:

Let $\{b_n\}$ be a sequence of complex numbers such that $b_0 \in R$ and $b_n\neq 0$ $\forall n$. Then for every $t\in U$ there exists a sequence of orthonormal polynomials over U, $\{\hat{Q}_n(z)\}$, such that $\hat{Q}_n(t)=b_n$ $\forall n$.

Proof:

Put $a_{n-1} = \dfrac{\bar{b}_{n-1}}{\bar{t}^{n-2}b_{n-1}}\left(\dfrac{1-\left(\dfrac{\bar{b}_n}{\bar{t}\,\bar{b}_{n-1}}\right)^2}{1+\left|\dfrac{b_n}{b_{n-1}}\right|^2}\right)$ $(n\geq 1)$. Then $a_{n-1} = \dfrac{\bar{b}_{n-1}}{\bar{t}^{n-1}b_{n-1}}\left(\bar{t}-(1-|a_{n-1}|^2)^{\frac{1}{2}}\cdot\dfrac{\bar{b}_n}{\bar{b}_{n-1}}\right)$ (3) and $|a_{n-1}|<1$ $(n\geq 1)$.

Hence there exists a sequence of monic orthogonal polynomials $\{\tilde{P}_n(z)\}$ such that $-\overline{\tilde{P}_n(0)} = a_{n-1}$ $(n\geq 1)$ verifying the well known recurrence formula. Since $\dfrac{e_n}{e_{n-1}} = 1-|\tilde{P}_n(0)|^2 = 1-|a_{n-1}|^2$, the corresponding orthonormal polynomial sequence satisfies the following relation:

$$(1-|a_{n-1}|^2)^{\frac{1}{2}}\hat{P}_n(z) = z\hat{P}_{n-1}(z)-\overline{a_{n-1}}\,z^{n-1}\overline{\hat{P}_{n-1}}(z).$$

Then $a_{n-1} = \dfrac{\overline{\hat{P}_{n-1}(t)}}{\bar{t}^{n-1}\hat{P}_{n-1}(t)}\left[\bar{t} - (1 - |a_{n-1}|^2)^{\frac{1}{2}}\dfrac{\overline{\hat{P}_n(t)}}{\overline{\hat{P}_{n-1}(t)}}\right]$, and as a particular situation we obtain $a_0 = \bar{t} - (1 -|a_0|^2)^{\frac{1}{2}}\dfrac{\overline{\hat{P}_1(t)}}{\overline{\hat{P}_0(t)}}$. By comparison with (3) $\hat{P}_1(t) = \dfrac{\hat{P}_0(t)}{b_0}b_1$, and by induction $\hat{P}_n(t) = \dfrac{\hat{P}_0(t)}{b_0}b_n$ $\forall n$.

If we define the inner product: $<P(z),Q(z)>_1 = \frac{\hat{P}_0^2(t)}{b_0^2}\cdot$ $\cdot<P(z),Q(z)>$ $\forall\, P(z),Q(z)\in\Pi$, where $<P(z),Q(z)>$ denotes the inner product whose corresponding orthonormal sequence is $\{\hat{P}_n(z)\}$, then the sequence $\{\hat{Q}_n(z)\} = \{\frac{b_0}{\hat{P}_0(t)}\hat{P}_n(z)\}$ satisfies $<\hat{Q}_n(z),\hat{Q}_m(z)>_1 = \delta_{nm}$ and $\hat{Q}_n(t) = b_n$ $\forall n$, and this completes the proof.

2. If we denote by $\{(E_n)^{-1/2}\}$ the sequence of the leading coefficients of the orthonormal polynomials $\{\hat{P}_n(z,d\tau)\}$ on the unit circle with respect to the measure $d\tau = |z-t|^2 d\alpha$, we study the sequences $e_n(\alpha)/e_n(\beta)$ and $E_{n-1}/e_n(\alpha)$. In case C, it is clear that their limits exist and are equal to 1, but in case D different situations appear.

Proposition 3:

i) If $\{\hat{P}_n(t,d\alpha)\}\in l^2$ or $\{\hat{P}_n(t,d\alpha)\}\notin l^2$ but $\exists \lim_{n\to\infty}|\hat{P}_n(t,d\alpha)|^2\in R$, then $\lim_{n\to\infty}\frac{e_n(\alpha)}{e_n(\beta)} = 1$.

ii) If $\exists \lim_{n\to\infty}|\hat{P}_n(t,d\alpha)|^2 = +\infty$ different situations appear.

iii) If $\{|\hat{P}_n(t,d\alpha)|^2\}$ has no limit, then $\overline{\lim}\frac{e_n(\alpha)}{e_n(\beta)} = 1$.

Proof:

i) Since $0\leq 1-\frac{e_n(\alpha)}{e_n(\beta)} = \frac{\varepsilon|\hat{P}_n(t,d\alpha)|^2}{1+\varepsilon K_n(t,t;d\alpha)}$ [see 1], it is clear that $\lim_{n\to\infty}\frac{e_n(\alpha)}{e_n(\beta)} = 1$.

ii) If $\lim_{n\to\infty}|\hat{P}_n(t,d\alpha)|^2=+\infty$ it is easy to give examples showing different situations. Let $\{\hat{P}_n(t,d\alpha)\}$ be a sequence of orthonormal polynomials over U defined by:

$\hat{P}_n(t,d\alpha)=n^{1/2}$ $n\geq 1$, $\hat{P}_0(t,d\alpha)=1$. In this case $\lim_{n\to\infty}\frac{e_n(\alpha)}{e_n(\beta)} =$

$$= \lim_{n\to\infty}\frac{1+\varepsilon K_{n-1}(t,t;d\alpha)}{1+\varepsilon K_n(t,t;d\alpha)} = \lim_{n\to\infty}\frac{1+\varepsilon[1+\frac{(n-1)n}{2}]}{1+\varepsilon[1+\frac{(n+1)n}{2}]} = 1.$$

$\hat{P}_n(t,d\alpha) = e^n$.

In this case $\lim\limits_{n\to\infty} \dfrac{e_n(\alpha)}{e_n(\beta)} = \lim\limits_{n\to\infty} \dfrac{1+\varepsilon \dfrac{(e^{2n}-1)}{e^2-1}}{1+\varepsilon \dfrac{(e^{2n+2}-1)}{e^2-1}} = \dfrac{1}{e^2}$

$\hat{P}_n(t,d\alpha) = (n!-(n-1)!)^{\frac{1}{2}}$

In this case $\lim\limits_{n\to\infty} \dfrac{e_n(\alpha)}{e_n(\beta)} = \lim\limits_{n\to\infty} \dfrac{1+\varepsilon(n-1)!}{1+\varepsilon n!} = 0$

$\hat{P}_{2n}(t,d\alpha) = n^{\frac{1}{2}}$ $n\geqslant 1$ $\hat{P}_0(t,d\alpha)=1$, $\hat{P}_{2n+1}(t,d\alpha)=e^n$.

In this case we have $K_{2n-1}(t,t;d\alpha) = 1+ \dfrac{n(n-1)}{2} + \dfrac{e^{2n}-1}{e^2-1}$,

$K_{2n}(t,t;d\alpha) = 1+ \dfrac{n(n+1)}{2} + \dfrac{e^{2n}-1}{e^2-1}$ and $K_{2n+1}(t,t;d\alpha) = 1+ \dfrac{n(n+1)}{2} + \dfrac{e^{2n+2}-1}{e^2-1}$

$\lim\limits_{n\to\infty} \dfrac{K_{2n-1}(t,t;d\alpha)}{K_{2n}(t,t;d\alpha)} = 1$, $\lim\limits_{n\to\infty} \dfrac{K_{2n}(t,t;d\alpha)}{K_{2n+1}(t,t;d\alpha)} = \dfrac{1}{e^2} \Longrightarrow \overline{\lim} \dfrac{e_n(\alpha)}{e_n(\beta)} = 1$ and

$\underline{\lim} \dfrac{e_n(\alpha)}{e_n(\beta)} > 0$

$\hat{P}_{2n}(t,d\alpha) = (n!-(n-1)!)^{\frac{1}{2}}$ $\hat{P}_{2n+1}(t,d\alpha) = n^{\frac{1}{2}}$.

In this case we have that $K_{2n-1}(t,t;d\alpha) = (n-1)! + \dfrac{n(n-1)}{2}$,

$K_{2n}(t,t;d\alpha) = n!+ \dfrac{n(n-1)}{2}$, $K_{2n+1}(t,t;d\alpha) = n!+ \dfrac{n(n+1)}{2}$

$\lim\limits_{n\to\infty} \dfrac{K_{2n-1}(t,t;d\alpha)}{K_{2n}(t,t;d\alpha)} = 0$, $\lim\limits_{n\to\infty} \dfrac{K_{2n}(t,t;d\alpha)}{K_{2n+1}(t,t;d\alpha)} = 1 \Longrightarrow \overline{\lim} \dfrac{e_n(\alpha)}{e_n(\beta)} = 1$ and

$\underline{\lim} \dfrac{e_n(\alpha)}{e_n(\beta)} = 0$.

iii) $\underline{\lim} |\hat{P}_n(t,d\alpha)|^2 \geqslant 0$ and $\underline{\lim} |\hat{P}_n(t,d\alpha)|^2 \neq \infty \Longrightarrow$

$\Longrightarrow \underline{\lim} \dfrac{\varepsilon |\hat{P}_n(t,d\alpha)|^2}{1+\varepsilon K_n(t,t;d\alpha)} = 0 \Longrightarrow \overline{\lim} \dfrac{e_n(\alpha)}{e_n(\beta)} = 1.$

<u>Remark</u> In case iii) it is also true that $\underline{\lim} \dfrac{e_n(\beta)}{e_n(\alpha)} = 1$, and it is not difficult to give an example showing that $\lim\limits_{n\to\infty} \dfrac{e_n(\alpha)}{e_n(\beta)} \neq 1$. Let $\{\hat{P}_n(z,d\alpha)\}$ be a sequence of orthonormal polynomials over U defined by their values at t: $\hat{P}_{2n}(t,d\alpha) = e^n$ and $\hat{P}_{2n-1}(t,d\alpha) = 1$. Since

$\lim_{n\to\infty} \frac{K_{2n}(t,t;d\alpha)}{K_{2n+1}(t,t;d\alpha)} = 1$ and $\lim_{n\to\infty} \frac{K_{2n-1}(t,t;d\alpha)}{K_{2n}(t,t;d\alpha)} = \frac{1}{e^2}$, $\{\frac{K_{n-1}(t,t;d\alpha)}{K_n(t,t;d\alpha)}\}$ has no limit, and the limit of $\{\frac{e_n(\alpha)}{e_n(\beta)}\}$ does not exist.

<u>Proposition 4:</u>

$$\text{i) } 1 - \frac{e_n(\alpha)}{e_n(\beta)} = \varepsilon\left(\frac{E_{n-1}}{e_n(\beta)} - 1\right)K_{n-1}(t,t;d\alpha) \qquad (4)$$

ii) $e_n(\alpha)<e_n(\beta)<E_{n-1}$

<u>Proof:</u>

i) It is a consequence of the relation

$\frac{e_n(\alpha)}{e_n(\beta)} = \frac{1+\varepsilon K_{n-1}(t,t;d\alpha)}{1+\varepsilon K_n(t,t;d\alpha)}$ [see 1] together with the relation

$$\frac{E_{n-1}}{e_n(\alpha)} = \frac{K_n(t,t;d\alpha)}{K_{n-1}(t,t;d\alpha)}$$

ii) Since $e_n(\beta)>e_n(\alpha)$, it follows from it that $E_{n-1}>e_n(\beta)$.

<u>Proposition 5:</u>

$$\lim_{n\to\infty} \frac{E_{n-1}}{e_n(\beta)} = 1$$

<u>Proof:</u> If $\{\hat{P}_n(t,d\alpha)\}\in l^2$, we can use proposition 3 together with (4) and the result follows.

If $\{\hat{P}_n(t,d\alpha)\}\notin l^2$, since $\{1 - \frac{e_n(\alpha)}{e_n(\beta)}\}$ is bounded and $\lim_{n\to\infty} \frac{1}{K_{n-1}(t,t;d\alpha)} = 0$, from (4) we have $\lim_{n\to\infty} \frac{E_{n-1}}{e_n(\beta)} = 1$.

<u>Corollary 1:</u>

i) If $\{\hat{P}_n(t,d\alpha)\}\in l^2$ or $\{\hat{P}_n(t,d\alpha)\}\notin l^2$ but $\lim_{n\to\infty}|\hat{P}_n(t,d\alpha)|^2\in R$, then $\lim_{n\to\infty} \frac{E_{n-1}}{e_n(\alpha)} = 1$.

ii) If $\lim_{n\to\infty}|\hat{P}_n(t,d\alpha)|^2=+\infty$, different situations appear.

iii) If $\{|\hat{P}_n(t,d\alpha)|^2\}$ has no limit, then $\overline{\lim}\ \frac{e_n(\alpha)}{E_{n-1}} = 1$.

Proof:

Since $\frac{e_n(\alpha)}{E_{n-1}} = \frac{e_n(\alpha)}{e_n(\beta)} \frac{e_n(\beta)}{E_{n-1}}$, i) and iii) follow from proposition 3 and 5.

If $\underline{\lim}_{n\to\infty}|\hat{P}_n(t,d\alpha)|^2=+\infty$, it is easy to give examples showing different situations. Let $\{\hat{P}_n(t,d\alpha)\}$ be a sequence of orthonormal polynomials over U defined as in proposition 3:

If $\hat{P}_n(t,d\alpha)=n^{\frac{1}{2}}$ $n\geq 1$, $\hat{P}_0(t,d\alpha)=1$, then $\lim_{n\to\infty} \frac{e_n(\alpha)}{E_{n-1}} = 1$.

If $\hat{P}_n(t,d\alpha)=e^n$, then $\lim_{n\to\infty} \frac{e_n(\alpha)}{E_{n-1}} = \frac{1}{e^2}$.

If $\hat{P}_n(t,d\alpha)=(n!-(n-1)!)^{\frac{1}{2}}$, then $\lim_{n\to\infty} \frac{e_n(\alpha)}{E_{n-1}} = 0$.

If $\hat{P}_{2n}(t,d\alpha)=n^{\frac{1}{2}}$ $n\geq 1$, $\hat{P}_0(t,d\alpha)=1$ and $\hat{P}_{2n+1}(t,d\alpha)=e^n$, then

$\underline{\lim}\ \frac{e_n(\alpha)}{E_{n-1}} = \underline{\lim}\ \frac{e_n(\alpha)}{e_n(\beta)} > 0$ and $\overline{\lim}\ \frac{e_n(\alpha)}{E_{n-1}} = 1$.

If $\hat{P}_{2n}(t,d\alpha) = (n!-(n-1)!)^{\frac{1}{2}}$, $\hat{P}_{2n+1}(t,d\alpha)=n^{\frac{1}{2}}$, then

$\underline{\lim}\ \frac{e_n(\alpha)}{E_{n-1}} = 0$ and $\overline{\lim}\ \frac{e_n(\alpha)}{E_{n-1}} = 1$.

Remark:

a) If $\{|\hat{P}_n(t,d\alpha)|^2\}$ has no limit, the above example shows that $\{\frac{e_n(\alpha)}{E_{n-1}}\}$ has no limit.

b) If $d\alpha>0$ a.e. on U, the preceding results follow from Rahmanov's result: $\lim_{n\to\infty} \frac{|\hat{P}_n(t,d\alpha)|^2}{K_n(t,t;d\alpha)} = 0$, which implies that

$\lim_{n\to\infty} \frac{e_n(\alpha)}{e_n(\beta)} = 1$ and $\lim_{n\to\infty} \frac{E_{n-1}}{e_n(\alpha)} = 1$.

Proposition 6:

$$\lim_{n\to\infty} E_{n-1} = \lim_{n\to\infty} e_n(\alpha)$$

Proof:

$$E_{n-1}-e_n(\beta) = e_n(\beta)\left(\frac{E_{n-1}}{e_n(\beta)} - 1\right)$$

Since $\lim\limits_{n\to\infty} \dfrac{E_{n-1}}{e_n(\beta)} = 1 \Longrightarrow \lim\limits_{n\to\infty} E_{n-1}-e_n(\beta) = 0 \Longrightarrow \lim\limits_{n\to\infty} E_{n-1} =$

$= \lim\limits_{n\to\infty} e_n(\beta)$.

As $\lim\limits_{n\to\infty} e_n(\alpha) = \lim\limits_{n\to\infty} e_n(\beta)$ (see [1]), then $\lim\limits_{n\to\infty} E_{n-1} = \lim\limits_{n\to\infty} e_n(\alpha)$

REFERENCES:

[1] Cachafeiro, A. y Marcellán, F. "Orthogonal polynomials and jump modifications". Communication to II International Symposium on Orthogonal Polynomials and their Applications. Segovia, 1.986

[2] Freud, G. "Orthogonal Polynomials". Pergamon Press. New York, 1.971.

[3] García Lazaro, P. "Modificaciones sobre la función de distribución". Tesina de Licenciatura. U.C. Madrid, 1.984.

[4] Geronimus, L. "Orthogonal Polynomials". Consultants Bureau. New York, 1.961.

[5] Nevai, P. "Orthogonal Polynomials". Mem. Amer. Math. Soc. 213, 1.979.

CONVERGENCE OF PADE APPROXIMANTS IN A NON-COMMUTATIVE ALGEBRA

André DRAUX
Université de Lille Flandres Artois
Laboratoire d'Analyse numérique-Bât M3
59655 VILLENEUVE D'ASCQ-CEDEX

Abstract

Properties of convergence of Padé approximants $[n+k+1/k]_f$ are proved by using theorems of convergence of non-commutative continued fractions.

1 INTRODUCTION

Some authors have already studied some convergence problems of matrix Padé approximants: Graffi and Grecchi [9] in the case where moments c_i are hermitian matrices, Von Sydow ([15] and [16]) for Stieltjes series, Delsarte and al [1] for a linear functional defined by a measure given by a hermitian matrix on the interval $[0,2\pi]$ for they study orthogonal polynomials on the unit circle.

As for us convergence theorems of non-commutative continued fractions will be used in this paper. Several authors have given such results: Wynn [17], Fair ([6] and [7]), Negoescu ([11] to [13]), Hayden [10], Peng and Hessel [14], Denk and Riederle [2], Field [8]. We will essentially used some theorems similar to Worpitzky theorem (Negoescu [11] and Field [8]) to give convergence theorems of Padé approximants.

Let $\mathcal{A}$ be a non-commutative Banach algebra with an unity element I on an infinite commutative field K.

Let f be a formal power series:

$$f(t) = \sum_{i=0}^{\infty} c_i t^i$$

for which $t \in K$, and $c_i \in \mathcal{A}$, $\forall i \in \mathbb{N}$.

Let $c^{(n)}$ $\forall n \in Z$ be left linear functionals acting on the set $\mathcal{P}$ of polynomials whose the coefficients belong to $\mathcal{A}$ and the variable to K. They are defined from their moments by:

$$c^{(n)}(\lambda x^r) = c_{n+r}\lambda \quad \forall r \in \mathbb{N},\ \forall n \in Z \text{ and } \forall \lambda \in \mathcal{A} \tag{1}$$

By convention $c_i=0$ if $i<0$.

Right linear functionals can be defined by the same way; in (1) $c_{n+r}\lambda$ is then replaced by λc_{n+r}. But in all the sequel left linear functionals and left expressions will only be used.

$M_k^{(n)}$ will denote the Hankel matrices $(c_{n+i+j})_{i=j=1}^{k-1}$ $\forall k \in \mathbb{N}$ and $\forall n \in Z$.

Definition 1

A polynomial P will be called monic if its leading coefficient is equal to I; P will be called quasi-monic if this coefficient has an inverse.

A left Padé approximant $[k\text{-}1/k]_f$ is the product of a polynomial $\tilde{Q}_k$ and the inverse of a polynomial $\tilde{P}_k$ such that:

$$f(t)-\tilde{Q}_k(t)(\tilde{P}_k(t))^{-1}=O(t^{2k})$$

where P_k is a monic polynomial of degree k exactly defined from $\tilde{P}_k$ by $\tilde{P}_k(t)=t^k P_k(t^{-1})$, and $Q_k(t)=t^{k-1}\tilde{Q}_k(t)$ is the left associated polynomial of P_k with respect to the left linear functional $c^{(0)}$ (see [3]), that is to say:

$$Q_k(t)=c^{(0)}((P_k(x)-P_k(t))(x-t)^{-1}) \tag{2}$$

Left Padé approximants with various degrees of the two polynomials can be defined (see[3]).

Let f_n be the formal power series $\sum_{j=0}^{\infty} c_{n+j}t^j$ $\forall n \in Z$. Then:

$$[k\text{-}1+n/k]_f= \sum_{i=0}^{n-1} c_i t^i +t^n [k\text{-}1/k]_{f_n} \text{ with } k\text{-}1+n\geqslant 0 \tag{3}$$

are left Padé approximants for which:

$$f(t)-[k\text{-}1+n/k]_f(t)=O(t^{2k+n}),$$

and $[k\text{-}1/k]_{f_n}$ is the left Padé approximant of f_n:

$$[k\text{-}1/k]_{f_n}(t)=\tilde{Q}_k^{(n)}(t)(\tilde{P}_k^{(n)}(t))^{-1}$$

Theorem 2 [3] and [4].

If $M_k^{(n)}$ has an inverse then the monic polynomial $P_k^{(n)}$ of degree k exactly is orthogonal with respect to $c^{(n)}$, and it is unique. $Q_k^{(n)}$ is the left associated polynomial of $P_k^{(n)}$ with respect to $c^{(n)}$ defined like in (2). The left Padé approximant $[n+k\text{-}1/k]_f$ also exists and is identical to the right Padé approximant. Moreover it is unique.

The Padé approximant $[k\text{-}1+n/k]_f$ will also be written:

$$[k\text{-}1+n/k]_f(t)=\tilde{Z}_k^{(n)}(t)(\tilde{P}_k^{(n)}(t))^{-1} \tag{4}$$

2 CONVERGENCE OF NON-COMMUTATIVE CONTINUED FRACTIONS.

A left non-commutative continued fraction will be written as follows:

$$b_0+a_1(b_1+(a_2(b_2+\ldots(\ldots)^{-1}\ldots)^{-1})^{-1})^{-1} \tag{5}$$

and the left n^{th} convergent:

$$E_n=b_0+a_1(b_1+a_2(b_2+\ldots+a_n(b_n)^{-1})^{-1}\ldots)^{-1} \tag{6}$$

can be computed thanks to the relations given by Wynn [18]:

$$G_0=b_n,\ G_{r+1}=b_{n-r-1}=a_{n-r}\ G_r^{-1}, \forall r \in \mathbb{N} \text{ such that } 0\leqslant r\leqslant n-1,\ G_n=E_n. \qquad (7)$$

Any convergent E_n can be written $Q_n(P_n)^{-1}$ where P_n and Q_n satisfy the same three-term recurrencerelation:

$$P_n=P_{n-1}b_n+P_{n-2}a_n,\quad \forall n \in \mathbb{N} \text{ such that } n\geqslant 1, \qquad (8)$$

with the initializations: $P_{-1}=0$, $P_0=I$, $Q_{-1}=I$ and $Q_0=b_0$.

Now a result of convergence proved by Negoescu [11] is given for a continued fraction (5) for which $b_0=0$ and $a_i=I$ $\forall i \in \mathbb{N}$ such that $i\geqslant 1$. The a_i's are assumed to depend on variables in a domain $\mathcal{D}$ within K.

Theorem 3 [11]

If in $\mathcal{D}$, $\|a_i\|\leqslant {}^1/_4$ $\forall i \in \mathbb{N}$ such that $i\geqslant 2$, then:

i) the continued fraction (5) converges uniformly to an element $F \in \mathcal{A}$.

ii) the values of the continued fraction (5) and of the convergents belong to the set:

$$V=\{X:\ X \in \mathcal{A} \ \text{ such that } \ \|a_1-X\|\leqslant\|X\|/2\}$$

iii) ${}^1/_4$ is the best constant such that $\|a_i\|\leqslant {}^1/_4$ involves the convergence of the continued fraction (5).

Field [8] gives another theorem for the convergence of (5), which can be extended to the case where the a_i's depend on variables in $\mathcal{D}$.

Theorem 4

If in $\mathcal{D}$, $\|a_{2n+1}\|\leqslant \upsilon^2$ and $\|(a_{2n})^{-1}\|\leqslant(1+\upsilon)^{-2}$ $\forall n \in \mathbb{N}$ such that $n\geqslant 1$, with $0<\upsilon<1$, then (5) converges uniformly to an element of $\mathcal{A}$.

A result of Wynn [18] can be generalized for a continued fraction (5).

Theorem 5

If

$$b_0+a_1\gamma_1(b_1\gamma_1+a_2\gamma_1\gamma_2(\gamma_1^{-1}b_2\gamma_1\gamma_2+\gamma_1^{-1}a_3\gamma_1\gamma_2\gamma_3((\gamma_1\gamma_2)^{-1}b_3\gamma_1\gamma_2\gamma_3+\ldots$$
$$\ldots+(\gamma_1\ldots\gamma_{n-2})^{-1}a_n\gamma_1\ldots\gamma_n((\gamma_1\ldots\gamma_{n-1})^{-1}b_n\gamma_1\ldots\gamma_n)^{-1})^{-1}\ldots)^{-1}=V_n(U_n)^{-1} \qquad (9)$$

$\forall n \in \mathbb{N}$ such that $n\geqslant 1$, and with γ_i invertible $\forall i \in \mathbb{N}$ such that $i\geqslant 1$, then

$$U_n=P_n\gamma_1\ldots\gamma_n \text{ and } V_n=Q_n\gamma_1\ldots\gamma_n, \quad \forall n\geqslant 1. \qquad (10)$$

Proof.

This result can be proved by recurrence.

$$V_1U_1=(b_0b_1+a_1)\gamma_1(b_1\gamma_1)^{-1}=Q_1\gamma_1(P_1\gamma_1)^{-1}.$$

The result is assumed to be satisfied up to the integer n-1. Thanks to (8) we have:

$$V_n=V_{n-1}(\gamma_1\dots\gamma_{n-1})^{-1}b_n\gamma_1\dots\gamma_n+V_{n-2}(\gamma_1\dots\gamma_{n-2})^{-1}a_n\gamma_1\dots\gamma_n$$

in which V_{n-1} and V_{n-2} are replaced by their expression given by (10).

Thanks to (8), (10) is satisfied for the integer n.

The proof for U_n is the same.

It can be interesting that any $(\gamma_1..\gamma_k)^{-1}b_{k+1}\gamma_1\dots\gamma_{k+1}$ in (8) is equal to I $\forall k \in \mathbb{N}$ with the convention that $(\gamma_1..\gamma_k)^{-1}=I$ for k=0.

Then it is easy to prove by recurrence that:

$$\gamma_1=(b_1)^{-1},\ \gamma_k=b_1\dots b_{k-1}(b_1\dots b_k)^{-1},\ \gamma_1..\gamma_k=(b_1\dots b_k)^{-1}. \tag{11}$$

Therefore the continued fraction has the following form:

$$b_0+a_1b_1^{-1}(I+a_2(b_1b_2)^{-1}(I+(b_1a_3(b_1b_2b_3)^{-1}(I+\dots(b_1\dots b_{n-2})a_n(b_1\dots b_n)^{-1}(I+\dots(\dots)^{-1})^{-1}\dots)^{-1} \tag{12}$$

It can also be interesting that any $(\gamma_1\dots\gamma_{n-2})^{-1}a_n\gamma_1\dots\gamma_n$ in (9) is equal to I $\forall n \in \mathbb{N}$ such that $n\geqslant1$ with the convention that $(\gamma_1\dots\gamma_{n-2})^{-1}=I$ for n=1 and n=2.

In this case it can be proved easily that:

$$\gamma_1=a_1^{-1},\ \gamma_1\gamma_2=a_2^{-1},$$

$$\gamma_1\gamma_2\dots\gamma_{2k}=(a_2a_4\dots a_{2k})^{-1},\ \gamma_1\gamma_2\dots\gamma_{2k+1}=(a_1a_3\dots a_{2k+1})^{-1} \tag{13}$$

and the continued fraction can be written:

$$b_0+I(b_1a_1^{-1}+I(a_1b_2a_2^{-1}+I(a_2b_3(a_1a_3)^{-1}+I(a_1a_3b_4(a_2a_4)^{-1}+\dots$$

$$\dots I(a_1a_3\dots a_{2k-1}b_{2k}(a_2a_4\dots a_{2k})^{-1}+I(a_2a_4\dots a_{2k}b_{2k+1}(a_1a_3\dots a_{2k+1})^{-1}+\dots)^{-1}\dots)^{-1} \tag{14}$$

A theorem proved by Field [8] can be extended without new proof to a Banach algebra for a continued fraction (5) in which $a_0=0$, $a_i=I$ $\forall i\geqslant1$, and the b_i's depend on variables in $\mathcal{D}$.

Theorem 6

If in $\mathcal{D}$, $\|(b_{2n-1})^{-1}\|\leqslant\upsilon$ and $\|(b_{2n})^{-1}\|\leqslant1-\upsilon$ $\forall n \in \mathbb{N}$ such that $n\geqslant1$ with $0<\upsilon<1$, then the continued fraction (4) converges uniformly to an element of $\mathcal{A}$.

3 CONVERGENCE OF PADE APPROXIMANTS

$M_i^{(0)}$ will be assumed to have an inverse $\forall i \in \mathbb{N}$. Then (see[4]) the left polynomials $P_k^{(0)}$ orthogonal with respect to $c^{(0)}$ and their left associated polynomials $Q_k^{(0)}$ satisfy the same three-term recurrence relation:

$$\tilde{P}_k^{(0)}(x)=\tilde{P}_{k-1}^{(0)}(x)(I+xB_k^{(0)})+x^2\tilde{P}_{k-2}^{(0)}(x)C_k^{(0)} \tag{15}$$

Thus the Padé approximant $[k-1/k]_f$, which is equal to $\tilde{Q}_k^{(0)}(\tilde{P}_k^{(0)})^{-1}$, is the kth left convergent of the following continued fraction:

$$C_1(I+B_1x+C_2x^2(I+B_2x+C_3x^2(I+B_3x+\dots(\dots)^{-1}\dots)^{-1} \tag{16}$$

Continued fraction (16) can be written in the same form as (11):

$$C_1(I+B_1x)^{-1}(I+x^2C_2((I+B_1x)(I+xB_2))^{-1}(I+\ldots$$
$$\ldots(I+(I+xB_1)\ldots(I+xB_{n-2})x^2C_n((I+xB_1)\ldots(I+xB_n))^{-1}(I+\ldots)^{-1}\ldots)^{-1} \quad (17)$$

Then the following result is proved thanks to theorem 3.

Theorem 7

If, $\forall x \in K$ such that $|x|\leqslant\rho$, $\|B_ix\|\leqslant\upsilon<1$ $\forall i \in \mathbb{N}$ such that $i\geqslant1$ and

$$\|C_i\|\leqslant(1-\upsilon)^i(4\rho^2(1+\upsilon)^{i-2})^{-1} \quad \forall i \in \mathbb{N} \text{ such that } i\geqslant2 \quad (18)$$

then $[k-1/k]_f$ converges uniformly to f when k tends to infinity.

Proof.

If in (7)

$$\|(I+xB_1)\ldots(I+xB_{n-2})x^2\,C_n((I+xB_1)\ldots(I+xB_n))^{-1}\|\leqslant{}^1/_4, \quad (19)$$

then the property holds thanks to theorem 3.

Since $\|B_ix\|\leqslant\upsilon<1$,

$$\|(I+xB_i)^{-1}\|=\|\sum_{j=0}^{\infty}(-xB_i)^j\|\leqslant\sum_{j=0}^{\infty}\upsilon^j=(1-\upsilon)^{-1}$$

Therefore if A is the left part of (18), we have:

$$A\leqslant(1+\upsilon)^{n-2}\|x^2C_n\|(1-\upsilon)^{-n}$$

By using (18) and $|x|\leqslant\rho$, (19) is satisfied.

From this theorem a result of convergence of $[n+k-1/k]_f$ can be deduced when n is fixed and k tends to infinity.

If $n \in \mathbb{N}$ the following continued fraction will be used:

$$\sum_{i=0}^{n-1}c_ix^i+x^nC_1^{(n)}(I+xB_1^{(n)}+x^2C_2^{(n)}(I+xB_2^{(n)}+x^2C_3^{(n)}(\ldots)^{-1}\ldots)^{-1} \quad (20)$$

If $n \in Z$, $n<0$, another form of continued fraction will be used, but to give its expression some preliminary results must be proved.

$M_i^{(n)}$ will be assumed to have an inverse $\forall i \in \mathbb{N}$ such that $i\geqslant1-n$. $M_{1-n}^{(n)}$ is a matrix whose all the coefficients over the main antidiagonal are zero. This antidiagonal only contains coefficients c_0, and therefore $M_{1-n}^{(n)}$ is invertible if and only if c_0 has an inverse.

Lemma 8

If $n<0$,

$$P^{(n)}_{-n+2}(x)=P^{(n)}_{-n+1}(x)(Ix+B^{(n)}_{-n+2})+C^{(n)}_{-n+2} \tag{21}$$

Proof.

By division of $P^{(n)}_{-n+2}$ by $P^{(n)}_{-n+1}$ we get:

$$P^{(n)}_{-n+2}(x)=P^{(n)}_{-n+1}(x)(Ix+B^{(n)}_{-n+2})+\overline{P}_{-n}(x)$$

with deg $\overline{P}_{-n}\leqslant -n$.

The orthogonality of $P^{(n)}_{-n+2}$ and $P^{(n)}_{-n+1}$ with respect to $c^{(n)}$ shows that :

$$c^{(n)}(x^j\overline{P}_{-n}(x))=0 \quad \forall j \in \mathbb{N} \text{ such that } 0\leqslant j\leqslant -n-1.$$

This linear system whose the unknowns are the coefficients of $\overline{P}_{-n}$ has an unique solution which is $\overline{P}_{-n}=b_0$ where b_0 is an invertible constant. Indeed

$$c^{(n)}(x^{-n}\,\overline{P}_{-n}(x))=c_0b_0=c^{(n)}(x^{-n+1}P^{(n)}_{-n+1}(x))$$

and the right part of this expression has an inverse (see [4]).

We will take $C^{(n)}_{-n+2}=b_0$.

Therefore (21) holds.

Lemma 8 obviously holds as soon as $M^{(n)}_{-n+2}$ and $M^{(n)}_{-n+1}$ have an inverse.

If $i>-n+2$ the left orthogonal polynomials satisfied the classical three-term recurrence relation (see[4]):

$$P^{(n)}_i(x)=P^{(n)}_{i-1}(x)(Ix+B^{(n)}_i)+P^{(n)}_{i-2}(x)C^{(n)}_i \tag{22}$$

with $C^{(n)}_i$ invertible.

We also have for the left associated polynomials:

Lemma 9

$$Q^{(n)}_{-n+1}(t)=c_0$$
$$Q^{(n)}_{-n+2}(t)=Q^{(n)}_{-n+1}(t)(It+B^{(n)}_{-n+2}) \tag{23}$$

Proof.

Since $c_i=0$ $\forall i<0$, (2) gives $Q^{(n)}_{-n+1}(t)=c^{(n)}(x^{-n}\,I)=c_0$.

(2) is also used from (21) to obtain:

$$Q^{(n)}_{-n+2}(t)=c^{(n)}(P^{(n)}_{-n+1}(x))+t\ c^{(n)}((P^{(n)}_{-n+1}(x)-P^{(n)}_{-n+1}(t))(x-t)^{-1})+Q^{(n)}_{-n+1}(t)B^{(n)}_{-n+2},$$

and (23) holds by using the orthogonality of $P^{(n)}_{-n+1}$.

If i>-n+2 the left associated polynomials satisfy the same relation as (22) (see[4]).

The transformed polynomials $\tilde{P}$ thus satisfy the following relation:

$$\tilde{P}^{(n)}_{-n+2}(x)=\tilde{P}^{(n)}_{-n+1}(x)(Ix+B^{(n)}_{-n+2})+C^{(n)}_{-n+2}x^{-n+2} \quad (24)$$

Moreover $\tilde{Q}^{(n)}_{-n+1}(t)=c_0t^{-n}$.

Therefore the Padé approximant $[n+k-1/k]_f$ which is equal to $x^n\,[k-1/k]_{f_n}$ for n<0 and $n+k-1\geqslant 0$ will be the $(k+n)^{th}$ left convergent of the following continued fraction:

$$x^n\tilde{Q}^{(n)}_{-n+1}(x)(\tilde{P}^{(n)}_{-n+1}(x)+x^{-n+2}C^{(n)}_{-n+2}(I+xB^{(n)}_{-n+2}+x^2C^{(n)}_{-n+3}(I+xB^{(n)}_{-n+3}+x^2C^{(n)}_{-n+4}(I+xB^{(n)}_{-n+4}+..)^{-1} \quad (25)$$

We will write :$P^{(n)}_{-n+1}(x)=I+\hat{P}(x)$ where $\hat{P}(0)=0$.

Then we have:

Corollary 10

Let n be a fixed integer belonging to Z.

If $\forall t\in K$ such that $|t|\leqslant\rho$ one of the two following properties holds:

i) If $n\in\mathbb{N}$, $\|B^{(n)}_i t\|\leqslant\upsilon<1$ $\forall i\in\mathbb{N}$ such that $i\geqslant 1$, and $\|C^{(n)}_i\|\leqslant(1-\upsilon)^i\;(4\rho^2\;(1+\upsilon)^{i-2})^{-1}$ $\forall i\in\mathbb{N}$ such that $i\geqslant 2$.

ii) If $n<0$, $\|\hat{P}(x)\|\leqslant\bar{\upsilon}<1$, $\|B^{(n)}_{-n+i}t\|\leqslant\upsilon<1$ $\forall i\in\mathbb{N}$ such that $i\geqslant 2$, $\|C^{(n)}_{-n+2}\|\leqslant(1-\bar{\upsilon})(1+\upsilon)(4\rho^{2-n})^{-1}$, and $\|C^{(n)}_{-n+i}\|\leqslant(1-\bar{\upsilon})(1-\upsilon)^{i-1}(4\rho^2(1+\bar{\upsilon})(1+\upsilon)^{i-3})^{-1}$ $\forall i\in\mathbb{N}$ such that $i\geqslant 3$,

then $[n+k-1/k]_f$ converges uniformly to f when k tends to infinity.

Proof.

i) By using theorem 7, (20) converges to $\sum_{i=0}^{\infty}c_i t^i+t^n f_n(t)=f(t)$.

ii) If n<0, by a similar proof as in theorem 7 it can be proved that $[k-1/k]_{f_n}$ converges uniformly to f_n when k tends to infinity.

By using theorem 4 another result of convergence is obtained.

Theorem 11

If, $\forall x\in K$ such that $0<\rho_1\leqslant|x|\leqslant\rho_2$, we have $\|B_k x\|\leqslant\mu<1$, $\|C^{-1}_{2k}\|\leqslant(1-\mu)^{2k-2}\rho_1^2(1+\mu)^{-2k}(1+\upsilon)^{-2}$ and $\|C_{2k+1}\|\leqslant(1-\mu)^{2k+1}(1+\mu)^{1-2k}\upsilon^2\rho_2^{-2}$ $\forall k\in\mathbb{N}$ such that $k\geqslant 1$, with $0<\upsilon<1$,

then $[k-1/k]_f$ converges uniformly to f when k tends to infinity.

Proof.

By using (17), if

$\|(I+xB_1)\ldots(I+xB_{2n})C_{2n}^{-1}x^{-2}((I+xB_1)\ldots(I+xB_{2n-2}))^{-1}\|\leqslant(1+\upsilon)^{-2}$ and

$\|(I+xB_1)\ldots(I+xB_{2n-1})C_{2n+1}x^2((I+xB_1)\ldots(I+xB_{2n+1}))^{-1}\|\leqslant\upsilon^2$,

then the property holds thanks to theorem 4.

It can easily be seen that these inequalities are satisfied with the assumptions and by using $\|(I+xB_i)^{-1}\|\leqslant(1-\mu)^{-1}$

For the Padé approximants $[n+k-1/k]_f$ the following corollary can be proved like corollary 10.

Corollary 12

Let n be a fixed integer belonging to Z.

If, $\forall x \in K$ such that $0<\rho_1\leqslant|x|\leqslant\rho_2$, one of the two following properties is satisfied:

i) If $n \in \mathbb{N}$, $\|B_i^{(n)}x\|\leqslant\mu<1$,

$\|(C_{2i}^{(n)})^{-1}\|\leqslant(1-\mu)^{2i-2}\rho_1^2(1+\mu)^{-2i}(1+\upsilon)^{-2}$ and

$\|C_{2i+1}^{(n)}\|\leqslant(1-\mu)^{2i+1}(1+\mu)^{1-2i}\upsilon^2\rho_2^{-2}$ $\forall i \in \mathbb{N}$ such that $k\geqslant1$, with $0<\upsilon<1$.

ii) If $n<0$, $\|\hat{P}(x)\|\leqslant\bar{\upsilon}<1$, $\|B_{-n+i}^{(n)}x\|\leqslant\mu<1$ $\forall i \in \mathbb{N}$ such that $i\geqslant2$,

$\|(C_{-n+2}^{(n)})^{-1}\|\leqslant(1+\bar{\upsilon})^{-1}(1+\upsilon)^{-2}(1+\mu)^{-1}\rho_1^{2-n}$, and

$\|(C_{-n+2k}^{(n)})^{-1}\|\leqslant(1-\upsilon)\rho_1^2(1+\bar{\upsilon})^{-1}(1+\upsilon)^{-2}(1-\mu)^{2k-3}(1+\mu)^{1-2k}$, and

$\|C_{-n+2k+1}^{(n)}\|\leqslant(1-\bar{\upsilon})\rho_2^{-2}(1+\bar{\upsilon})^{-1}(1-\mu)^{2k}(1+\mu)^{2-2k}\upsilon^2$, $\forall k \in \mathbb{N}$ such that $i\geqslant1$ with $0<\upsilon<1$

then $[n+k-1/k]_f$ converges uniformly to f when k tends to infinity.

There exists another three-term recurrence relations between the left orthogonal polynomials, which can be used to prove another results of convergence.

Let n be a fixed integer belonging to Z.

$M_{i+1}^{(n)}$ and $M_{j+1}^{(n+1)}$ are assumed to have an inverse $\forall i$ and $j \in \mathbb{N}$ such that $n+i\geqslant0$ and $n+1+j\geqslant0$.

If $n \in \mathbb{N}$, then $\tilde{P}$ satisfies the two following three-term recurrence relation (see [4]):

$$\tilde{P}_{k+1}^{(n)}(x)=\tilde{P}_k^{(n+1)}(x)-x\tilde{P}_k^{(n)}(x)q_{k+1}^{(n)}, \tag{26}$$

$$\tilde{P}_{k+1}^{(n+1)}(x)=\tilde{P}_{k+1}^{(n)}(x)-x\tilde{P}_k^{(n+1)}(x)e_{k+1}^{(n)}, \tag{27}$$

where $q_{k+1}^{(n)}$ and $e_{k+1}^{(n)}$ have an inverse.

By using (2) as well as the orthogonality of polynomials P and relations satisfied by P deduced from (26) and (27), relations satisfied by $\tilde{Q}$ can be proved.

$$\tilde{Q}_{k+1}^{(n)}(t)=t\tilde{Q}_k^{(n+1)}(t)-t\tilde{Q}_k^{(n)}(t)q_{k+1}^{(n)}+c_n\tilde{P}_k^{(n+1)}(t) \tag{28}$$

$$t\tilde{Q}_{k+1}^{(n+1)}(t)=\tilde{Q}_{k+1}^{(n)}(t)-c_n\tilde{P}_{k+1}^{(n)}(t)-t^2\tilde{Q}_k^{(n+1)}(t)e_{k+1}^{(n)}. \tag{29}$$

Thanks to these two last relations and (3) and (4), it is easy to see that $\tilde{Z}_{k+1}^{(n)}$ and $\tilde{Z}_{k+1}^{(n+1)}$ satisfied the same three-term recurrence relations as (26) and (27) respectively.

Then $[n+k/k+1]_f$ and $[n+k+1/k+1]_f$ are respectively the $2(k+1)^{th}$ and $(2k+3)^{th}$ left convergent of the following continued fraction:

$$\tilde{Z}_0^{(n)}(x) + c_n x^n (I - xq_1^{(n)}(I - xe_1^{(n)}(I - xq_2^{(n)}(I - xe_2^{(n)}(I\ldots)^{-1}\ldots)^{-1} \qquad (30)$$

In the case where $n<0$ the following result is proved:

Theorem 13

If $n \in Z$ such that $n<0$, then:

$$xP_{-n}^{(n+1)}(x) = P_{-n+1}^{(n)}(x) + q_{-n+1}^{(n)}$$

with $q_{-n+1}^{(n)} = -P_{-n+1}^{(n)}(0)$.

$$Q_{-n}^{(n+1)}(x) = Q_{-n+1}^{(n)}(x).$$

Proof.

Let us denote by P the polynomial $P_{-n+1}^{(n)}(x) + q_{-n+1}^{(n)}$ where $q_{-n+1}^{(n)} = -P_{-n+1}^{(n)}(0)$.

Then $c^{(n)}(x^r \overline{P}(x)) = 0 \ \forall r \in \mathbb{N}$ such that $0 \leqslant r \leqslant n-1$, thanks to the orthogonality of $P_{-n+1}^{(n)}$ and by using the fact that $c^{(n)}(x^r q_{-n+1}^{(n)}) = c_{n+r} q_{-n+1}^{(n)} = 0$ if $n+r<0$.

$x^{-1}\overline{P}(x)$ which is of degree $-n$, is orthogonal with respect to $c^{(n+1)}$. Thus it is identical to $P_{-n}^{(n+1)}$.

On the other hand:

$$Q_{-n+1}^{(n)}(t) = c^{(n)}(x(P_{-n}^{(n+1)}(x) - P_{-n}^{(n+1)}(t))(x-t)^{-1}) + c^{(n)}(P_{-n+1}^{(n)}(t))$$

$$= Q_{-n}^{(n+1)}(t) + c_n P_{-n}^{(n+1)}(t) = Q_{-n}^{(n+1)}(t), \text{ since } c_n = 0.$$

Therefore

$$\tilde{P}_{-n+1}^{(n)}(x) = \tilde{P}_{-n}^{(n+1)}(x) - x^{-n+1} q_{-n+1}^{(n)}$$

$$x\tilde{Q}_{-n}^{(n+1)}(x) = \tilde{Q}_{-n+1}^{(n)}(x).$$

and the corresponding continued fraction is:

$$x^{n+1}\tilde{Q}_{-n}^{(n+1)}(x)(\tilde{P}_{-n}^{(n+1)}(x) - x^{-n+1} q_{-n+1}^{(n)}(I - xe_{-n+1}^{(n)}(I - xq_{-n+2}^{(n)}(I - xe_{-n+2}^{(n)}(I-\ldots)^{-1}\ldots)^{-1} \qquad (31)$$

$[j/-n+j]_f$ and $[j/-n+j+1]_f$ are respectively the $(2j+1)^{th}$ and $(2j+2)^{th}$ left convergent of (31) $\forall j \in \mathbb{N}$.

(31) will be written as:

$$x^{n+1}\tilde{Q}_{-n}^{(n+1)}(x)(\tilde{P}_{-n}^{(n+1)}(x))^{-1}(I - x^{-n+1} q_{-n+1}^{(n)}(\tilde{P}_{-n}^{(n+1)}(x))^{-1}(I - x\tilde{P}_{-n}^{(n+1)}(x)e_{-n+1}^{(n)}(\tilde{P}_{-n}^{(n+1)}(x))^{-1}$$

$$(I - x\tilde{P}_{-n}^{(n+1)}(x)q_{-n+2}^{(n)}(\tilde{P}_{-n}^{(n+1)}(x))^{-1}(I - x\tilde{P}_{-n}^{(n+1)}(x)e_{-n+2}^{(n)}(\tilde{P}_{-n}^{(n+1)}(x))^{-1}(I-\ldots)^{-1}\ldots)^{-1} \qquad (32)$$

$\tilde{P}_{-n}^{(n+1)}(x)$ will be written as $I + \hat{P}(x)$ where $\hat{P}(0) = 0$.

Then theorems 3 and 4 give results of convergence along two adjacent diagonals n and n+1.

Theorem 14

Let n a fixed integer belonging to Z.

$M_{i+1}^{(n)}$ and $M_{j+1}^{(n+1)}$ are assumed to have an inverse $\forall i$ and $j \in \mathbb{N}$ such that $n+i\geq 0$ and $n+1+j\geq 0$.

If $\forall x \in K$ such that $|x|\leq\rho$ one of the two following properties is satisfied:

i) If $n \in \mathbb{N}, \|q_i^{(n)}\|\leq(4\rho)^{-1}$ and $\|e_i^{(n)}\|\leq(4\rho)^{-1}$, $\forall i \in \mathbb{N}$ such that $i\geq 1$.

ii) If $n<0$, $\|\hat{P}(x)\|\leq\bar{\upsilon}<1$, $\|q_{-n+1}^{(n)}\|\leq(1-\bar{\upsilon})(4\rho^{-n+1})^{-1}$,

$\|q_{-n+2+j}^{(n)}\|\leq(1-\bar{\upsilon})(4\rho(1+\bar{\upsilon}))^{-1}$ and $\|e_{-n+1+j}^{(n)}\|\leq(1-\bar{\upsilon})(4\rho(1+\bar{\upsilon}))^{-1}$ $\forall j \in \mathbb{N}$.

then $[n+k/k+1]_f$ and $[n+k+1/k+1]_f$ converge uniformly to f when k tends to infinity.

Theorem 15

Let n a fixed integer belonging to Z.

$M_{i+1}^{(n)}$ and $M_{j+1}^{(n+1)}$ are assumed to have an inverse $\forall i$ and $j \in \mathbb{N}$ such that $n+i\geq 0$ and $n+1+j\geq 0$.

If $\forall x \in K$ such that $0<\rho_1\leq|x|\leq\rho_2$ one of the two following properties is satisfied:

i) If $n \in \mathbb{N}$, $\|(q_k^{(n)})^{-1}\|\leq\rho_1(1+\upsilon)^{-2}$ and $\|e_k^{(n)}\|\leq\rho_2^{-1}\upsilon^2$ $\forall k \in \mathbb{N}$ such that $k\geq 1$ with $0<\upsilon<1$.

ii) If $n<0$, $\|\hat{P}(x)\|\leq\bar{\upsilon}<1$, $\|(q_{-n+1}^{(n)})^{-1}\|\leq(1+\bar{\upsilon})^{-1}(1+\upsilon)^{-2}\rho_1^{-n+1}$,

$\|(q_{-n+2+j}^{(n)})^{-1}\|\leq(1-\bar{\upsilon})(1+\upsilon)^{-2}\rho_1(1+\bar{\upsilon})^{-1}$ and $\|e_{-n+1+j}^{(n)}\|\leq(1-\bar{\upsilon})\rho_2^{-1}\upsilon^2(1+\bar{\upsilon})^{-1}$ $\forall j \in \mathbb{N}$ with $0<\upsilon<1$

then $[n+k/k+1]_f$ and $[n+k+1/k+1]_f$ converge uniformly to f when k tends to infinity.

On the other hand if $M_k^{(n)}$ and $M_k^{(n+1)}$ have an inverse $\forall k \in \mathbb{N}$ such that $k\geq 1$ with $n+k-1=s$ fixed belonging to $\mathbb{N}$, we have (see [5]):

$$\tilde{P}_{k+1}^{(n-1)}(x)=\tilde{P}_k^{(n)}(x)(I+x\beta_{k+1}^{(n-1)})+x\tilde{P}_{k-1}^{(n+1)}(x)\Gamma_{k+1}^{(n-1)} \tag{33}$$

with $\beta_{k+1}^{(n-1)}$) and $\Gamma_{k+1}^{(n-1)}$ invertible.

By using (2) as well as the orthogonality of polynomials P, it can easily be proved that the associated polynomials $\tilde{Q}$ satisfied the following recurrence relation:

$$tQ_{k+1}^{(n-1)}(t)=Q_k^{(n)}(t)(It+\beta_{k+1}^{(n-1)})+Q_{k-1}^{(n+1)}(t)\Gamma_{k+1}^{(n-1)}+c_n P_{k-1}^{(n+1)}(t)\Gamma_{k+1}^{(n-1)}+c_{n-1}P_{k+1}^{(n-1)}(t) \tag{34}$$

with $Q_{-1}^{(n)}=I$ and $Q_0^{(n)}=0$.

Moreover thanks to (3), (4) and (34) the same three-term recurrence relation as (33) is satisfied by the polynomials $\tilde{Z}$.

From (33) the following continued fraction can be deduced:

$$\tilde{Z}_0^{(n+k)}(x)+x^{n+k}\Gamma_1^{(n+k-1)}(I+x\beta_1^{(n+k-1)}+x\Gamma_2^{(n+k-2)}(I+x\beta_2^{(n+k-2)}+ x\Gamma_3^{(n+k-3)}(I+x\beta_3^{(n+k-3)}+\ldots)^{-1}\ldots)^{-1} \tag{35}$$

We have:

$$\tilde{Z}_0^{(n+k)}(x)=\sum_{i=0}^{n+k-1} c_i x^i.$$

and the kth left convergent is $[n+k-1/k]_f=\tilde{Z}_k^{(n)}(x)(\tilde{P}_k^{(n)}(x))^{-1}$.

Let us determine β_1 and Γ_1. We have:

$$P_1^{(n+k-1)}(x)=Ix+\beta_1^{(n+k-1)} \text{ and } c^{(n+k-1)}(P_1^{(n+k-1)}(x))=0.$$

Therefore $\beta_1^{(n+k-1)}= -(c_{n+k-1})^{-1}c_{n+k}$.

Moreover, from (34) written for $tQ_1^{(n+k-1)}(t)$, we get:

$$tQ_1^{(n+k-1)}(t)=c_{n+k-1}P_1^{(n+k-1)}(t)+\Gamma_1^{(n+k-1)}.$$

Thus $\Gamma_1^{(n+k-1)}= -c_{n+k-1}P_1^{(n+k-1)}(0)=c_{n+k-1}P_0^{(n+k-1)}(0)q_1^{(n+k-1)}=c_{n+k}$, by using (26) and the value of q_1 (see[4]).

(36) then can be transformed as:

$$\tilde{Z}_0^{(n+k)}(x)+x^{n+k}\Gamma_1^{(n+k-1)}(I+x\beta_1^{(n+k-1)})^{-1}(I+x\Gamma_2^{(n+k-2)}((I+x\beta_1^{(n+k-1)})(I+x\beta_2^{(n+k-2)}))^{-1}(I+ (I+x\beta_1^{(n+k-1)})..(I+x\beta_{i-2}^{(n-i+k+2)})x\Gamma_i^{(n+k-i)}((I+x\beta_1^{(n+k-1)})\dots(I+x\beta_i^{(n+k-i)}))^{-1}(I+\dots)^{-1}\dots)^{-1}$$

By a similar proof a theorem of convergence like theorem 7 can be obtained.

Theorem 16

Let n+k-1 be a fixed integer $s \in \mathbb{N}$, $\forall k \in \mathbb{N}$.

If $\forall x \in K$ such that $|x|\leqslant\rho$, we have:

$$\|\beta_i^{(n+k-i)}x\|\leqslant\upsilon<1, \quad \forall i \in \mathbb{N} \text{ such that } i\geqslant 1, \text{ and}$$

$$\|\Gamma_i^{(n+k-i)}\|\leqslant(1-\upsilon)^i(4\rho(1+\upsilon)^{i-2})^{-1} \quad \forall i \in \mathbb{N} \text{ such that } i\geqslant 2,$$

then $[n+k-1/k+1]_f$ converges uniformly to f when k tends to infinity.

Another result of convergence is also obtained like theorem 11.

Theorem 17

Let n+k-1 be a fixed integer $s \in \mathbb{N}$, $\forall k \in \mathbb{N}$.

If $\forall x \in K$ such that $0<\rho_1\leqslant|x|\leqslant\rho_2$, we have:

$$\|\beta_i^{(n+k-i)}x\|\leqslant\mu<1, \quad \|(\Gamma_{2i}^{(n+k-2i)})^{-1}\|\leqslant(1-\mu)^{2i-2}\rho_1(1+\mu)^{-2i}(1+\upsilon)^{-2} \text{ and}$$

$$\|\Gamma_{2i+1}^{(n+k-2i-1)}\|\leqslant(1-\mu)^{2i+1}\rho_2^{-1}(1+\mu)^{1-2i}\upsilon^2 \quad \forall i \in \mathbb{N} \text{ such that } i\geqslant 1 \text{ with } 0<\upsilon<1,$$

then $[n+k-1/k+1]_f$ converges uniformly to f when k tends to infinity.

Bibliography.

[1] DELSARTE P., GENIN Y. and KAMP Y.
Orthogonal polynomial matrices on the unit circle.
IEEE Trans. Circuits and Systems. CAS.25, (1978), p.149-160.

[2] DENK H. and RIEDERLE M.
A generalization of a theorem of Pringsheim.
J. Approx. Th., 35 (1982), p.355-363.

[3] DRAUX A.
The Padé approximants in a non-commutative algebra and their applications
in "Padé Approximation and its applications. Bad Honnef 1983". Proceedings, H.Werner and H.J.Bünger eds., Lecture Notes in Mathematics 1071, Springer Verlag, Berlin 1984, p.117-131.

[4] DRAUX A.
Formal orthogonal polynomials and Padé approximants in a non-commutative algebra.
in "Mathematical Theory of Networks and Systems", Proceedings of the MTNS 83 International Symposium, Beer Sheva, Israel, June 20-24,1983. Lecture Notes in Control and Information Sciences 58, Springer Verlag, Berlin 1984, p.278-292.

[5] DRAUX A.
On semi-orthogonal polynomials (submitted)

[6] FAIR W.
Non commutative continued fractions.
SIAM J. Math. Anal., 2 (1971), p.226-232.

[7] FAIR W.
A convergence theorem for non commutative continued fractions.
J. Approx. Th., 5, (1972), p.74-76.

[8] FIELD D.A.
Convergence theorems for matrix continued fractions.
SIAM J. Math. Anal., 15, (1984), p.1220-1227.

[9] GRAFFI S. and GRECCHI V.
Matrix moment methods in perturbation theory, Boson Quantum field models, an anharmonic oscillators.
Commun Math. Phys., 35 (1974), p.235-252.

[10] HAYDEN T.L.
Continued fractions in Banach spaces.
Rocky Mountains J. Math., 4 (1974), p.367-370.

[11] NEGOESCU N.
Un théorème de convergence pour les fractions continues non commutatives.
C.R. Acad. Sc. Paris, 278A (1974), p.689-692.

[12] NEGOESCU N.
Sur les fractions continues non commutatives.
Proceedings of the Institute of Mathematics, Iasi, Romania, (1976), p.137-143.

[13] NEGOESCU N.
Convergence theorems on non-commutative continued fractions.
Rev. Anal. Numer. Theor. Approx., 5 (1976), p.165-180.

[14] PENG S.T. and HESSEL A.
Convergence on non-commutative continued fractions.
SIAM J. Math. Anal., 6 (1975), p.724-727.

[15] SYDOW Von B.
Padé approximation of matrix-valued series of Stieltjes.
Ark. Mat., 15 (1977), p.199-210.

[16] SYDOW Von B.
Matrix-valued Padé approximation and Gaussian quadrature.
Det Kongelige Norske Videnskabers Selskab. 1, (1983), p.117-127.

[17] WYNN P.
A note on the convergence of certain non-commutative continued fractions.
Mathematics Research Center, Madison, report 750, (1967).

[17] WYNN P.
Continued fractions whose coefficients obey a non-commutative law of multiplication.
Arch. Rat. Mech. Anal., 12 (1963), p.273-312.

SUBSETS OF UNICITY IN UNIFORM APPROXIMATION

Charles B. Dunham

Department of Computer Science
The University of Western Ontario
London, Canada N6A 5B7

Uniqueness of best uniform approximation on open covers of the set of maximal points of the error is shown equivalent to uniqueness on a single open cover for families with the betweeness property. An example is given of an asymptotically convex family (hence a sun) for which the equivalence fails. That example is used to show that asymptotic convexity (being a sun) is not preserved under constraints on values.

Let X be a compact Hausdorff space. Let Y be a vector space with norm $|\ |$. Let $C(X)$ be the space of continuous functions from X into Y. For $g \in C(X)$ and W a non-empty subset of X, define

$$||g||_W = \sup\{|g(x)| : x \in W\}.$$

Let $\mathcal{G}$ be a subset of $C(X)$ with the betweeness property. The approximation problem is, given $f \in C(X)$ and W a non-empty subset of X, to find G^* minimizing $||f-G||_W$ over $G \in \mathcal{G}$. Such an element G^* is called a best approximation to f on W.

DEFINITION: $\mathcal{G}$ has the betweeness property if for any two elements G and F of $\mathcal{G}$, there exists a <u>λ-set</u> $\{H_\lambda\} \subset \mathcal{G}$ such that

(a) $H_0 = G, H_1 = F$.

(b) For $\lambda \in [0,1]$, $H_\lambda(x)$ is on the line joining $G(x)$ and $F(x)$.

(c) If $G(x) = F(x)$, then $H_\lambda(x) = G(x)$, $0 < \lambda < 1$, and if $G(x) \neq F(x)$, $|G(x) \ H_\lambda(x)|$ is a strictly monotonic continuous function of λ, $0 \leq \lambda \leq 1$.

Families with betweeness include linear families (finite and infinite dimensional) and (generalized) rational families with positive denominators. Real families with betweeness retain this property under certain transformations (Dunham, 1977). Other examples are given in the papers of the author in the references.

Define $M(f,G) = \{x : |f(x) - G(x)| = \|f-G\|_X\}$.

By compactness of X and continuity of $|f-G|$, $M(f,G)$ is a non-empty closed set.

LEMMA: Let $\mathcal{G}$ have the betweeness property. Let G be best to f on X , then G is best to f on any subset W containing $M(f,G)$.

Proof: Suppose there is $F \in \mathcal{G}$ with

$$\|f-F\|_W < \|f-G\|_W = \|f-G\|_X = \|f-G\|_{M(f,G)}$$

then

$$\|f-F\|_{M(f,G)} < \|f-G\|_{M(f,G)} \quad .$$

Apply the characterization theorem for betweeness to get a contradiction.

THEOREM: Let $\mathcal{G}$ have the betweeness property. The following are equivalent:

(i) G is a unique best approximation to f on X .
(ii) For each subset W containing an open cover of $M(f,G)$, G is the unique function on X of best approximation to f on W.
(iii) G is the unique function on X of best approximation to f on some subset Q containing an open cover of $M(f,G)$.

Proof: Suppose that (iii) holds (we note that (i) is a special case of (iii)) and (ii) fails for $W=V$. By the preceding lemma G is best on V , with $F \neq G$ also best to f on V .

By uniqueness of G on Q , H_λ is not best on Q for any $\lambda \in (0,1]$ and so there is $x_k \in Q$ such that

(1) $$|f(x_k) - H_{1/k}(x_k)| > \|f-G\|_Q = \|f-G\|_X .$$

As X is compact, $\{x_k\}$ has a cluster point x (that is, every neighbourhood of x contains infinitely many points of the sequence). Suppose that

(2) $$|f(x) - G(x)| < \|f-G\|_X \quad .$$

As $\{H_{1/k}\}$ converges uniformly to G on X , we must have for all k sufficiently large

$$|f(x_k) - H_{1/k}(x_k)| < \|f-G\|_X \quad .$$

But this contradicts (1), hence (2) is false and $x \in M(f,G)$. V is a neighbourhood of x and so contains an element x_k of $\{x_k\}$. As H_λ lies between F and G for $0<\lambda<1$, H_λ is best to f on V for $0<\lambda<1$. But this contradicts (1). We have proven that (iii) or (i) implies (ii). It is obvious that (ii) implies (i) and (iii), and the theorem is proven.

The equivalence of (i) and (iii) were shown by Johnson (Theorem 5) for linear approximation and Q open.

NOTE: If the restriction of G to Q is the unique function of best approximation on Q to f, we can conclude nothing.

EXAMPLE 1: Let $X = \{0,1\}$, $\mathcal{G} = \{ax\}$ and $f(0) = 1$, $f(1) = 0$. $M(f,0) = \{0\}$ and 0 is the unique best approximation on $M(f,0)$ to f, but 0 is not unique on X.

NOTE. Best approximations need not be unique on $M(f,G)$ to be unique on X. Consider the approximation of $f(x) = x^2-1$ by $\{ax\}$ on $[-1,1]$, in which 0 is uniquely best on $[-1,1]$ (Cheney, 1966, p.82), but not uniquely best on $M(f,0) = \{0\}$. This example shows that we may not have uniqueness if the subset merely contains $M(f,G)$. On the other hand, if the best approximation on $M(f,G)$ is unique, we trivially have uniqueness on all subsets containing $M(f,G)$.

The situation when $\mathcal{G}$ does not have the betweeness property is also of interest. Terminology for this situation will be taken from the paper of Braess. In case $\mathcal{G}$ is not a sun, the theorem need not hold (in particular the lemma need not hold).

EXAMPLE 2: Let $X = [0,3/2]$ and $\mathcal{G}$ be a family of constants including $\{0,1\}$ but excluding $(0,1)$. Let $f(x) = x-1/2$. As $f(3,2) = 1$, $\|f\| = 1$, and $|f(0)-c| \geq 3/2$ for $c \geq 1$, 0 is uniquely best to f on X. $M(f,0) = \{3/2\}$. But 1 is uniquely best to f on any non-empty subset of $(1,3/2]$.

If $\mathcal{G}$ is a sun, the lemma holds because it is implied by the Kolmogorov characterization, but not necessarily the theorem.

EXAMPLE 3: Let $X = [0,1]$ and $\mathcal{G} = \{0\} \cup \{x\} \cup$ {first degree polynomials >0 on $X\}$. We claim that $\mathcal{G}$ is asymptotically convex as defined by Braess, who gives a relation numbered (2.5) in his definition. Asymptotic convexity implies being a Kolmogorov set of the second kind and being regular (which is equivalent to being a sun). We set Braess' g equal to 1. All but the case $\{v,v_0\} = \{0,x\}$ are covered by $v_t = (1-t)v_0 + tv$, which makes (2.5) equal to zero. The case where $v_0 = x$, $v=0$ is handled by $v_t = (1-t)x+t^2$ and (2.5) becomes

$$\|(1-t)x-(1-t)x + t^2\| = \|t^2\| .$$

The case where $v_0 = 0$, $v = x$ is handled by $v_t = tx + t^2$ and (2.5) becomes

$$\|tx - tx + t^2\| = \|t^2\| .$$

Now let $f(x) = x^{1/2}/2-1$, then $f(0) = -1$ and $f(1) = -1/2$. $\|f\| = 1$ and $M(f,0) = \{0\}$, from which it can be shown that 0 is uniquely best on X. But 0 and x are best on $[0,\alpha]$ for α small.

In the case X is finite, $M(f,G)$ is an open cover of itself. The theorem may still not be true if $\mathcal{G}$ does not have the betweeness property on X. In Example 2, change X to any subset of $[0,3/2]$ containing the endpoints. In Example 3, change X to any subset of $[0,1]$ containing the endpoints.

APPENDIX: Properties under Constraints

A generic question which occurred to the author recently is the following: if $\mathcal{G}$ is an approximating family with a (desirable) property, does $\mathcal{G}$ still have that property under constraints? In the case of both Lagrange-type interpolatory constraints (Chalmers and Taylor, 1979, p. 50ff), or of restricted range of values (Chalmers and Taylor, 1979, p. 62ff; Dunham, 1974), the answer is yes for betweeness. For asymptotically convex families (and suns) the answer is no: in Example 3, interpolate at zero or require approximants to be ≤ 0 at 0 respectively. The requirement that the restricted approximating family satisfy the weak betweeness property in the paper of Smarzewski is thus necessary.

REFERENCES

D. BRAESS (1974), Geometrical Characterizations for nonlinear uniform approximation, J. Approximation Theory 11, 260-274.

B.L. CHALMERS and G.D. TAYLOR (1979), Uniform approximation with constraints, Jber. Deutsch Math. Verein, 81, 49-86.

E.W. CHENEY (1966), Introduction to Approximation Theory, McGraw-Hill, New York.

C.B. DUNHAM (1969), Chebyshev approximation by families with the betweeness property, Trans. Amer. Math. Soc. 136, 131-137.

C.B. DUNHAM (1974), Chebyshev approximation with restricted ranges by families with the betweeness property, J. Approximation Theory 11, 254-259.

C.B. DUNHAM (1975), Chebyshev Approximation by Families with the Betweeness Property, II, Indiana Univ. Math. J. 24, 727-732.

C.B. DUNHAM (1977), Transformed rational Chebyshev approximation, J. Approximation Theory 19, 200-204.

C.B. DUNHAM (1980), Uniqueness in restricted range approximation with betweeness, J. Approximation Theory 30, 157-159.

C.B. DUNHAM (1983), Monotone approximation by reciprocals, in C. Chui, L. Schumaker and J. Ward, Approximation Theory IV, Academic Press, New York, 441-444.

C.B. DUNHAM (1985), Chebyshev approximation by restricted rationals, Approx. Theory Appl. 1, 111-118.

L.W. JOHNSON (1970), Unicity in the Uniform Approximation of Vector-valued Functions, Bull. Austral. Math. Soc. 3, 193-198.

R. SMARZEWSKI (1979), Nonlinear Chebyshev approximations having restricted ranges. J. Approximation Theory 25, 56-64.

ON QUALITATIVE KOROVKIN THEOREMS WITH A-DISTANCE

J.L.Fernández, Havana University, Cuba.

In a paper of the author [1] it is obtained qualitative Korovkin theorems for a certain class of complex linear operators on complex function soaces over a topological space.

These results are based in two fundamental theorems posed below. The slight modifications which have been introduced in their reformulation are not relevant.

DEFINITION 1 Let X be a Hausdorff topological space. Let F(X) be a linear space of complex functions on X, E a subspace of F(X) that contains the constant 1. $\emptyset \neq Y \subseteq X$, $(L_n)_{n\in\mathbb{N}}$ a sequence of linear operators, $L_n: E \longrightarrow F(Y)$. We shall say that the sequence $(L_n)_{n\in\mathbb{N}}$ belongs to the class $\tilde{R}$ in E if for every $\varepsilon>0$, and every $f\in E$ with $\mathrm{Re}f\geq 0$ there exists $N(\varepsilon,f)$ such that for every $n\geq N(\varepsilon,f)$, $\mathrm{Re}L_n f>-\varepsilon$.

THEOREM 1 Let B(X) be the space of complex and bounded functions on X and E a subspace of B(X). Then for every $f\in E$, the sequence $L_n f$ converges uniformly to zero on Y if and only if

1) $(L_n)_{n\in\mathbb{N}}\in\tilde{R}$ in E

2) $\mathrm{Re}L_n 1$ converges uniformly to zero on Y.

THEOREM 2 Let E be a subspace of B(X), $H_n: E\longrightarrow B(Y)$, $n\in\mathbb{N}$, a sequence of linear operators such that for every $f\in E$, $H_n f$ converges uniformly to f in Y. Then for every $f\in E$, the sequence $L_n f$ converges uniformly to f in Y if and only if:

1) $(L_n - H_n)_{n\in\mathbb{N}} \in \tilde{R}$ in E.

2) $\mathrm{Re}\, L_n 1$ converges uniformly to 1 in Y.

Remark If there exists a sequence (H_n) which satisfies condition 1, then every other sequence under the same conditions also satisfies it.

We obtain qualitative Korovkin results of Korovkin type as corollaries of these theorems in the forementioned paper [1]. The fundamental role in the proofs is played by the following concept of test family:

DEFINITION 2 (test family) Let $\emptyset \neq Y \subseteq X$. A family of functions $(f_x)_{x \in \overline{Y}}$ is a test family in E if

1) For every $x \in \overline{Y}$, $f_x \in E$ and the function $(x,t) \longrightarrow f_x(t)$ is continuous on $\overline{Y} \times X$.

2) $\operatorname{Re} f_x(x)=0$

3) For every $t \in X$, $t \neq x$, $\operatorname{Re} f_x(t) > 0$

We want to generalize theorems 1 and 2 in the A-distance sense, based on paper [2]. It includes the uniform case.

DEFINITION 3 Let X be a non-empty set, F(X) a linear space of complex functions on X, $\emptyset \neq Y \subseteq X$, $1 \in F(X)$. We shall say that a function $\|\cdot\|_A : F(X) \longrightarrow \mathbb{R}^+$ is a A-norm on F(X) for Y, if

1) $\|\lambda\, g\|_A = |\lambda|\ \|g\|_A$; $\lambda \in \mathbb{C}$, $g \in F(X)$

2) $\|f+g\|_A \leq \|f\|_A + \|g\|_A$; $f,g \in F(X)$

3) Let $f,g \in F(X)$. If for real functions $\phi, \theta \in F(X)$, $c \in \mathbb{R}$ and for every $x \in Y$ we have

$$\phi(x) \leq \operatorname{Re} f(x) \leq \theta(x)$$

$$\phi(x) \leq \operatorname{Im} f(x) \leq \theta(x)$$

$$\phi(x)-c \leq \operatorname{Re} g(x) \leq \theta(x)+c$$

$$\phi(x)-c \leq \operatorname{Im} g(x) \leq \theta(x)+c$$

then

$$\|f-g\|_A \leq \|\phi-\theta\|_A + |c|.$$

4) For every constant function $c \in \mathbb{R}$, $\|c\|_A = 0$ if and only if $c=0$.

NOTE If we define the convergence of a sequence as usual, then uniform convergence implies A-norm convergence.

EXAMPLES

1) $F(X)=B(X)$ (Bounded complex functions on X) $\|f\|_\infty = \sup\limits_{x \in X} |f(x)|$

2) Let μ be a positive and finite measure on X,

$$\|f\|_p = \left(\frac{1}{\mu(X)} \int_X |f|^p \, d\mu \right)^{\frac{1}{p}}, \quad p \geq 1.$$

THEOREM 3 Let X be a non-empty set , and let B(X) be the linear space of bounded and complex functions on X, E is a subspace of B(X), $1\in E$,, $\|.\|_A$ a A-norm on E for Y, $\emptyset\neq Y\subseteq X$, $L_n:E\longrightarrow B(Y)$, $n\in\mathbb{N}$, a sequence of linear operators.

Then for every $f\in E$ the sequence $\|L_nf\|_A$ converges to zero if

1) $\left[L_n\right]_{n\in\mathbb{N}} \in \tilde{R}$ in E

2) $\|ReL_n1\|_A\longrightarrow 0$

Proof. Let $f\in E$, $M=\sup\limits_{x\in X} |f(x)|$, $\phi_1=M-f$, $\phi_2=M+f$.

We have $Re\phi_1\geq 0$, $Re\phi_2\geq 0$, then for $\varepsilon>0$, there exists $No\in\mathbb{N}$ such that if $n\geq No$, $ReL_n\phi_1>-\varepsilon$, $ReL_n\phi_2>-\varepsilon$.

From here we have

$$-M\ Re\ L_n1-\varepsilon < Re\ L_nf < M\ Re\ L_n1+\varepsilon$$

$$-M\ Re\ L_n1-\varepsilon < Im\ L_nf < M\ Re\ L_n1+\varepsilon$$

Also there exists $N_1\in N$, such that if $n\geq N_1$, we obtain $Re\ L_n1 > -\frac{\varepsilon}{M}$, then

$$-M\ Re\ L_n1-\varepsilon < 0 < M\ Re\ L_n1+\varepsilon$$

From the convergence of $\|Re\ L_n1\|_A$ to zero, there exists $N_2\in\mathbb{N}$ such that if $n\geq N_2$, $\|Re\ L_n1\|_A < \varepsilon$.

Let $N=\max\left\{N_o,N_1,N_2\right\}$, From the properties of the A-norm we obtain for $n\geq N$

$$\|L_nf\|_A \leq 2\ \|M\ Re\ L_n1+\varepsilon\|_A \leq 2\ M\ \|Re\ L_n1\| + 2\ \varepsilon\ \|1\|_A < \left[2M + 2\|1\|_A\right]\varepsilon$$

The proof is completed.

THEOREM 4 Let $\left[H_n\right]_{n\in\mathbb{N}}$ be a sequence of linear operators, $H_n:E\longrightarrow B(Y)$ such that for every $f\in E$, $\|H_nf-f\|_A\longrightarrow 0$. Then for every $f\in E$, $\|L_nf-f\|_A\longrightarrow 0$ if

1) $\left[L_n-B_n\right]_{n\in\mathbb{N}} \in \tilde{R}$ in E

2) $\|Re\ L_n1-1\|_A\longrightarrow 0$.

Proof. It is a consequence of theorem 3.

REFERENCES

1) Fernández Muñiz, José Luís, Qualitative theorems of Korovkin type for sequences of operators of the class $\tilde{R}$, Revista Ciencias Matemáticas, Vol III, No 1, 1982. (in spanish).

2) Sendov, Bl. Convergence of sequences of monotonic operators in A-distance, Comptes Rendus de l'Académie Bulgare des Sciences, Tome 30, No 5, 1977.

RELATIVE ASYMPTOTICS OF ORTHOGONAL POLYNOMIALS WITH RESPECT TO VARYING MEASURES II

René Hernández Herrera   Instituto Sup.Ped. Enrique J. Varona
Guillermo López Lagomasino   Havana University

1. INTRODUCTION

In the past few years, relative asymptotics of orthogonal polynomials nas brought considerable attention because of different applications. Suppose we have appropiate information regarding a measure $d\sigma$ and the associated orthogonal polynomials, and we know that $d\mu$ can be expressed in terms of $d\sigma$ as $d\mu = g\, d\sigma$ (where g is a reasonably well behaved function), then in certain cases we can deduce information on the orthogonal polynomials associated with $d\mu$.

In this work, we will extend some results obtained by E.A.Rahmanov [18] and Maté, Nevai and Totik [11] to the case of varying measures, i.e., when the associated measures vary with n in a special way.

Let $d\sigma$ be a positive Borel measure on the interval $[0,2\pi]$, whose support consists of an infinite set of points. By $d\theta$ we denote Lebesgue's measure on $[0,2\pi]$ and $\sigma' = \sigma'(\theta) = d\sigma/d\theta$ is the Radon-Nykodym derivative of $d\sigma$ with respect to $d\theta$.

Let $\{W_n\}, n \in \mathbb{N}$, be a sequence of polynomials such that for each $n \in \mathbb{N}$, W_n has degree n ($\deg W_n = n$) and all its zeros $(w_{n,i})$, $1 \le i \le n$, lie in $[|z| \le 1]$.

Set

$$d\sigma_n(\theta) = \frac{d\sigma(\theta)}{|W_n(z)|^2}, \qquad z = e^{i\theta} \tag{1.1}$$

<u>Definition 1.1</u> Let k be a fixed integer. We say that $(\sigma, \{W_n\}, k)$ is admissible if:

i) $\sigma' > 0$ almost everywhere on $[0,2\pi]$

ii) $\|d\sigma_n\| = \int_0^{2\pi} d\sigma_n(\theta) < +\infty, n \in \mathbb{N}$

iii) $\int_0^{2\pi} \prod_{i=1}^{-k} |z - w_{n,i}|^{-2} \, d\sigma_n(\theta) \leq M < +\infty$, $z = e^{i\theta}$, for $k = -1, -2, \ldots$

iv) $\lim_n \sum_{i=1}^{n} (1 - |w_{n,i}|) = +\infty$

In case that iv) doesn't hold, but $0 \leq w_{n,1} \leq \ldots \leq w_{n,n} \leq 1$, then in place of iv) we suppose that

iv') $$\lim_n \sum_{m=1}^{n} \left(\sqrt[2m]{c_{n,m}} \right)^{-1} = +\infty \quad ,$$

where

$$c_{n,m} = \int |W_{n,m}(z)|^{-2} \, d\sigma(\theta) \quad \text{and} \quad W_{n,m} = \prod_{i=1}^{m} (z - w_{n,i}) \, .$$

We note that if $(\sigma, \{W_n\}, k)$ is admissible then $(\sigma, \{W_n\}, k+1)$ is also admissible.

Condition ii) guarantees that for each pair (n,m) of natural numbers we can construct a polynomial

$$\phi_{n,m}(d\sigma_n, z) = \phi_{n,m}(z) = \alpha_{n,m}(d\sigma_n) z^m + \ldots$$

that is uniquely determined by the conditions

$$\frac{1}{2\pi}\int_0^{2\pi} \bar{z}^i \phi_{n,m}(d\sigma_n, z) \, d\sigma_n(\theta) = 0 \quad , \ i = 0, 1, \ldots, m-1 \ , \ z = e^{i\theta}, \tag{1.2}$$

$$\frac{1}{2\pi}\int_0^{2\pi} |\phi_{n,m}(d\sigma_n, z)|^2 \, d\sigma_n(\theta) = 1 \quad , \ \deg \phi_{n,m}(d\sigma_n, z) = m \ ,$$
$$\alpha_{n,m}(d\sigma_n) > 0 \, .$$

The monic polynomials $\Phi_{n,m}(d\sigma_n, z)$ are defined by

$$\Phi_{n,m}(d\sigma_n, z) = \alpha_{n,m}^{-1}(d\sigma_n) \, \phi_{n,m}(d\sigma_n, z) \quad .$$

We set

$$\Phi^*_{n,m}(d\sigma_n, z) = z^m \bar{\Phi}_{n,m}(d\sigma_n, z^{-1}) \quad .$$

By $\Gamma = \{ z : |z| = 1 \}$ and $\Gamma_\varepsilon(z) = \{ e^{i\varphi} : |\theta - \varphi| < \varepsilon \}, z = e^{i\theta}$, we will denote the unit circle and an open circular neighborhood of a point $z \in \Gamma$ respectively. $\overline{\Gamma_\varepsilon(z)} = \{ e^{i\theta} : |\theta - \varphi| \leq \varepsilon \}$ is the corresponding closed neighborhood.

Let g be a non-negative function on Γ, $\log g \in L^1(\Gamma)$ and $D(g,z)$, $|z|\geq 1$, be Szegö's exterior function corresponding to g

$$D(g,z) = \exp\left\{ \frac{1}{4\pi}\int_0^{2\pi} \frac{\xi + z}{\xi - z} \log g(\xi)\, d\theta \right\}, \quad |z| < 1 \quad , \tag{1.3}$$

$$D(g,z) = \lim_{r \longrightarrow 1-0} D(g,rz) \quad , \quad |z| = 1 \quad .$$

The problem of relative asymptotics consists in finding conditions for which in some region of values of z the following relation takes place

$$\lim_{n \longrightarrow \infty} \frac{\phi_{n,n+k}(d\mu_n,z)}{\phi_{n,n+k}(d\sigma_n,z)} = \overline{D(g^{-1},\bar{z}^{-1})} \tag{1.4}$$

with $d\mu_n(z) = g(z)\, d\sigma_n(z)$.

Formulas of type (1.4) have many different applications in the theory of orthogonal polynomials. It seems that the first result about relative asymptotics of type (1.4) for sufficiently extensive classes of measures was obtained by A.A.Gonchar [3]. Recently, E.A. Rahmanov [18] and Mate,Nevai and Totik [10,11] have obtained, independently, similar results in this direction for the classical case, when the measure doesn't vary with n .

In this work we use the course of arguments in [18] extended to the case of varying measures.

The purpose of this paper is to prove the following result about relative asymptotics of orthogonal polynomials with respect to varying measures. This is an analogous version of theorem 2 of E.A. Rahmanov appearing in [18]. In its proof, we use techniques and results of that paper.

<u>THEOREM 1.1</u>

Let $d\mu_n = g\, d\sigma_n$, where $(\sigma, \{W_n\}, k)$ is admissible for each k and g is a positive continuous function on Γ such that

$$L_g(z) = \max_{\xi \in \Gamma} \frac{|g(z) - g(\xi)|}{|z - \xi|} \leq M < \infty \, , \quad z \in \gamma \quad ,$$

where γ is an arbitrary subset of Γ. Then relation (1.4) takes place uniformly with respect to $z \in \gamma$.

We remind that $L_g(z)$ denotes Lipschitz's constant of function g at point z (in the following it is convenient to consider that $L_g(z)$ is defined everywhere, taking possibly the value $+\infty$).

In [4] we have obtained relations of type (1.4) for $|z|\geq r>1$. The case when $|z|=1$ needs to be treated independently because it has additional difficulties.

We wish to express our gratitude to Paul Nevai and E.A.Rahmanov for their encouragement and useful discussions.

2.AUXILIARY RESULTS

Before we prove theorem 1.1, we need to establish several auxiliary results.

In [6], the following was proved:

Lemma. Let $(\sigma, \{W_n\}, k)$ be admissible on $[0,2\pi]$. Then

$$\lim_n \Phi_{n,n+k+1}(0) = 0 \tag{2.1}$$

and

$$\lim_n \int_0^{2\pi} f(\theta)\ |\Phi_{n,n+k}(d\sigma_n,z)|^2 d\sigma_n(\theta) = \int_0^{2\pi} f(\theta)d\theta \tag{2.2}$$

$z = e^{i\theta}$, for every bounded and Borel-measurable function f .

From (2.2) we can deduce, taking f as the characteristic function, that for every measurable subset $\gamma \subseteq \Gamma$ we have

$$\lim_n \int_\gamma |\phi_{n,n+k}(d\sigma_n,z)|^2\ d\sigma_n(\theta) = \text{mes}\ (\gamma) \quad , \tag{2.3}$$

where mes$(\cdot)$ denotes Lebesgue's measure of the measurable set $(\cdot)$. The following theorem is an analogous version of a result appearing in [18] (lemma 4) .

Lemma 2.1

Let $d\mu_n = g\ d\sigma_n$, and $(\mu, \{W_n\}, k)$, $(\sigma, \{W_n\}, k)$ be admissible. Suppose that $g(e^{i\theta}) = t_1(\theta)\ /\ t_2(\theta)$, where t_1, t_2 are non-negative trigonometrical polynomials. Then relation (1.4) takes place uniformly on compacts in $\{z : |z|\geq 1, z \neq e^{i\beta_j}, j = 1,\ldots,k \}$ where $\left\{\beta_j\right\}_{j=1}^k$ is the set of all zeros of the trigonometrical polynomials t_1 , t_2 .

Proof

In virtue of a known theorem on the representation of non-negative trigonometrical polynomials ([19], Th.1.2.1), there exist algebraical polynomials P_1 , P_2 with zeros in the circle $|z| \le 1$ such that $|P_j(e^{i\theta})|^2 = t_j(\theta)$, $j = 1,2$. Furthermore, Szegö's function is multiplicative, i.e. , $D(g_1 \cdot g_2, z) = D(g_1,z)D(g_2,z)$, so it is sufficient to consider two cases : $t_1(\theta) = |e^{i\theta} - z_0|^2$, $t_2(\theta) \equiv 1$ and $t_1(\theta) \equiv 1$, $t_2(\theta) = |e^{i\theta} - z_0|^2$, where $|z_0| \le 1$; the second of these cases reduces to the first interchanging measures $d\mu_n$ and $d\sigma_n$ and so it is sufficient to prove the statement of the theorem for the case

$$d\mu_n = |\ z - z_0\ |^2\ d\sigma_n \quad . \tag{2.4}$$

Put $z_0^* = 1 / \bar{z}_0$; for $z_0 = 0$ the statement of the theorem is trivial so we consider that $z_0 \ne 0$, $z_0^* \ne \infty$.

Let us consider the the polynomial

$$Q_{n+k+2}(z) = (z-z_0)(z-z_0^*)\Phi_{n,n+k}(d\mu_n,z) - \gamma_{n,n+k}\Phi^*_{n,n+k}(d\sigma_n,z) \ ,$$

where

$$\gamma_{n,n+k} = z_0^* \frac{\alpha^2_{n,n+k}(d\sigma_n)}{\alpha^2_{n,n+k}(d\mu_n)}\Phi_{n,n+k+1}(d\mu_n,0) \quad ,$$

and let us prove that it satisfies the following orthogonality relation with respect to the measure $d\sigma_n$:

$$\int_\Gamma Q_{n+k+2}(z)\ \bar{z}^m\ d\sigma_n(z) = 0 \quad , \quad m = 0,1,\dots,,n+k \quad . \tag{2.5}$$

We have

$$\int_\Gamma \Phi^*_{n,n+k}(d\sigma_n,z)\ \bar{z}^m\ d\sigma_n(z) = \int_\Gamma \overline{\Phi_{n,n+k}}(d\sigma_n,\bar{z})\ z^{n+k}\ \bar{z}^m\ d\sigma_n(z)$$

$$= \int_\Gamma \overline{\Phi_{n,n+k}(d\sigma_n,z)\ \bar{z}^{(n+k)}\ z^m}\ d\sigma_n(z) = \int_\Gamma \overline{\Phi_{n,n+k}(d\sigma_n,z)\ \bar{z}^{(n+k-m)}}\ d\sigma_n(z)$$

$$= \begin{cases} 0 & ,m=1,\dots,n+k \\ \dfrac{2\pi}{\alpha^2_{n,n+k}(d\sigma_n)} & ,m=0 \end{cases}$$

On the other hand, since $(z-z_0)(z-z_0^*) = -z\, z_0^*\, |z-z_0|^2$, $z \in \Gamma$, we obtain

$$\int_\Gamma (z-z_0)(z-z_0^*)\Phi_{n,n+k}(d\mu_n,z)\ \bar{z}^m\ d\sigma_n(z) =$$

$$= -z_0^*\int_\Gamma \Phi_{n,n+k}(d\mu_n,z)\ \bar{z}^m z\ |z-z_0|^2 d\sigma_n(z) =$$

$$-z_0^*\int_\Gamma \Phi_{n,n+k}(d\mu_n,z)\ \bar{z}^{(m-1)} d\mu_n(z) = 0 \quad , \quad m = 1,2,\ldots,n+k \ .$$

For $m = 0$, using the recurrence formula

$$z\,\Phi_{n,n+k}(d\mu_n,z) = \Phi_{n,n+k+1}(d\mu_n,z) - \Phi_{n,n+k+1}(d\mu_n,0)\ \Phi^*_{n,n+k}(d\mu_n,z)$$

(see [1,10,18]) we obtain

$$-z_0^*\int_\Gamma z\ \Phi_{n,n+k}(d\mu_n,z)\ d\mu_n(z) =$$

$$= -z_0^*\int_\Gamma \Phi_{n,n+k+1}(d\mu_n,z)d\mu_n(z) +$$

$$+ z_0^*\ \Phi_{n,n+k+1}(d\mu_n,0)\int_\Gamma z^{n+k}\ \overline{\Phi_{n,n+k}}(d\mu_n,z)d\mu_n(z)$$

$$= z_0^*\ \Phi_{n,n+k+1}(d\mu_n,0)\ \int_\Gamma \overline{z^{-n+k}\ \Phi_{n,n+k}(d\mu_n,z)}\ d\mu_n(z) =$$

$$= z_0^*\ \Phi_{n,n+k+1}(d\mu_n,0)\ \frac{2\pi}{\alpha^2_{n,n+k}(d\mu_n)} =$$

$$= \frac{2\pi}{\alpha^2_{n,n+k}(d\sigma_n)} z_0^* \frac{\alpha^2_{n,n+k}(d\sigma_n)}{\alpha^2_{n,n+k}(d\mu_n)}\ \Phi_{n,n+k+1}(d\mu_n,0) =$$

$$= \frac{2\pi}{\alpha^2_{n,n+k}(d\sigma_n)}\gamma_{n,n+k}$$

Adding these results together we obtain (2.5).

In turn, from (2.5) follows that the expansion of the polynomial $Q_{n+k+2}(z)=z^{n+k+2}+\ldots$ with respect to the system $\left\{\Phi_{n,n+i}(d\sigma_n,z)\right\}_{i=-n}^{k+2}$

only contains the two leading terms. Hence we obtain the representation

$$(z-z_0)(z-z_0^*)\Phi_{n,n+k}(d\mu_n,z) = \tag{2.6}$$

$$= \Phi_{n,n+k+2}(d\sigma_n,z) + \beta_{n,n+k}\Phi_{n,n+k+1}(d\sigma_n,z) + \gamma_{n,n+k}\Phi^*_{n,n+k}(d\sigma_n,z) .$$

From (2.4) it follows that $d\mu_n \le 4\, d\sigma_n$, consequently

$$\alpha^2_{n,n+k}(d\sigma_n) \le 4\, \alpha^2_{n,n+k}(d\mu_n)$$

and

$$|\gamma_{n,n+k}| \le 4\, |z_0^*|\, |\Phi_{n,n+k+1}(d\mu_n,0)| \longrightarrow 0 \text{ for } n \longrightarrow \infty$$

(using (2.1)).

Further we divide (2.6) by $\Phi_{n,n+k+1}(d\sigma_n,z)$ and placing $z = z_0^*$ we obtain

$$\beta_{n,n+k} = - \frac{\Phi_{n,n+k+2}(d\sigma_n,z_0^*)}{\Phi_{n,n+k+1}(d\sigma_n,z_0^*)}\, \gamma_{n,n+k}\frac{\Phi^*_{n,n+k}(d\sigma_n,z_0^*)}{\Phi_{n,n+k+1}(d\sigma_n,z_0^*)} .$$

Letting now $n \longrightarrow \infty$, using the theorem in [6] (see also theorem in [5]) on the asymptotics of the ratio and the obvious bound $|\Phi^*_{n,n+k}(d\sigma_n,z)| \le |\Phi_{n,n+k}(d\sigma_n,z)|$ for $|z| \ge 1$, we obtain

$$\beta_{n,n+k} \longrightarrow - z_0^* .$$

We note that in [6] the theorem on the asymptotics of the ratio is actually for admissibility of type (i)-(iv). When (iv) is substituted by (iv') the same result holds and the proof follows the same arguments .

Now let us divide (2.6) by $\Phi_{n,n+k}(d\sigma_n,z)$ and we obtain

$$(z-z_0)(z-z_0^*)\frac{\Phi_{n,n+k}(d\mu_n,z)}{\Phi_{n,n+k}(d\sigma_n,z)} = \frac{\Phi_{n,n+k+2}(d\sigma_n,z)}{\Phi_{n,n+k+1}(d\sigma_n,z)}\ \frac{\Phi_{n,n+k+1}(d\sigma_n,z)}{\Phi_{n,n+k}(d\sigma_n,z)}$$

$$+ \beta_{n,n+k}\ \frac{\Phi_{n,n+k+1}(d\sigma_n,z)}{\Phi_{n,n+k}(d\sigma_n,z)} + \gamma_{n,n+k}\ \frac{\Phi^*_{n,n+k}(d\sigma_n,z)}{\Phi_{n,n+k}(d\sigma_n,z)} ,$$

and taking limit as $n \longrightarrow \infty$, we have

$$\lim_n (z-z_0)(z-z_0^*)\frac{\Phi_{n,n+k}(d\mu_n,z)}{\Phi_{n,n+k}(d\sigma_n,z)} = z^2 - z\, z_0^* = z\,(z - z_0^*)$$

$|z| \geq 1$. Therefore , uniformly on closed subsets of $\left\{ z : |z|\geq 1 ,\ z \neq z_0^* \right\}$ (if $|z| < 1$, uniformly on the closed disk $|z| \geq 1$) we have the relation

$$\lim_n \frac{\Phi_{n,n+k}(d\mu_n,z)}{\Phi_{n,n+k}(d\sigma_n,z)} = \frac{z}{z - z_0} = \overline{D(g,^{-1}z)} \ , \ g(z) = |z-z_0|^2 \ . \quad (2.7)$$

It rests to show that $\lim_n \frac{\alpha_{n,n+k}(d\mu_n)}{\alpha_{n,n+k}(d\sigma_n)} = 1$. For $|z_0| < 1$ this is

an immediate consequence of the definition of $\alpha^{-2}_{n,n+k}(d\mu_n) =$

$= \frac{1}{2\pi}\int_\Gamma |\Phi_{n,n+k}(d\mu_n,z)|^2 d\mu_n(z)$ (the same for $\alpha_{n,n+k}(d\sigma_n)$) and (2.7).

When $z_0 \in \Gamma$ it is necessary to add relation (2.2). On account of (2.2) from (2.6) it follows that for each $\varepsilon > 0$ and all sufficiently large n we have the inequalities

$$1 - 3\varepsilon \leq \frac{1}{2\pi}\int_{\Gamma\setminus\Gamma_\varepsilon(z_0)} |\phi_{n,n+k}(d\mu_n,z)|^2 d\mu_n(z)$$

$$= \frac{\alpha^2_{n,n+k}(d\mu_n)}{2\pi}\int_{\Gamma\setminus\Gamma_\varepsilon(z_0)} |\Phi_{n,n+k}(d\mu_n,z)|^2 d\mu_n(z)$$

$$\leq (1+\varepsilon)\frac{\alpha^2_{n,n+k}(d\mu_n)}{2\pi}\int_{\Gamma\setminus\Gamma_\varepsilon(z_0)} |\Phi_{n,n+k}(d\sigma_n,z)|^2 d\sigma_n(z)$$

$$\leq (1+\varepsilon)\frac{\alpha^2_{n,n+k}(d\mu_n)}{\alpha^2_{n,n+k}(d\sigma_n)} \ .$$

Interchanging $d\mu_n$ and $d\sigma_n$ we obtain the additional estimate

$$1 - 3\varepsilon \le (1+\varepsilon)\,\frac{\alpha^2_{n,n+k}(d\sigma_n)}{\alpha^2_{n,n+k}(d\mu_n)}\,.$$

Since $\varepsilon > 0$ is arbitrary , then from this follows the statement we need. The proof is complete.

We note that with the aid of standard approximation techniques the statement of lemma 2.1 can be extended to more general classes of functions g .

Lemma 2.2

Let $d\mu_n(z) = g(z)\, d\sigma_n(z)$ be a positive measure on Γ and $\|d\mu_n\|$, $\|d\sigma_n\| < +\infty$ for each $n \in \mathbb{N}$. For $z \in \Gamma$, the following inequality holds :

$$\left| \frac{\phi_{n,n+k}(d\mu_n,z)}{\phi_{n,n+k}(d\sigma_n,z)} - \frac{\alpha_{n,n+k}(d\mu_n)}{\alpha_{n,n+k}(d\sigma_n)} \right| \le$$

$$\le \frac{1}{\pi\, g(z)} \int_\Gamma \left| \frac{g(z) - g(\xi)}{z - \xi} \right| |\phi_{n,n+k}(d\mu_n,z)|\,|\phi_{n,n+k}(d\sigma_n,z)|\,d\sigma_n(z) \qquad (2.8)$$

Proof

Let $K_{n,n+k}(d\sigma_n,\xi,z) = \sum_{i=-n}^{k-1} \overline{\phi_{n,n+i}(d\sigma_n,\xi)}\ \phi_{n,n+i}(d\sigma_n,z)$ be the Szegö's kernel with respect to the measure $d\sigma_n$. The following identity holds (it can be checked inmediately on the basis of the orthogonality relations) :

$$\phi_{n,n+k}(d\mu_n,z) - \frac{\alpha_{n,n+k}(d\mu_n)}{\alpha_{n,n+k}(d\sigma_n)}\phi_{n,n+k}(d\sigma_n,z) =$$

$$= \frac{1}{2\pi}\int_\Gamma \phi_{n,n+k}(d\mu_n,\xi)K_{n,n+k}(d\sigma_n,\xi,z)\ \left[\, d\sigma_n(\xi) - \frac{1}{g(z)}d\mu_n(\xi)\,\right] ,$$

$z \in \Gamma$. It rests to divide both sides of this equality by $\phi_{n,n+k}(d\sigma_n,z)$, then

$$\frac{\phi_{n,n+k}(d\mu_n,z)}{\phi_{n,n+k}(d\sigma_n,z)} - \frac{\alpha_{n,n+k}(d\mu_n)}{\alpha_{n,n+k}(d\sigma_n)} =$$

$$= \frac{1}{2\pi g(z)}\int_\Gamma \frac{\phi_{n,n+k}(d\mu_n,z)}{\phi_{n,n+k}(d\sigma_n,z)}\ K_{n,n+k}(d\sigma_n,\xi,z)\ [g(z) - g(\xi)]d\sigma_n(\xi)$$

and using the estimate

$$| K_{n,n+k}(d\sigma_n,\xi,z) | \le 2 |\xi - z|^{-1} |\phi_{n,n+k}(d\sigma_n,\xi)||\phi_{n,n+k}(d\sigma_n,z)|$$

ξ , $z \in \Gamma$, which follows from the Christoffel-Darboux formula [1,18].

The lemma has been proved.

Lemma 2.3

Let $d\mu_n(z) = g(z)\, d\sigma_n(z)$, $\|\sigma_n\| < \infty$, for each $n \in \mathbb{N}$, and let

$$1 - \varepsilon \le g(z) \le 1 + \varepsilon \ , \ \varepsilon \in (0,1/2) \ .$$

The following relations hold :

$$1 - \varepsilon \le \frac{1}{1 + \varepsilon} \le \frac{\alpha_{n,n+k}(d\mu_n)}{\alpha_{n,n+k}(d\sigma_n)} \le 1 + \varepsilon \ , \quad n+k = 0,1,2,... \tag{2.9}$$

$$\left| \frac{\phi_{n,n+k}(d\mu_n,z)}{\phi_{n,n+k}(d\sigma_n,z)} - 1 \right| \le C \sqrt{\varepsilon} \left[1 + L_g(z) \right] \ , \tag{2.10}$$

$$, \ z \in \Gamma \ , \ n \ge n(\varepsilon,d\mu,d\sigma)$$

Proof

From the known representation of the leading coefficient of an orthogonal polynomial

$$\alpha^{-2}_{n,n+k}(d\nu_n) = \inf \left\{ \frac{1}{2\pi}\int_\Gamma |P_{n+k}(\xi)|^2 d\nu_n(\xi) \ : \ P_{n+k}(\xi) = \xi^{n+k}+... \right\}$$

it follows that if $\nu_{n,1}$, $\nu_{n,2}$ are positive measures and $\nu_{n,1} \le M\, \nu_{n,2}$ where $M > 0$, then

$$\alpha_{n,n+k}(d\nu_{n,1}) \ge M^{-1/2} \alpha_{n,n+k}(d\nu_{n,2})$$

In our case $d\mu_n \le (1 + \varepsilon)\, d\sigma_n$ and $d\sigma_n \le (1 - \varepsilon)^{-1}\, d\mu_n$, where $\varepsilon \in (0,1/2)$. Now

$$\alpha_{n,n+k}(d\mu_n) \ge (1 + \varepsilon)^{-1/2} \alpha_{n,n+k}(d\sigma_n) > (1 + \varepsilon)^{-1} \alpha_{n,n+k}(d\sigma_n)$$

and

$$1 - \varepsilon < \frac{1}{1 + \varepsilon} \le \frac{\alpha_{n,n+k}(d\mu_n)}{\alpha_{n,n+k}(d\sigma_n)}$$

On the other hand $d\sigma_n \le (1 - \varepsilon)^{-1} d\mu_n$, then

$$\alpha_{n,n+k}(d\sigma_n) \ge (1 - \varepsilon)^{1/2} \alpha_{n,n+k}(d\mu_n)$$

and

$$\frac{\alpha_{n,n+k}(d\mu_n)}{\alpha_{n,n+k}(d\sigma_n)} \le (1-\varepsilon)^{-1/2} \le (1+\varepsilon) \quad .$$

Then (2.9) is proved.

Let us prove (2.10). Now we make use of lemma 2.2 . Let us divide the integral which figures in the estimate of this theorem into two terms corresponding to the integration over $\Gamma \setminus \Gamma_{\sqrt{\varepsilon}}(z)$ and $\Gamma_{\sqrt{\varepsilon}}(z)$. Using the inequality

$$\left| \frac{g(z) - g(\xi)}{z - \xi} \right| \le \begin{cases} L_g(z), \xi \in \Gamma_{\sqrt{\varepsilon}}(z) \\ \pi\sqrt{\varepsilon}\ , \xi \in \Gamma \setminus \Gamma_{\sqrt{\varepsilon}}(z) \end{cases}$$

and then the Cauchy - Schwarz inequality we obtain

$$\left| \frac{\phi_{n,n+k}(d\mu_n,z)}{\phi_{n,n+k}(d\sigma_n,z)} - \frac{\alpha_{n,n+k}(d\mu_n)}{\alpha_{n,n+k}(d\sigma_n)} \right| \le$$

$$\le \frac{2\pi\sqrt{\varepsilon}}{g(z)} \left[\frac{1}{2\pi} \int_{\Gamma\setminus\Gamma_{\sqrt{\varepsilon}}(z)} |\phi_{n,n+k}(d\mu_n,\xi)|^2 d\sigma_n(\xi) \right]^{\frac{1}{2}} \cdot$$

$$\cdot \left[\frac{1}{2\pi} \int_{\Gamma\setminus\Gamma_{\sqrt{\varepsilon}}(z)} |\phi_{n,n+k}(d\sigma_n,\xi)|^2 d\sigma_n(\xi) \right]^{\frac{1}{2}} +$$

$$+ \frac{2\,L_g(z)}{g(z)} \left[\frac{1}{2\pi} \int_{\Gamma_{\sqrt{\varepsilon}}(z)} |\phi_{n,n+k}(d\mu_n,\xi)|^2 d\sigma_n(\xi) \right]^{\frac{1}{2}} \cdot$$

$$\cdot \left[\frac{1}{2\pi} \int_{\Gamma_{\sqrt{\varepsilon}}(z)} |\phi_{n,n+k}(d\sigma_n,\xi)|^2 d\sigma_n(\xi) \right]^{\frac{1}{2}} .$$

Since $d\sigma_n \le 2\,d\mu_n$ and $1/2 \le g(z)$ ($d\mu_n(z) = g(z)\,d\sigma_n(z)$ and $1 - \varepsilon \le g(z) \le 1 + \varepsilon$ for $\varepsilon \in (0,1/2)$) we obtain :

$$\left| \frac{\phi_{n,n+k}(d\mu_n,z)}{\phi_{n,n+k}(d\sigma_n,z)} - \frac{\alpha_{n,n+k}(d\mu_n)}{\alpha_{n,n+k}(d\sigma_n)} \right| \le$$

$$\leq \frac{2\pi\sqrt{\varepsilon}\,\sqrt{2}}{\frac{1}{2}}\left[\frac{1}{2\pi}\int_{\Gamma\setminus\Gamma_{\sqrt{\varepsilon}}(z)}|\phi_{n,n+k}(d\mu_n,\xi)|^2 d\mu_n(\xi)\right]^{\frac{1}{2}}\cdot$$

$$\cdot\left[\frac{1}{2\pi}\int_{\Gamma\setminus\Gamma_{\sqrt{\varepsilon}}(z)}|\phi_{n,n+k}(d\sigma_n,\xi)|^2 d\sigma_n(\xi)\right]^{\frac{1}{2}}$$

$$+\frac{2\,L_g(z)}{\frac{1}{2}}\left[\frac{1}{2\pi}\int_{\Gamma_{\sqrt{\varepsilon}}(z)}|\phi_{n,n+k}(d\mu_n,\xi)|^2 d\mu_n(\xi)\right]^{\frac{1}{2}}\cdot$$

$$\cdot\left[\frac{1}{2\pi}\int_{\Gamma_{\sqrt{\varepsilon}}(z)}|\phi_{n,n+k}(d\sigma_n,\xi)|^2 d\sigma_n(\xi)\right]^{\frac{1}{2}}$$

$$\leq 6\pi\sqrt{\varepsilon} + 6\,L_g(z)\left[\frac{1}{2\pi}\int_{\Gamma_{\sqrt{\varepsilon}}(z)}|\phi_{n,n+k}(d\mu_n,\xi)|^2 d\mu_n(\xi)\right]^{\frac{1}{2}}\cdot$$

$$\cdot\left[\frac{1}{2\pi}\int_{\Gamma_{\sqrt{\varepsilon}}(z)}|\phi_{n,n+k}(d\sigma_n,\xi)|^2 d\sigma_n(\xi)\right]^{\frac{1}{2}}$$

Using now (2.3), we obtain that for large enough n we have

$$\left|\frac{\phi_{n,n+k}(d\mu_n,z)}{\phi_{n,n+k}(d\sigma_n,z)}-\frac{\alpha_{n,n+k}(d\mu_n)}{\alpha_{n,n+k}(d\sigma_n)}\right| \leq 6\pi\sqrt{\varepsilon} + 6\,L_g(z)(3\sqrt{\varepsilon})$$

$$= 6\pi\sqrt{\varepsilon} + 18\,L_g(z)$$

$$\leq C\sqrt{\varepsilon}\left[1 + L_g(z)\right]$$

and using (2.9) we obtain (2.10).

Let us introduce the following notation. Let n be a natural even number, $m = 1 + \frac{n}{2}$. Set

$$J_n(z) = \frac{1}{2\pi m(2m^2+1)}\left|\frac{z^m - 1}{z - 1}\right|^4 ,$$

$$J_n(z,g) = \int_\Gamma J_n(\xi).g(\xi z)\, d\theta \quad ,$$

where $J_n(z)$ represents Jackson's kernel, $J_n(z,g)$ Jackson's sum of the integrable function $g(z)$ on Γ. Note that $J_n(e^{i\theta},g)$ is a trigonometrical polynomial on θ of degree n , which converges uniformly to g on Γ as $n \longrightarrow \infty$ if g is a continuous function on Γ .

Now we will write two lemmas of E.A.Rahmanov, whose proofs are in [18].

Lemma 2.4

Let $g \in L^1(\Gamma)$, $g_n(z) = J_n(z,g)$. The following inequality holds:

$$L_{g_n}(z) \le C\, L_g(z) \quad , \quad z \in \Gamma \quad , \quad n = 2,4,6,... \qquad (2.11)$$

Lemma 2.5

Let g be a continuous positive function on Γ and $L_g(z) \le C$, $z \in \gamma \le \Gamma$, where γ is an arbitrary subset of Γ . Let $g_n(z) = J_n(z,g)$. Then

$$\lim_n D(g_n,z) = D(g,z) \qquad (2.12)$$

uniformly with respect to $z \in \gamma$.

3.PROOF OF THE MAIN RESULT

Now we are able to prove theorem 1.1 , the main result of this work . Set

$$g_1(z) = J_1(z,g) \quad , \quad d\mu_n^1 = g_1\, d\sigma_n = \frac{g_1}{g} d\mu_n \quad ,$$

here and in the following l is a natural even number. For any l we have the identity

$$\frac{\phi_{n,n+k}(d\mu_n,z)}{\phi_{n,n+k}(d\sigma_n,z)} - D(g,z) = \left[\frac{\phi_{n,n+k}(d\mu_n,z)}{\phi_{n,n+k}(d\mu_n^1,z)} - 1\right]\frac{\phi_{n,n+k}(d\mu_n^1,z)}{\phi_{n,n+k}(d\sigma_n,z)} +$$

$$+ \left[\frac{\phi_{n,n+k}(d\mu_n^1,z)}{\phi_{n,n+k}(d\sigma_n,z)} - D(g_1,z)\right] + \Big[D(g_1,z) - D(g,z)\Big] . \qquad (3.1)$$

Let $\varepsilon \in (0,1/2)$ be arbitrary. Let us choose and fix l such that the following inequalities hold

$$1 - \varepsilon \le \frac{g(z)}{g_1(z)} \le 1 + \varepsilon \quad , \quad z \in \Gamma \qquad (3.2)$$

and

$$| D(g_1,z) - D(g,z) | \leq \varepsilon \quad , \quad z \in \gamma \tag{3.3}$$

(see lemma 2.5) . From lemma 2.4 it follows that

$$L_{g/g_1}(z) \leq K_1 \, L_g(z) \leq K_2 \quad , \quad z \in \gamma$$

Using now lemma 2.3 , we obtain that

$$\overline{\lim_n} \left| \frac{\phi_{n,n+k}(d\mu_n,z)}{\phi_{n,n+k}(d\mu_n^1,z)} - 1 \right| \leq K_3 \sqrt{\varepsilon} \tag{3.4}$$

uniformly with respect to $z \in \gamma$. On the other hand , from lemma 2.1 it follows that

$$\lim_n \frac{\phi_{n,n+k}(d\mu_n^1,z)}{\phi_{n,n+k}(d\sigma_n,z)} = D(g_1,z) \tag{3.5}$$

(since $d\mu_n^1 = g_1 \, d\sigma_n$, and g_1 is a trigonometrical polynomial) uniformly with respect to Γ. Now from (3.1), (3.3), (3.4) and (3.5) we obtain that uniformly with respect to $z \in \Gamma$ takes place the relation

$$\overline{\lim_n} \left| \frac{\phi_{n,n+k}(d\mu_n,z)}{\phi_{n,.n+k}(d\sigma_n,z)} - D(g,z) \right| \leq K_4 \sqrt{\varepsilon}$$

Hence the theorem has been proved.

REFERENCES

[1] Freud, Geza : Orthogonal Polynomials . New York. Akademiai Kiado Pergamon Press.Oxford-New York-Toronto-Sidney-Braunschweig, 1971.

[2] Geronimus, A.L. : Orthogonal Polynomials on the circle and the segment. Moskow : Gosizdat Mat. Fizmatguiz Literature , 1958 .

[3] Gonchar, A.A. : On convergence of Padé approximants for some classes of meromorphic functions. Mat.Sb. 97 (139) (1975) , 607 - 629 .

[4] Hernández H., René ; López L., Guillermo : Relative asymptotics of orthogonal polynomials with respect to varying measures . Ciencias Matemáticas vol. VII (1987) , no. 2 .

[5] López L., Guillermo : On the asymptotics of the ratio of orthogo-

nal polynomials and convergence of multipoint Padé approximants. Mat. Sb. 128 (1985) , 216 - 229 .

[6] López L., Guillermo : Asymptotics of orthogonal polynomials with respect to varying measures . To appear in Constructive Approximation .

[7] López L., Guillermo : Szegö's theorem for polynomials orthogonal with respect to varying measures. To appear .

[8] Maté, A.; Nevai, P.; Totik, V. : What is beyond Szegö's theory of orthogonal polynomials . Rational Approximation and interpolation (P.R.Graves - Morris, E.B. Saff and R.S. Varga , eds). Lecture Notes in Mathematics , vol 1105 . Berlin , Heidelberg , New York , Tokyo : Springer Verlag (1984) , 502 - 510 .

[9] Maté, A.;Nevai, P.; Totik, V. : Asymptotics for the ratio of leading coefficients of orthogonal polynomials on the unit circle. Constructive Approximation 1 (1985) , 63 - 69 .

[10] Maté, A.;Nevai, P.;Totik, V. : Extensions of Szegö's theory of orthogonal polynomials II. Constructive Approximation 3 (1987) , 51 - 72 .

[11] Maté, A.;Nevai, P.;Totik, V. : Extensions of Szegö's theory of orthogonal polynomials III. Constructive Approximation 3 (1987) , 73 - 96 .

[12] Maté, A.;Nevai, P.;Totik, V. : Strong and weak convergence of orthogonal polynomials. To appear in Amer. J. Math.

[13] Nevai, P. : Orthogonal Polynomials . Mem. Amer. Math. Soc. 213 (1979) , 1 - 185 .

[14] Nevai, P. : Extensions of Szegö's theory of orthogonal polyno - mials . Orthogonal Polynomials and their Applications (C . Brezinski, A. Draux, A.P. Magnus, P. Maroni, A. Ronveaux , eds.) Lecture Notes in Mathematics, vol 1171 Berlin : Springer Verlag (1985) , 230 -238 .

[15] Nevai,P. : G. Freud , orthogonal polynomials and Christoffel functions. A case study. J. Approx. Theory, 48 (1986) , 3 - 167 .

[16] Rahmanov, E.A. : On the asymptotics of the ratio of orthogonal polynomials . Mat. Sb. 103 (145) (1977) , 237 - 252 .

[17] Rahmanov, E.A. : On the asymptotics of the ratio of orthogonal polynomials II . Mat. Sb. 118 (160) (1982) , 104 - 117 .

[18] Rahmanov, E.A. : On the asymptotics properties of polynomials, orthogonal on the circle with respect to weigths that do not satisfy Szegö's condition. Mat. Sb. 130 (172) (1986) , 151 - 169 .

[19] Szegö, G. : Orthogonal Polynomials . Amer. Math. Soc. Coll. Publ. vol 23 . Providence R.I. (1939) . Amer. Math. Soc. 4th ed. 1975 .

ON THE RATIONAL APPROXIMATION OF H^p FUNCTIONS IN THE $L_p(\mu)$ METRIC

J. Illán
Havana University, Fac. of Mathematics and Cybernetics
Vedado, Havana City, Cuba

ABSTRACT Upper estimates of the best $L_p(\mu)$ rational approximation of H^p functions are found when μ is a Carleson measure on the interval $(-1,1)$.

1. INTRODUCTION. Let $R_n(f,[a,b])$ be the least uniform deviation from f to the set F_n of rational functions with order not greater than n, $n\in\mathbb{N}$. In the last two decades some important results have been proved in relation to the degree of convergence to zero of the sequence $R_n(f,[a,b])$, when f is analytic on [a,b] except at a finite number of points. At present it is well known that the existence of singularities of the function f on the approximation interval [a,b], has a strong connection with the rate of convergence of $R_n(f,[a,b])$ (see [1-3,6,9]). An interesting situation occurs when f is analytic in an open region U which contains the interval [a,b]. In such case the minimal speed is given by $\exp(-nc)$, where c is a positive constant which only depends on the Green capacity of the condenser $(\partial U,[a,b])$ (see [2,7,8]). If the function f has a real singularity in the points a or b, it can be thought that the capacity of the above condenser is infinite. Then the typical speed becomes $\exp(-d\sqrt{n})$, $d>0$. (see [1,2,6,9]).

For the $L_p(\mu)$ rational approximation of the function $f\in H(U)$, $U\supseteq(a,b)$, it is easy to choose a measure μ in order to appreciate that the behaviour of the sequence $R_n(f,[a,b])_p:=\inf_{r\in F_n}\|f-r\|_{L_p(\mu)}$ is in general different from that of the above mentioned uniform norm case. Besides, it is expected to obtain different exact rates of convergence of the sequence $R_n(f,[a,b])_p$, according to the nature of μ.

The purpose of this note is to find general and uniform estimates of

$$R_n(f)_p:=R_n(f,(-1,1))_p,$$

where $f\in H^p$, $1\le p\le\infty$, and μ is a Carleson measure on $(-1,1)$.

Here it is proved that the upper estimate can be improved when f fullfills a Hölder condition and « μ is thin at the ends of the interval», although the capacity of the condenser (supp μ, $|z|=1$) can be infinite.

Some estimates obtained in this paper (theorems 3.1, 3.2) represent a generalization of classical results of A. A. Gonchar [2] (see also [4]).

2. PRELIMINARIES.

Henceforth μ denotes a Carleson measure on the interval $(-1,1)$; c, c_o, $c_1, \ldots$ represent absolute and positive constants and $f_h(x)=f((1-h)x)$.

2.1 DEFINITION The integral modulus of continuity of order p, $1\le p\le\infty$, associated to the measure μ, of the function f, $f\in H^p$, is given by

$$\omega_p(f,\mu,\delta):=\sup\left\{ \|f-f_h\|_{L_p(\mu)};\ 0<h<\delta \right\},$$

where $0<\delta<1$.

2.2 DEFINITION Let $1\le p\le\infty$, and $\alpha>0$.

$$H(p,\mu,\alpha):=\left\{ f;\ f\in H^p,\ \omega_p(f,\mu,\delta)\le c\delta^\alpha \right\},$$

where c is fixed.

2.3 DEFINITION Let M_δ, $\delta>0$, be the class of Carleson measures μ on $(-1,1)$, such that

$$\int_{-1}^{1}(1-x^2)^{-\delta}d\mu(x)<\infty.$$

2.4 LEMMA (Gonchar [2]) For every h, $0<h<e^{-1}$, $n\in\mathbb{N}$, there exists a rational function r_n

$$r_n(x)=\prod_{k=1}^{n}\left(\frac{x-b_k}{x+b_k}\right),$$

where $b_k=b_k(n,h)\in[h,1[$, such that

$$|r_n(x)|\le e\ \exp\left\{-\frac{n}{\log\left(\frac{1}{h}\right)}\right\},$$

$x\in[h,1]$.

2.5 LEMMA (Ganelius [6]) For every $r>0$ and $n\in\mathbb{N}$, there exists a rational function l_n

$$l_n(x)=\prod_{k=1}^{n}\left(\frac{x-a_k}{x+a_k}\right),$$

where $a_k=a_k(n,r)\in(0,1)$ such that

$$\max_{0\le x\le 1} x^r|l_n(x)|\le C_r\ \exp(-\pi\sqrt{nr}),$$

and $\sup\left\{C_r,\ 0\le r\le r_o\right\}<\infty$, $r_o>0$.

2.6 LEMMA Let $f\in L_p(\lambda)$, where λ is a positive and finite measure on the Borel sets of $[-b,b]$, $b>0$. Let also I_1 and I_2 be respectively the intervals $[-b,a]$ and $[-a,b]$, $0<a<b$.

Now let us suppose that r_i $i=1,2$, are two rational functions such that

$$\|(f-r_i)1_I\|_{L_p}\le\beta_1, \tag{1}$$

$$\|r_i\|_{L_p}\le A, \tag{2}$$

$$\deg r_i \le k_i, \tag{3}$$

i=1,2, where $\beta_1>0$, $A>0$, $k_i\ge 0$, $i=1,2$, are given, and $1_{I_i}(x)=\begin{cases}0 & x\notin I_i\\ 1 & x\in I_i\end{cases}$.

For each $\beta_2>0$ there exists a rational function r such that

$$\|f-r\|_{L_p} \le 6\,(\beta_1+\beta_2), \tag{4}$$

$$\|r\|_{L_p} \le 2A, \tag{5}$$

$$\deg r \le k_1+k_2+c_o \log\left(e+\frac{b}{a}\right)\log\left(e+2\left(\|f\|_{L_p}+A\right)\beta_2^{-1}\right), \tag{6}$$

$c_o>1$.

PROOF Let us make $\beta_2>0$, and $k>0$. From lemma 2.4 we have that there exists a rational function ρ such that

$$|\rho(x)| \le k\ \beta_2 \quad x\in[-b,-a],$$
$$|1-\rho(x)| \le k\ \beta_2 \quad x\in[a,b],$$
$$0\le\rho(x)\le 1 \quad x\in(-a,a),$$
$$\deg \rho \le c_o \log(e+b/a)\log(e+1/k\beta_2)\ ,\ c_o>1.$$

Let $k=\left[2^{(p-1)/p}\left(\|f\|_{L_p}+A\right)\right]^{-1}$, and let r be the rational function defined by $r=(1-\rho)\,r_1 + \rho\, r_2$. It is easy to prove the inequalities (5) and (6). In order to obtain (3) let us observe that

$$\|f-r\|^p_{L_p} \le \|(f-r)1_{I_1}\|^p_{L_p}+\|(f-r)1_{[-a,a]}\|^p_{L_p}+\|(f-r)1_{I_2}\|^p_{L_p}.$$

We have from condition (1)

$$\|(f-r)1_{I_i}\|^p_{L_p} \le 2^{p-1}\left[\beta_1{}^p + \beta_2{}^p\, 2^{p-1}\, k^p\left[\|f\|^p_{L_p} + A^p\right]\right] \le$$

$$\le 2^{p-1}\left[\beta_1^p + \beta_2^p\right], \tag{7}$$

i=1,2. Besides

$$\|(f-r)1_{[-a,a]}\|^p_{L_p} \le 2^p\,\beta_1^p \le 2\left[\beta_1^p + \beta_2^p\right] \tag{8}$$

From (7) and (8), and bearing in mind the known inequality

$$\left[\beta_1^p + \beta_2^p\right]^{1/p} \le \beta_1 + \beta_2,$$

(4) is obtained.

3. MAIN RESULTS

The next result generalizes theorem 1 of [2].

3.1. THEOREM The following estimate holds for every $f\in H^p$, $1\le p\le\infty$

$$R_n(f)_p=O\left[\inf_{t\ge 1}\left\{\omega_p(f,\mu,e^{-t})+t\exp\left(-\frac{nc_1}{t}+\frac{t}{p}\right)\right\}\right],$$

where $0<c_1<1$.

PROOF We shall follow the method suggested by Gonchar in [2]

Let r_n be the rational function associated to h, $0<h<e^{-1}$, and $n\in\mathbb{N}$, according to lemma 2.4.

Let us define $g_n(z)=r_n(x(z))$, where $x(z)$ is the rational function given by

$$x(z)=\frac{h(2(z+1)-h)}{(4-3h)(2(1-z)-h)}$$

Below we pose some properties of $x=x(z)$ which we shall use

$$\frac{h^2}{16}<x(-1+h)\le x(z)\le x(1-h)=1,\ z\in\mathbb{R},\ |z|\le(1-h), \tag{9}$$

$$\Gamma_h:=\left\{s;\ |s|=(1-h/2)\right\}=\left\{s;\ \mathrm{Re}\ x(s)=0\right\}, \tag{10}$$

$$|g_n(z)|\le e\ \exp\left\{-\frac{nc_1}{\log(\frac{1}{h})}\right\}, \tag{11}$$

for $z\in\mathbb{R}$, $|z|\le(1-h)$, and $c_1=(2\log 4e)^{-1}$;

$$|g_n(z)|=1,\ z\in\Gamma_h. \tag{12}$$

Let $f\in H^p$, $1\le p\le\infty$, and $z\in\mathbb{R}$, $|z|\le(1-h)$. We define

$$\rho_n(z)=\frac{1}{2\pi i}\int_{\Gamma_h}\frac{g_n(s)s-g_n(z)z}{g_n(s)s(s-z)}\ f(s)\ ds \tag{13}$$

The rational function ρ_n has order n and interpolates to f in the zeros of g_n and in z=0.

From the properties (9-12) and the definition (13) of ρ_n, we obtain for $z\in\mathbb{R}$, $|z|<(1-h)$, the following inequality

$$|f(z)-\rho_n(z)|\le$$

$$\le c_2\|f\|_p\log(1/h)\exp\left(\frac{-nc_1}{\log(1/h)}+\frac{\log(1/h)}{p}\right)=m(n,h,p) \tag{14}$$

Thus

$$\|f-(\rho_n)h\|_{L_p}\le\|f-f_h\|_{L_p}+m(n,h,p)\ \|\mu\|^{1/p}, \tag{15}$$

where $\|\mu\|=\mu(-1,1)$.

From (15) we have for $0<h<e^{-t}$, $t\ge1$, the estimate

$$R_n(f)_p\le c\ \max\left\{1,\|f\|_p\|\mu\|^{1/p}\right\}\left[\omega_p(f,\mu,e^{-t})+t\ \exp\left(-\frac{nc_1}{t}+\frac{t}{p}\right)\right]$$

The theorem is so proved.

REMARK When $p=\infty$ it follows the classical theorem of Gonchar.

From (14) we can draw out the next uniform estimate

3.2. THEOREM Let a be such that $1-e^{-1}<a<1$. Then

$$\lim_n\sup\left(\sup_{(f,\mu)}R_n(f,[-a,a])_p\right)^{1/n}\le\exp(-c_1/\log(1/(1-a))),$$

where the pair (f,μ) ranges over all $f\in H^p_?$ and μ such that

$$\|f\|_p\ \mu[-a,a]^{1/p}\le c_2.$$

REMARK Observe that

$$\exp\left(-\frac{nc_1}{t}+\frac{t}{p}\right)\le\exp\left(-\frac{nc_1}{t}+t\right),$$

$1\le p<\infty$. Furthermore, this term does not depend on the measure μ. In order to obtain more information on μ we may focus our attention on the functions of Lipschitz class and M_δ measures.

3.3 THEOREM Let $\mu\in M_\delta$, $\delta>0$. Then

$$\sup_{\substack{f\in H(p,\mu,\alpha)\\ \|f\|_p\le1}} R_n(f)_p=O\left(\exp\left(-\alpha\pi\frac{\sqrt{n\phi p\delta}}{\alpha p+1}\right)\right),$$

where $\alpha>0$, $1\le p<\infty$, and $0<\phi<1/2$.

PROOF Let $l_n(x)=l_n(x,r)$, $r>0$, be the corresponding rational function given by lemma 2.5

Let us define

$$x_k(z)=\frac{(1-h/2+(-1)^{k+1}z)}{2(3-h)(\frac{3}{4}+(-1)^k z)},\quad k=1,2.$$

Let us observe that $x_k((-1)^k(1-h/2))=0$, $k=1,2$. Besides

$$\frac{h}{21}<x_k\left[(-1)^k(1-h)\right]\le x_k(z)\le x_k\left[\frac{(-1)^{k+1}}{2}\right]=1,$$

for z between $(-1)^k(1-h)$ and $\frac{(-1)^{k+1}}{2}$, $k=1,2$.

Let $I_1=[-1,1/2]$, $I_2=[-1/2,1]$, and let $\Gamma_{h,k}$ be the circle centered at the point $((-1)^k\frac{(1-2h)}{8};0)$ with radius $\frac{(7-2h)}{8}$, $k=1,2$. We can easily prove that

$$\Gamma_{h,k}=\left\{s;\operatorname{Re} x_k(s)=0\right\},\quad k=1,2.$$

For $f\in H^p$, $1\le p<\infty$, $n\in\mathbb{N}$, $k=1,2$, let us consider the rational function

$$R_{n,k}(z)=\frac{1}{2\pi i}\int_{\Gamma_{h,k}}\frac{\rho_{n,k}(s)(s-(-1)^k\frac{(1-2h)}{8})-\rho_{n,k}(z)(z-(-1)^k\frac{(1-2h)}{8})}{(\rho_{n,k}(s)(s-z)(s-(-1)^k\frac{(1-2h)}{8}))}f(s)\,ds,$$

where $\rho_{n,k}(z)=l_n(x_k(z))$, $z\in[-1+h,1/2]$ if $k=1$, and $z\in[-1/2,1-h]$, for $k=2$.

We have

$$\|(f-R_{n,k})_h\,1_{I_k}\|_{L_p}\le M\exp(-\pi\sqrt{nr})\,C(z,k,h,n,p,r),$$

where

$$C(z,k,h,n,p,r)^p=\int_{I_k}\left(|x_k((1-h)z)|^{-r}\int_{\Gamma_{h,k}}|f(s)B^{-1}||ds|\right)^p d\mu(z),$$

and $B=\rho_{n,k}(s)\left[s-(1-h)z\right]\left[s-(-1)^k\frac{(1-2h)}{8}\right]$.

Hence

$$\|(f-R_{n,k})_h 1_{I_k}\|_{L_p}\le M\|f\|_p\, h^{-1/p}\exp(-\pi\sqrt{nr})\log(1/h)\; K,$$

where $K=\left(\int_{I_k}|x_k((1-h)z)|^{-pr}d\mu(z)\right)^{1/p}$.

The above inequalities are true because

$$\int_{\Gamma_{h,k}}|s-(1-h)z|^{-1}|ds|\le c_3\log(1/h),$$

$$|\rho_{n,k}(s)|=1,\ \text{if } s\in\Gamma_{h,k},$$

and lemma 2.5.

According to the preceeding statements we obtain

$$\|(f-R_{n,k})_h 1_{I_k}\|_{L_p}\le$$

$$\le c_4\|f\|_p\log(1/h)K_k(h,s,p)\exp(-\pi\sqrt{nr}+p^{-1}\log(1/h)) \tag{16}$$

where $\delta=rp$ and

$$K_k(h,s,p)^p=\int_{I_k}|(1-h/2)+(-1)^{k+1}(1-h)x|^{-\delta}d\mu(x).$$

From the inequality

$$K_k(h,s,p)\le 2^{\delta/p}\left(\int_{-1}^{1}|(1-h/2)^2-(1-h)^2x^2|^{-\delta}d\mu(x)\right)^{\frac{1}{p}}, \tag{17}$$

$k=1,2$; and from (16) and (17) we obtain

$$\|(f-R_{n,k})_h 1_{I_k}\|_{L_p}\le$$

$$\le c_4\|f\|_p\log\left(\frac{1}{h}\right)\left(\int_{-1}^{1}\left|\left(1-\frac{h}{2}\right)^2-(1-h)^2x^2\right|^{-\delta}d\mu(x)\right)^{1/p}\exp\left(-\pi\sqrt{nr}+\frac{\log\left(\frac{1}{h}\right)}{p}\right)=$$

$$=\beta(n,p,h,\delta).$$

Thus

$$\|(f-(R_{n,k})_h)1_{I_k}\|_{L_p}\le\|f-fh\|_{L_p}+\beta(n,p,h,\delta).$$

If $s\in\Gamma_{h,k}$; $k=1,2$ and $z\in\begin{cases}[-1+h,1/2] & \text{if } k=1\\ [-1/2,1-h] & \text{if } k=2\end{cases}$, then $|s-z|\ge h/2$, $(h<e^{-1})$.

Therefore $\|(R_{n,k})_h\|_{L_p}\le c_5\|f\|_p\dfrac{\|\mu\|^{1/p}}{h}$, $k=1,2$.

Let us put $h_n=e^{-t\sqrt{n}}$ and $\beta_n=e^{-d\sqrt{n}}$, $t,d>0$ with n sufficiently large.

If $0<a<b<1$, we can find n_o such that for $n>n_o$ the following inequality holds

$$n^{-1}c_o\log(e+2)\log(e+2\beta_n^{-1}\|f\|_p(c_6+c_5\|\mu\|^{1/p}h_n^{-1}))<$$
$$< a < b < 1-\frac{1}{2n},$$

where $\|f\|_{L_p}\le c_6\|f\|_p$ (μ is a Carleson measure).

Let m be defined by

$$m=\left[\frac{n-2-c_o\log(e+2)\log\left(e+2\beta_n^{-1}\|f\|_p\left(c_6+c_5\|\mu\|^{1/p}h_n^{-1}\right)\right)}{2}\right]+1,$$

where $n>n_o$, and $[x]$ denotes here the greatest integer smaller than x.

From lemma 2.6 there exists a rational function r such that

$$\|f-r\|_{L_p}\le 6\left(\beta(m,p,h_m,\delta)+\beta_n+\|f-f_{h_m}\|_{L_p}\right).$$

Furthermore, according to the definition of m we have for $n>n_o$

$$\deg r\le 2m+c_o\log(e+2)\log\left(e+2\beta_n^{-1}\|f\|_p\left(c_6+c_5\|\mu\|^{1/p}h_n^{-1}\right)\right)\le n,$$

where we have had in mind that $n>m$. On the other hand

$$\|(R_{m,k})_{h_m}\|_{L_p}\le c_5\|f\|_p\|\mu\|^{1/p}e^{t\sqrt{n}}.$$

An estimate of the L_p norm of r follows from lemma 2.6

$$\|r\|_{L_p}\le 2\,c_5\|f\|_p\|\mu\|^{1/p}e^{t\sqrt{n}}\quad\text{(condition (2))}$$

For $n>n_o$ we have

$$R_n(f)_p\le 6\|f-f_{h_m}\|_{L_p}+ \tag{18}$$
$$+6c_7\|f\|_p\log(1/h_m)K(h_m,\delta,p)\exp\left(-\pi\sqrt{mr}+\frac{t\sqrt{m}}{p}\right)+6e^{-d\sqrt{n}}$$

where $K(h,\delta,p)=\left(\int_{-1}^{1}|(1-h/2)^2-(1-h)^2x^2|^{-\delta}d\mu(x)\right)^{1/p}.$

Let σ be small $(0<\sigma<1-\sqrt{2}/2)$. Then

$$\left|\left(1-\frac{h_m}{2}\right)^2-\left(1-h_m\right)^2x^2\right|^{-\delta}\le v(\delta,x),$$

$m\in\mathbb{N}$, where

$$v(\delta,x)=\begin{cases}|1-x^2|^{-\delta} & \text{if } x\in[-1,-\sigma-\sqrt{2}/2]\cup[\sqrt{2}/2+\sigma,1]\\ M & \text{if } x\in(-\sigma-\sqrt{2}/2,\sqrt{2}/2+\sigma)\end{cases},$$

and M is some positive constant.

The condition $\mu\in M_\delta$ implies that $v(\delta,z)\in L_1(\mu)$. From the dominated convergence theorem we can say that the limit $\lim_n K(h_n,\delta,p)$ exists and hence $\sup_n K(h_n, \delta, p) = K(\delta, p) < \infty$.

For n sufficiently large $m \ge n(b-a)/2 = n\phi$

Let $f \in H(p, \mu, \alpha)$. From (18) we have for $\pi\sqrt{\frac{\delta}{p}} > \frac{t}{p}$ the following inequalities

$$R_n(f)_p \le M_{\delta,\phi}\left[e^{-\alpha t\sqrt{m}} + e^{-d\sqrt{n}} + \|f\|_p \sqrt{n}\, \exp\left(-\pi\sqrt{\frac{m\delta}{p}} + \frac{t\sqrt{m}}{p}\right)\right] \le$$

$$\le M_{\delta,\phi}\left[e^{-\alpha t\sqrt{n\phi}} + e^{-d\sqrt{n}} + \|f\|_p \sqrt{n}\, \exp\left(-\sqrt{n\phi}\left(\pi\sqrt{\frac{\delta}{p}} - \frac{t}{p}\right)\right)\right].$$

If t and d satisfy

$$\sqrt{\phi}\left[\pi\sqrt{\frac{\delta}{p}} - \frac{t}{p}\right] = d = \alpha t\sqrt{\phi},$$

then we shall have

$$R_n(f)_p \le M'(\delta, \phi_o) \max\left\{1, \|f\|_p\right\} \exp\left(-\alpha\pi\frac{\sqrt{np\delta\phi_o}}{\alpha p+1}\right),$$

where $0<\phi_o<\phi$, n is sufficiently large, and $M'(\delta, \phi_o)$ only depends on δ and ϕ_o. The theorem is so proved.

3.4 COROLLARY Let $\mu \in \bigcap\left\{Mp, \delta;\ 0<\delta<\delta_o\right\}$, $1 \le p < \infty$. Then

$$\lim\sup_n \left(\sup_{\substack{f\in H(p,\mu,\alpha)\\ \|f\|_p \le 1}} R_n(f)_p\right)^{1/\sqrt{n}} \le \exp\left(-\frac{\alpha\pi}{(\alpha p+1)}\sqrt{\frac{p\delta_o}{2}}\right)$$

3.5 COROLLARY Let $\mu \in \bigcap\left\{Mp, \delta;\ \delta>0\right\}$, $1 \le p < \infty$. Then

$$\lim_n \left(\sup_{\substack{f\in H(p,\mu,\alpha)\\ \|f\|_p \le 1}} R_n(f)_p\right)^{1/\sqrt{n}} = 0$$

REFERENCES

1. Newman D.J., Rational approximation to $|x|$, Michigan, Math. Journal 11 (1964), 11-14, MR 30 # 1344.

2. Gonchar A.A., On the speed of rational approximation to continuous functions with characteristic singularities, Mat. Sbornik T 73 (115) No 4 (1967) (in russian).

3. ___________, The rate of rational approximation and the property of single-valuedness of an analytic function in the neighborhood of an isolated singular point, Math. USSR Sbornik Vol. 23 (1974) No 2.

4. Illán J., Estimates for the best rational approximation of H^p functions, Comptes Rendus de l'Académie Bulgare des Sciences, Tome 39, No 3, 1986.

5. Viacheslavov N.S., On the uniform and rational approximation to $|x|$, Dokl. A.N.SSSR, 220(1975) p.512-515. (in russian).

6. Ganelius T., Rational approximation to x^{α} on [0,1], Analysis Mathematica, 5 (1979), 19-33.
7. __________, Rational approximation in the complex plane and on the line, Annales Academic Scientarum Fennicae, Series A, I. Mathematica. Vol. 2, 1976, p.129-145.

8. Widom H., Rational approximation and n-dimensional diameter, J. Approximation Theory, 5 (1972) 343-361.

9. Stenger F,, Polynomial, sinc and rational function methods for approximating analytic functions, Lecture Notes in Mathematics, Rational approximation and interpolation, Proceedings, Tampa, Florida 1983. Springer Verlag.

ON THE TRAJECTORIES OF INCLINED OIL-WELLS

M. A. Jiménez Pozo. University of Havana

ABSTRACT

The author presents two mathematical models for representing the trajectories of inclined oil-wells. The main tool is parametric quadratic splines approximation with free knots. Formulas for calculating the solutions are obtained explicitly and both models are compared.

INTRODUCTION

Inclined oil-wells are designed by engineers taking into account special conditions of the ground, economic aspects of the problem and technical possibilities that consider mathematical parameters such as the curvature of a path. For a large introduction to the subject, see [1].

However, inclined oil-wells in real situations follow trajectories which depend not only on the technical plan but also on random conditions that appear in practice.

Very often, polygonal curves are used to describe such trajectories, but the development of computer science gives us the possibility of improving the description. In fact, here we present the mathematical aspects of two models for representing these trajectories. The first one has been used with success in industry. The main tool will be the use of parametric quadratic splines with free knots.

STATEMENT OF THE MATHEMATICAL PROBLEM

Denote by Γ the path of the oil-well and consider that a system of rectangular three dimensional coordinates whose origen coincides with the origen of the oil-well is given. Let (P_k) be a finite sequence of ordered points in Γ and denote by l_k the length of the arc joining P_k and P_{k+1}, $k=0,1,..$ Using physical measures, the spherical coordinates of the unit tangent vectors (v_k) to the path at (P_k) are obtained.

The data (l_k) and (v_k) are given and $P_o=(0,0,0)$. Then the problem consists in defining by induction a function whose graph approximates Γ between P_k and P_{k+1} and that allows us to determine P_{k+1}. Using a translation of coordinates at P_k as origen of the system for $k>0$, we simplify the problem as follows:

Denote $P:=P_k=(0,0,0)$, $P_1:=P_{k+1}:=(x_1,y_1,z_1)$, $L=l_k$ and let l be the length of arc as variable. Then the path is described by the function

$$z = Z(l), \quad y = Y(l), \quad x = X(l) \qquad l \in [0,L] \tag{1}$$

From the physical meaning of the problem we can state that z represents the depth and it is a strictly increasing function. Also that the y-axis is in the direction of the petrolic stratum. So y is a nondecreasing function. From the first remark there exists the inverse function

$$l = Z^{-1}(z) \qquad z \in [0,z_1] \tag{2}$$

and then y and x are also functions of z.

Denote by y' and x' the derivatives of those functions. Since the unit tangent vectors at P and P_1 are known, we can deduce the values of y' and x' at $z=0$ and $z=z_1$.

In [2], from physical considerations, we have supposed that

$$y = Az^2 + Bz \qquad x = Cz^2 + Dz \tag{3}$$

Then, we proved that there exists one and only one system of values for A,B,C,D,x_1,y_1,z_1 which satisfies all of the prescribed conditions on derivatives and the length of arc.

Later, the computing program and practice in real oil-wells showed that there is only a small difference between the values of z when we calculate them by this model and by the simple model which approximates the path by a polygonal curve.

Since the variable z in our model is the independent variable, we decided to prepare a second model traying to avoid any privilege with respect to any particular variable.

So we have supposed that

$$z = al^2 + bl \quad y = cl^2 + dl \quad x = el^2 + fl \qquad l \in [0,L] \tag{4}$$

Later, the computing programs showed us that there is no significant difference between the results of our two models. So we have compared them in theoretical form.

FIRST MODEL

As we have already mentioned, it is going to appear in [2]. However, since we want to compare later the results on both models, we give the formulas here.

If $y'(0)=y'(z_1)$ and $x'(0)=x'(z_1)$, then the curve joining P and P_1 is a straight line. So, we will consider only the main case when at least one of the above equalities does not hold. This gives us a nonlinear problem on spline functions.

Without loss of generality we may suppose that

$$y'(0) \neq y'(z_1) \tag{5}$$

Define

$$\alpha = (x'(z_1)-x'(0)) \,/\, (y'(z_1)-y'(0)) \tag{6}$$

$$\beta = x'(0)- \alpha\, y'(0) \tag{7}$$

Then

$$A = \frac{1+\alpha^2+\beta^2}{2+L\,(1+\alpha^2)^{3/2}} \int_{\frac{(1+\alpha^2)\, y'(0)\, +\, \alpha\beta}{(1+\alpha^2+\beta^2)^{1/2}}}^{\frac{(1+\alpha^2)\, y'(z_1)\, +\, \alpha\beta}{(1+\alpha^2+\beta^2)^{1/2}}} (1+u)^{1/2}\, du \tag{8}$$

$$B = y'(0) \qquad C = \alpha A \qquad D = x'(0) \tag{9}$$

$$z_1 = (y'(z_1) - y'(0)) \,/\, 2A \tag{10}$$

while y_1 and x_1 now follow from (3).

SECOND MODEL

Denote by $\dot{z}$, $\dot{y}$, $\dot{x}$, the derivatives in (4). From elementary analysis it follows that

$$\dot{y} = y'\,\dot{z} \qquad \dot{x} = x'\,\dot{z} \tag{11}$$

$$\dot{x}^2+\dot{y}^2+\dot{z}^2 = 1 \tag{12}$$

Thus, from (11) and (12)

$$\dot{z} = 1 \,/ \sqrt{1+x'^2+y'^2} \tag{13}$$

Equation (13) allows us to know $\dot{z}(0)$ and $\dot{z}(L)$ and, substituting these values in (11), we obtain $\dot{y}(0)$, $\dot{y}(L)$, $\dot{x}(0)$, $\dot{x}(L)$.

Now, taking derivatives in (4), it follows that

$$b = \dot{z}(0) \qquad a = (\dot{z}(L)-\dot{z}(0)) \,/\, 2L \tag{14}$$

$$d = \dot{y}(0) \qquad c = (\dot{y}(L)-\dot{y}(0)) \,/\, 2L \tag{15}$$

$$f = \dot{x}(0) \qquad e = (\dot{x}(L)-\dot{x}(0)) \,/\, 2L \tag{16}$$

and evaluating (4) at $l = L$ we obtain

$$z_1 = (\dot{z}(L)+\dot{z}(0))\, L \,/\, 2 \tag{17}$$

$$y_1 = (\dot{z}(L)\, y'(z_1) + \dot{z}(0)\, y'(0))\, L \,/\, 2 \tag{18}$$

$$x_1 = (\dot{z}(L)\, x'(z_1) + \dot{z}(0)\, x'(0))\, L \,/\, 2 \tag{19}$$

COMPARISON BETWEEN BOTH MODELS

The variable x denotes the error in the direction of the oil-well. Since there exists a systematic control on the trajectory to correct this error during its perforation, it follows that no significant difference for this parameter may hold for any z. However we will

consider its influence in the estimations of z and y.
Denote the coordinates of P_1 by (x_1, y_1, z_1) when we calculate them by the first model and by $(\bar{x}_1, \bar{y}_1 \bar{z}_1)$ in the second one.
The derivative $y'(z)$ represents two different functions depending on the model. However, these functions have equal values at P and P_1. A similar remark holds for $x'(z)$. Thus, sometimes we will write $y'(P_1)$, $x'(P_1)$, and so on.
On the other hand, observe that $y_1 = \bar{y}_1$, $x_1 = \bar{x}_1$, $z_1 = \bar{z}_1$ in the linear case. So we must only compare the results in the case given by (5).
The formula (8) has been deduced from

$$L = \int_0^{z_1} \sqrt{1 + y'(t)^2 + x'(t)^2}\,dt \tag{20}$$

Thus, from the mean value theorem for integrals of continuos functions, there exists

$$\lambda \in \Gamma\ (0, P_1) \tag{21}$$

where $\Gamma\ (0, P_1)$ denotes the arc of Γ joining $P \equiv 0$ and P_1, such that

$$z_1 = L / \sqrt{1 + y'(\lambda)^2 + x'(\lambda)^2} \tag{22}$$

On the other hand, making use of (17) and (13), we obtain that

$$\bar{z}_1 = \frac{L}{2}\left[\frac{1}{(1+y'(P_1)^2 + x'(P_1)^2)^{1/2}} + \frac{1}{(1+y'(0)^2+x'(0)^2))^{1/2}}\right]$$

Since $f(t) = 1 / (1 + y'(t)^2 + x'(t)^2)^{1/2}$ is a continuous function, there exists

$$\mu \in \Gamma\ (0, P_1) \tag{23}$$

such that

$$\bar{z}_1 = L / (1 + y'(\mu)^2 + x'(\mu)^2)^{1/2} \tag{24}$$

Since y' and x' only have small variations on $\Gamma(0, P_1)$, and it does not depend on the model it follows from formulas (21) and (24) that

$$z_1 \sim \bar{z}_1 \tag{25}$$

The mean value theorem of Lagrange in the particular case of functions such as (3), allows us to write

$$y_1 = y'(z_1/2)\ z_1$$

Thus, from (22)

$$y_1 = L\ y'(z_1/2) / (1 + y'(\lambda)^2 + x'(\lambda)^2)^{1/2} \tag{26}$$

Now, from (18) and (13), we obtain that

$$\overline{y}_1 = \left[\frac{y'(z_1)}{(1 + y'(\bar{z}_1)^2 + x'(\bar{z}_1)^2)^{1/2}} + \frac{y'(0)}{(1 + y'(0)^2 + x(0)^2)^{1/2}} \right] L/2$$

Using the continuity of the function

$$f(t) = y'(t)/(1 + y'(t)^2 + x'(t)^2)^{1/2}$$

we deduce from the last expression that there exists

$$\nu \in \Gamma\,(0, P_1) \tag{27}$$

such that

$$\overline{y_1} = L\; y'(\nu)/\,(\,1 + y'(\nu)^2 + x'(\nu)^2)^{1/2} \tag{28}$$

Once again we point out that y' and x' only have small variations on $\Gamma(0, P_1)$. Then, from (26) and (28) we can observe that in practice

$$y_1 \sim \overline{y}_1 \tag{29}$$

Remark

The coincidence of results of these two models and the experience in practice make them available for use. However, the second one has the advantage that algorithms are simpler and also it allows us to evaluate directly the coordinates at any point in the path for a given length. From the mathematical point of view, the first model gives us a nonlinear problem on splines with fixed length of arc and fixed derivatives at the end points. It is not hard to prove that this last problem also has solution in m-dimensions.

BIBLIOGRAPHY

[1] Mavliutov, M. R. "Tecnologia de perforación de pozos profundos" MIR, URSS, 1986 (from the original book in russian edited by Nedra, 1982).

[2] Jiménez Pozo, M. A. "Modelación de las trayectorias de pozos inclinados de petroleo mediante splines paramétricos cuadráticos con extremos libres" To appear in Ciencias Matemáticas, Univ. of Havana, in spanish.

ON SOME CONTRIBUTIONS OF HALASZ TO THE TURAN POWER–SUM THEORY

Lee Lorch and Dennis Russell

York University
4700 Keele Street
North York, Ontario, Canada M3J 1P3

1. Background.

Turán's power–sum theory, to which he devoted his remarkable talent for a major portion of his scientific life, is presented in his posthumous book [2]. He left in finished form the fundamental analysis and a great variety of applications. The published text includes also elaborations of many applications along lines he had indicated. This was accomplished thanks to years of caring and creative work by his pupils Gabor Halász and Janos Pintz.

A number of specific results due to them, some inserted already by Turán, are incorporated.

Here we examine a few due to Halász and establish sharper lower bounds in his inequalities without serious alteration of method.

These inequalities, typical of the theory, deal with sums of the type $g(\nu)$ defined by (4) and localized lower estimates for them. Special cases of such problems and methods are found in H. Bohr's study of the Riemann zeta function and related topics in analysis and analytic number–theory [2, Introduction]. This motivated Turán to construct a greatly generalized and deepened theory with a surprisingly vast field of applications.

2. Some Preliminaries.

The improvements presented here rest chiefly on a more precise analysis of the properties of the function $U(x)$ than is made in [2, Appendix A, pp. 555–556]. This function is defined by

(1) $$U(x) = \frac{1}{\sin^2 x} - \frac{1}{x^2}\,,\; U(0) = U(0+) = \tfrac{1}{3}\,,\; x \neq n\pi\,.$$

The properties used are:

Lemma 1. *The function* $U(x)$ *defined by* (1)

(i) *increases for* $0 \le x < \pi$,

(ii) *is convex everywhere, i.e.,* $U''(x) > 0$, $x > 0$, *in fact,* $U^{(2m)}(x) > 0$, $m = 1,2,\ldots$,

(iii) *has one and only one extremum (a minimum) in* $n\pi < x < (n+1)\pi$, *say at* x_n, $n = 1,2,\ldots$, $x_0 = 0$,

(iv) *the successive minima* $\{U(x_n)\}$, $n = 0,1,\ldots$, *increase, and*

(v) $\tfrac{1}{3} < U(x) < U(\kappa x)$, *for* $\kappa > 1$, $0 < x < \tfrac{1}{2}\pi$, $\kappa x \neq n\pi$.

Proof: By differentiating the well–known partial fraction expansion for $\cot x$ [1, p.207], we obtain

(2) $$U(x) = \sum_{n=1}^{\infty} \left\{ \frac{1}{(n\pi - x)^2} + \frac{1}{(n\pi + x)^2} \right\},\; x \neq \pm\pi, \pm 2\pi, \ldots\,.$$

Hence

$$U'(x) = \sum_{n=1}^{\infty} \left\{ \frac{2}{(n\pi - x)^3} - \frac{2}{(n\pi + x)^3} \right\} > 0 \text{ for } 0 < x < \pi\,,$$

and so (i) follows. For (ii) we have

(3) $$U^{(2m)}(x) = \sum_{n=1}^{\infty} \left\{ \frac{(2m+1)\,!}{(n\pi + x)^{2m+2}} + \frac{(2m+1)\,!}{(n\pi - x)^{2m+2}} \right\} > 0\,.$$

This proves (iii), as well. For (iv) it suffices to note that

$$U(x_n) \le U(x_{n+1} - \pi) = \frac{1}{\sin^2 x_{n+1}} - \frac{1}{(x_{n+1} - \pi)^2} < U(x_{n+1})\,.$$

Clearly, (v) follows from (i) when $0 < \kappa x < \pi$. For $\kappa x > \pi$, (iv) and (i) imply

$$U(\kappa x) \ge U(x_1) = .955 > 1 - 4\pi^{-2} = U(\tfrac{1}{2}\pi) > U(x) > \tfrac{1}{3}\,.$$

This proves the lemma.

Remark. It is clear that $U(x_n) < 1$, $n = 0,1,2,\ldots$, so that the sequence $\{U(x_n)\}$ has a limit not greater than 1. Actually, the limit is exactly one, a result not required in this paper, although we do use, in (v), the numerical result $U(x_1) = .955$, a value already fairly close to 1, contrasting with $U(x_0) = \tfrac{1}{3}$.

Lemma 1(v) implies an improvement by an order of magnitude in [2, (A.1,4), p.556], an inequality which plays a crucial role. It will be shown that

$$\text{(A.1.4')} \qquad 1-\frac{\sin^2\kappa\beta}{\kappa^2\sin^2\beta} > \left[1-\frac{1}{\kappa^2}\right]\left[1-\frac{\sin^2\kappa\beta}{\kappa^2\beta^2}\right],$$

$0<\beta<\frac{1}{2}\pi$, $\kappa>1$.

For $\kappa\beta=n\pi$, this is obvious. For $\kappa\beta\neq n\pi$, it is equivalent to

$$U(\kappa\beta)=\frac{1}{\sin^2\kappa\beta}-\frac{1}{\kappa^2\beta^2}>\frac{1}{\sin^2\beta}-\frac{1}{\beta^2}=U(\beta),$$

which, under the assumptions, is precisely the inequality established in Lemma 1(v).

With (A.1,4′) established, we can pass to the main results after recording the appropriate notations.

3. Notations.

Hereafter $\kappa=k+1$. Following Turán, we define the *generalized power sum*

$$\text{(4)} \qquad g(\nu)=\sum_{j=1}^{n} b_j z_j^\nu ,$$

where b_j, z_j are complex numbers and ν is an integer. The *norm* of the polynomial $P_k(z)=\Sigma_0^k\,\mu_\nu z^\nu$ is defined to be

$$\|P_k(z)\|=\|\sum_{\nu=0}^{k}\mu_\nu z^\nu\|=\sum_{\nu=0}^{k}|\mu_\nu| .$$

4. Results.

With the exception of Corollaries 2 and 3 below (closely related to the other results) we confine ourselves to sharpening estimates in a lemma and some theorems and corollaries all due to G. Halász and published first in [2]. To facilitate identification, the new estimates will be denoted by adding the prime symbol (′) to the designations of the corresponding references in [2]. Thus, associated with Lemma 5.8 of [2, p.52], we begin with a result in which only the primed property 4′ is new:

Lemma 5.8′. *For each* $k \geq 2$, *there is a polynomial* $\overset{*}{\pi}_k(z)$ *of degree* k *such that (recalling that* $\kappa = k+1$) ,

1. $\overset{*}{\pi}_k(z) \neq 0$, $|z| < 1$,
2. $\overset{*}{\pi}_k(0) = 1$
3. $\overset{*}{\pi}_k(1) = 0$

4′. $\rho_k = \max_\theta |\overset{*}{\pi}_k(e^{i\theta})| < \exp\{3/(2\kappa)\}$.

The polynomials $\overset{*}{\pi}_k(z)$ here will be precisely the ones introduced in [2, p.52]. Hence, properties 1,2, and 3 above, coinciding as they do with the corresponding properties in Lemma 5.8 [2, p.52], are automatically satisfied. Only 4′ above represents a sharpening from [2] where the exponent is $2/k$. Indeed, our exponent can be reduced still further, as can be seen from the proof.

From Lemma 5.8′ (4′) there follows a corresponding sharpening of the Corollary in [2, §5.8, pp.53–54]. The new version reads:

Corollary′. *The polynomial* $\overset{*}{\pi}_k(z)$ *of Lemma 5.8′ satisfies the inequality*

(5.8.1′)
$$\frac{1}{|\overset{*}{\pi}_k(z)|} \leq \exp\left\{\frac{3r}{\kappa(1-r)}\right\} , \; |z| \leq r < 1 \, , \; \kappa = k+1 \, .$$

Together with the Lemma this implies an increase in the lower bound in Halász's Theorem 7.1 [2, p.73], namely:

Theorem 7.1′. *For* $m = 0,1,2,\ldots,k = 2,3,\ldots,$ *and arbitrary complex numbers* b_j, z_j, *with* $\min_j |z_j| = 1$ *and* $g(\nu)$ *defined by* (4), *we have*

(5)
$$\max|g(\nu)| \geq |g(0)|(kn+1)^{-1/2}\, 2^{-m-1} \exp\{-9n/(2\kappa)\} ,$$

where the maximum is taken over $\nu = m+1,\ldots,m+kn$, *and* $\kappa = k+1$.

This replaces Halász's exponent $6n/k$ by $9n/(2\kappa)$.

In turn, this permits improving his Corollary [2, p.75] to the following.

Corollary″. *For* $\min_j |z_j| = 1$ *and arbitrary complex* b_j, *we have*

(7.1.11′)
$$\max|g(\nu)| \geq \tfrac{1}{2}\, n^{-1}\, e^{-9/2}|g(0)| ,$$

where the maximum is taken over $\nu = 1,2,\ldots,n(n-1)$.

Remark. In [2, (7.1.11), p.75] either the strict inequality should be replaced by $\geq$, or it should be assumed that not all b_j vanish. In the latter case, $\geq$ can be replaced by $>$ also in (7.1.11′), which provides a lower bound $2e^{3/2} = 8.963$ times as large as the one in (7.1.11).

The above improvements lead to a corresponding sharpening in the application Halász made to ordinary differential equations [2, Theorem 19.2, p.216]. In it, the function $y(t)$ is any solution of

$$y^{(n)}(t) + a_1 y^{(n-1)}(t) + \ldots + a_n y(t) = 0$$

with constant coefficients, expressed so that

$$y(t) = \sum_{j=1}^{n} b_j \exp(i\alpha_j t) \, , \; \min_j \operatorname{Re}(\alpha_j) = 0 \, . \tag{6}$$

Theorem 19.2′. *For* $y(t)$ *as defined in* (6) *and* $0 < a < 2d/(3 \log 2)$, *we have*

$$\begin{aligned} \max|y(t)| &\geq |y(0)|(14n)^{-1/2} (a/d)^{1/4} \exp\left\{-\tfrac{25}{7}n(a/d)^{1/2}\right\} \\ &\geq |y(0)|(14n)^{-1/2} (a/d)^{1/4} \exp\{-4n(a/d)^{1/2}\} \, , \end{aligned} \tag{7}$$

where the maximum is taken over $a \leq t \leq a + d$.

The smaller lower bound in (7) is larger than the one in Theorem 19.2 by a factor of $5(14)^{-1/2} \exp\{2n(a/d)^{1/2}\}$.

Consequently, the Corollary in [2, p.217] can be improved to give a lower bound more than 15 times as large; namely

Corollary 1. *Under the same hypotheses on* $y(t)$ *we have*

$$\max|y(t)| \geq (14)^{-1/2} e^{-25/7} n^{-1} |y(0)| \, ,$$

where the maximum is taken over $n^{-2} \leq t \leq 1 + n^{-2}$.

With an interval of the same unit length, but with end–points chosen for optimal effect, a far larger lower bound can be achieved, one eight times as large as in Corollary 1 and more than 120 times the one in [2, p.217]:

Corollary 2. *With the same hypotheses on* $y(t)$, *we have*

$$\max|y(t)| \geq \frac{3}{50} n^{-1} |y(0)| \, ,$$

provided the maximum is taken over $(8n)^{-2} \leq t \leq 1 + (8n)^{-2}$.

Moreover, this conclusion can be increased further by a more careful calculation to yield

Corollary 3. *Again with the same hypotheses on* $y(t)$ *we have*

$$\max|y(t)| \geq \frac{1}{16}n^{-1}|y(0)| ,$$

where the maximum is taken over $(8n)^{-2} \leq t \leq 1 + (8n)^{-2}$.

5. Proof of Lemma 5.8′.

For this we use the same polynomials as in the proof of Lemma 5.8 [2, pp.52–53], attributed there to Q.I. Rahman and F. Stenger. Accordingly, properties 1., 2. and 3. are already verified.

As in [2, p.53],

$$\rho_k = \exp\left\{\frac{1}{2\pi}\int_0^\pi \log t_0(\theta)\,d\theta\right\} ,$$

where

$$t_0(\theta) = 1 - \left[\frac{\sin\frac{1}{2}\,\kappa\theta}{\kappa\sin\frac{1}{2}\theta}\right]^2 ,$$

and, using (A.1.4′) to majorize ρ_k, we find

$$\rho_k < \left[\exp\left\{\frac{1}{2\pi}\int_0^\pi \log\left[1 - \frac{1}{\kappa^2}\right] d\theta\right\}\right] \left[\exp\left\{\frac{1}{\pi\kappa}\int_0^{\pi\kappa/2} \log\left[1 - \frac{\sin^2 x}{x^2}\right] dx\right\}\right]$$

$$= \left[1 + \frac{1}{\kappa^2-1}\right]^{1/2} \exp\left\{\frac{1}{\kappa}\,I_\kappa\right\} < \exp\left\{\frac{1}{2(\kappa^2-1)} + \frac{I_\kappa}{\kappa}\right\} < \exp\left\{\frac{1}{2(\mu^2-1)} + \frac{I}{\mu}\right\} .$$

Here,

$$I_\kappa = -\frac{1}{\pi}\int_0^{\pi\kappa/2} \log\left[1 - \frac{\sin^2 x}{x^2}\right] dx \; < \; I = -\frac{1}{\pi}\int_0^\infty \log\left[1 - \frac{\sin^2 x}{x^2}\right] dx < 1.278,$$

where the upper bound has been determined by numerical integration and is very close to the actual value. But

$$\frac{1}{2(\kappa^2-1)} + \frac{1.278}{\kappa} < \frac{1.4655}{\kappa} < \frac{3}{2\kappa} ,\; \kappa = 3,4, \ldots ,$$

completing the proof of property 4′ and hence of the lemma.

Remark. The above argument actually establishes more, since

$$\frac{1}{\kappa^2-1} = \frac{1}{\kappa^2}\frac{\kappa^2}{\kappa^2-1} < \frac{9}{8\kappa^2}, \ \kappa = 3,4,\ldots,$$

namely

4″ $$\rho_k < \exp\left[\frac{9}{16\kappa^2} + \frac{\Gamma}{\kappa}\right], \ \kappa = 3,4,\ldots,$$

an inequality which can be strengthened further if need be. However, Lemma 5.8′ is sufficiently precise to lead to a proof of Theorem 7.1′.

6. Proof of Theorem 7.1′.

This requires amending various steps in the proof of Theorem 7.1 [2, pp. 73–75] but not altering the line of reasoning. We adopt the various notations found in [2, loc. cit.].

Employing (5.8.1′) sharpens (7.1.2) [2, p.74] to

(7.1.2′) $$\|F_1\| \le (kn+1)^{1/2} \exp\{3n/(2\kappa)\},$$

and also (7.1.4) to

$$\sum_{\nu=0}^{m} |d^{(2\nu)}| < 2^{m+1} \exp(3n/\kappa),$$

so that

$$\|s_m(1/F_1)\| < 2^{m+1} \exp(3n/\kappa).$$

Using this and (7.1.2′) in (7.1.10) [2, p.75] increases therein the lower bound to the value asserted in Theorem 7.1′ and the proof is complete.

7. Proof of Corollary″.

This is done simply by putting in Theorem 7.1′, $m = 0$, $\kappa = n \ge 2$. Hence, $k = n - 1$, and so

$$(kn+1)^{1/2} = (n^2-n+1)^{1/2} < n.$$

8. Proof of Theorem 19.2′.

The lower bound provided by Theorem 7.1′ above, when used instead of [2, p.216 (19.4.1)], permits the strengthening of (19.4.2) [2, p.216] to

(19.4.2′) $$\max|y(t)| \geq \tfrac{1}{2}(kn+1)^{-1/2}\ |y(0)|\exp\left\{-\left[\tfrac{a}{d}k\log 2+\tfrac{9}{2\kappa}\right]n\right\}.$$

As in [2, p.216], let (with $[x]$ the usual integer–valued bracket function)

(8) $$k = \left[\left[\frac{6d}{a\ \log\ 2}\right]^{1/2}\right],$$

a value at least 3. Then, since $\kappa = k+1$,

$$\tfrac{a}{d}k\log 2 + \frac{9}{2\kappa} < \tfrac{a}{d}(\log 2)\left[\frac{6d}{a\ \log\ 2}\right]^{1/2} + \tfrac{9}{2}\left[\frac{a\ \log\ 2}{6d}\right]^{1/2}$$

$$= \tfrac{7}{4}(6\log 2)^{1/2}(a/d)^{1/2} < \tfrac{25}{7}(a/d)^{1/2} < 4(a/d)^{1/2}.$$

To complete the proof of (7), we need to show that

$$4(kn+1) \leq 14n(d/a)^{1/2}$$

with k as defined in (8). For this value of k,

$$4(kn+1) \leq 4n\left[\frac{6d}{a\ \log\ 2}\right]^{1/2} + 4$$

and the right side of this inequality will be less than or equal to $14n(d/a)^{1/2}$ if

$$4 \leq 2n(d/a)^{1/2}\{7 - 2(6/\log 2)^{1/2}\}.$$

But this is the case already for $n = 2$, since $d/a \geq \frac{3}{2}\log 2$ and so the proof is complete.

9. Proofs of Corollaries 1, 2 and 3.

Corollary 1 is the special case of (7) in Theorem 19.2′ in which $d = 1$ and $a = n^{-2}$.

Corollary 2 is the special case of (7) in which $d = 1$ and $a = (8n)^{-2}$. This value is selected because the function $x^{1/2}\exp(-\alpha nx)$ achieves its unique maximum when $x = (2\alpha n)^{-1}$. In Theorem 19.2′, the lower bound is of this form with

$$\alpha = 4,\ x = (a/d)^{1/2} = (8n)^{-1}.$$

These choices for d and a provide the largest lower bound attainable without reference to improvements in the estimate of $(kn+1)^{1/2}$ from the inequality in Theorem 19.2′.

Corollary 3 is a consequence of a closer estimate of $(kn+1)^{1/2}$,

since it can be shown that

$$\left[\frac{d}{a}\right]^{1/4}\left\{4\left[\frac{6}{\log 2}\right]^{1/2}+\frac{4}{n}\left[\frac{a}{d}\right]^{1/2}\right\}^{1/2} n^{1/2} \geq 2(kn+1)^{1/2},$$

with k defined as above. This estimate increases also the constant in Corollary 1 from $(14)^{-1/2}$ to $(13)^{-1/2}$.

10. Conclusion.

Still more precise results can be obtained. These require more detailed arguments which will be presented elsewhere. In this paper we have increased noticeably the lower bounds in some of Halász's contributions to Turán's power–sum theory. In doing so, our objective has been to follow the book's methodology, to provide estimates of the same form and of the same simplicity.

11. Acknowledgements.

Our work has been supported by the Natural Sciences and Engineering Research Council of Canada. We are pleased to have had the opportunity to present it to the Seminar on Approximation and Optimization, University of Havana, Cuba.

References

1. Konrad Knopp, *Theory and Application of Infinite Series*, English translation (second ed.), Blackie, London, 1966.

2. Paul Turán, *On a New Method of Analysis and its Applications*, John Wiley & Sons, New York, 1984.

ON THE M-TH ROW OF NEWTON TYPE (α,β)-PADE TABLES AND SINGULAR POINTS

Andrei Martínez Finkelshtein,
Dept. Theory of Functions, Havana University

Abstract An extension is obtained of inverse type results for rows of Padé approximants on the location of singularities on the boundary of the m-th meromorphic extension of a formal power series to Newton-type Padé approximants.

§ 1. INTRODUCTION

1. Newton type tables play an important role in generalized Padé approximation (see [1,2]). Such tables are constructed according to a system $\alpha=(\alpha_n)$ of interpolation nodes, and the system $\beta=(\beta_n)$ of fixed poles (in each table some terms can be repeated) such that $\alpha \cap \beta = \emptyset$. Besides, we assume that $\alpha \subseteq E$, $\beta \subseteq F$, where E and F are compact sets of the extended plane $\hat{\mathbb{C}}$, $E \cap F = \emptyset$, $\infty \notin E$, $0 \notin F$.

Using the same notation as in [11] we take

$$a_n(z)=\prod_{k=1}^{n}\left(z-\alpha_k\right),\quad b_n(z)=\prod_{k=1}^{n}\left(1-\frac{z}{\beta_k}\right),\quad n=1,2,\ldots .$$

Given a holomorphic function f on E ($f\in H(E)$), for an arbitrary pair (n,m) of nonnegative integer numbers, there exist two polynomials P and Q such that

$$\deg P \le n,\ \deg Q \le m,\ \frac{Qb_{n-m}f-P}{a_{n+m+1}} \in H(E) \tag{1}$$

which define the unique rational function

$$r_{n,m}=\frac{P}{Q\,b_{n-m}}$$

called (α,β)-Padé approximant of type (n,m) associated with the function f.

In order to study the convergence of the rows of an (α,β)-Padé table, we suppose that (E,F) is a regular condenser with the above mentioned properties, ω is a continuous function on $\hat{\mathbb{C}}$ and harmonic on the connected region $G=\hat{\mathbb{C}}\setminus(E\ \cup\ F)$ which satisfies $\omega|_F=0$, $\omega|_E=1$, and $\phi=\exp\left(\frac{\omega}{c}\right)$, where $c=c(E,F)$ is the capacity of the condenser (E,F) (see [21]).

The canonical regions associated with (E,F) are defined as follows

$$E_\rho=\left\{z:\ \phi(z)<\rho\right\},\quad 1<\rho\le\exp(1/c)$$

According to [11], we consider the family $W(E,F)$ of pairs of newtonian tables (α,β), which satisfies for each n

$$0<c_1(K)\le\frac{\left|\frac{a_n(z)}{b_n(z)}\right|}{\tau^n\ \phi^n(z)}\le c_2(K)<\infty,\ z\in K. \tag{2}$$

where K is a compact set of G, $\tau\in\mathbb{R}$ is fixed, and $c_i(K)$, $i=1,2$, only depend on K. For examples of such pairs (α,β) (see [11]).

2.Let $f\in H(E)$, $m\in\mathbb{N}$ and E_R be the maximal canonical region such that f has meromorphic extension with no more than m poles.

Henceforth, the rational function r_n will be denoted by

$$r_n=r_{n,m}=\frac{P_n}{Q_n b_{n-m}}$$

where P_n and Q_n do not have common roots and

$$Q_n(z)=\prod_{|\eta_i|\le1}\left(z-\eta_i\right)\prod_{|\eta_i|>1}\left(\frac{z}{\eta_i}-1\right) \tag{3}$$

We have the relation

$$P_{n+1}(z)Q_n(z)-P_n(z)Q_{n+1}(z)\left(1-\frac{z}{\beta_{n+m+1}}\right)=A_n\hat{a}_{n+m+1}(z) \tag{4}$$

where $\hat{a}_{n+m+1}(z)$ is different from $a_{n+m+1}(z)$ in no more than m factors, and

$$(r_{n+1}-r_n)(z)=A_n\frac{\hat{a}_{n+m+1}(z)}{b_{n-m+1}(z)}\ \frac{1}{(Q_nQ_{n+1})(z)},\ n=0,1,2,\ldots$$

Hence, the convergence of the sequence $\left\{r_n(z)\right\}$ is equivalent to that of the series

$$\sum_{n\ge0}A_n\frac{\hat{a}_{n+m+1}(z)}{b_{n-m+1}(z)}\ \frac{1}{(Q_nQ_{n+1})(z)}.$$

We know that if in particular $(\alpha,\beta)\in W(E,F)$ then the following formula takes place (see [11])

$$\overline{\lim_{n}} |A_n|^{1/n} = \frac{1}{\tau R}. \qquad (5)$$

3. Let $P_n = \left\{\zeta_{n,1}, \ldots, \zeta_{n,m_n}\right\}$, $0 \le m_n \le m$, be the set of poles of r_n which belong to G (in particular, no one element from β belongs to P_n), where each pole is written as many times as its multiplicity.

Let $\zeta \in G$, $\zeta \ne \infty$, and let us suppose that for $(\alpha,\beta) \in W(E,F)$ we have

$$\operatorname{dist}(\zeta, P_n) = \min_{1 \le j \le m_n} |\zeta - \zeta_{n,j}| \longrightarrow 0, \quad n \longrightarrow \infty. \qquad (6)$$

In [11] we proved that under the above conditions $\phi(\zeta) \le R$, and if in addition $\phi(\zeta) < R$, then $z=\zeta$ is a pole of the function f.

In the present paper we prove the following result corresponding to the case $\phi(\zeta)=R$.

THEOREM Let us suppose that (6) takes place for $(\alpha,\beta) \in W(E,F)$, and

$$\overline{\lim_{n}} \frac{\log |\tau^n R^n A_n|}{\log n} > -\infty \qquad (7)$$

and $\phi(z)=R$, then $z=\zeta$ is a singular point of f.

The proof of this theorem is based on a method of M.Riezs's method used in [5] (see also [7,8]), which generalizes the known scheme given for the case $m=0$ (see [7,8]).

An additional difficulty in the case considered here comes up from the general form of the level curves, because it makes it difficult to obtain the necessary estimates, and therefore we had to find certain general condition which would contain a sufficiently large class of tables whose level curves behaved "nice enough". Because of the reasons, above, in the proof of our theorem it was necessary to introduce in Riesz's method the use of new mathematical tools. from the geometric theory of functions and topology.

§ 2. Proof of the theorem

Let us see briefly the proof of the theorem stated above .

If the coefficients $\{A_n\}$ given in (4) satisfy (7), and $B_n = A_n \tau^n R^n n^p$, then we can choose a value of p ,$p\in N$,such that

$$\overline{\lim_{n}}\ |B_n|^{1/n}=1,\quad \overline{\lim_{n}}\ |B_n|=+\infty \tag{8}$$

Under these conditions we know from [7] and [8] that there exists a concave function $C(x)$, $x\geq 0$, which fulfils the following conditions

1) $C(x)\geq 0$ for $x\geq 0$; $C(0)=0$;

2) $C(n)\geq \log |B_n|$, $n\geq 1$;

3) For any other concave function $C_1(x)$, $x\geq 0$, satisfying 1) and 2), we have that $C_1(x) \geq C(x)$.

From condition 2) we obtain that

$$\lim_{x\longrightarrow\infty} \frac{C(x)}{x}=0$$

and there exists an infinite sequence $\Lambda\subseteq N$ such that $C(n)=\log|B_n|$, $n\in\Lambda$.

Let $E'_R=E_R\setminus[f=\infty]$. From the above asumptions the following result takes place

LEMMA If $f\in H(\mathcal{D})$, $\left(E'_R\setminus E_R\right) \subseteq D \subseteq G$, then for an arbitrary point $\zeta\in\mathcal{D}$ there exists a neighbourhood U of this point such that

$$\max_{z\in U}\left|\frac{\left(Q_n b_{n-m} f-P_n\right)(z)}{d_n R^{-n}\ n^{-p}\phi^n(z)\ b_{n-m}(z)}\right|=O(1),\ n\longrightarrow\infty \tag{9}$$

where $d_n=\exp C(n)$.

In sections 2–5 we shall outline the proof of this fundamental lemma.

2. Let us fix the point $\zeta\in G$, $\zeta\neq\infty$. Our immediate purpose is to construct a conformal mapping of G which so as to regularize the convergence region and the level curves. Although we can consider a univalent branch of that mapping on some neighbourhood of the point ζ, we shall define it on a larger region, because, among other reasons, it has interest in itself due to the topological results used here.

We are going to use the following obvious proposition.

PROPOSITION Let (E,F) be a regular condenser, with the properties stated at 1.1. Then we can assert that there exist two compact sets $\mathcal{E}$ and $\mathcal{F}$ of the extended complex plane $\hat{\mathbb{C}}$, such that:

i) Both $\mathcal{E}$ and $\mathcal{F}$ are simply connected;

ii) $E \subseteq \mathcal{E}$, $F \subseteq \mathcal{F}$, $\mathcal{E} \cap \mathcal{F} = \emptyset$;

iii) $\mathcal{E}$ is compact of $\hat{\mathbb{C}}$;

iv) $\zeta \notin \mathcal{E} \cup \mathcal{F}$.

Having in mind this result, let us denote by $G_1=\hat{\mathbb{C}}\setminus(\mathcal{E} \cup \mathcal{F})$, where $\mathcal{E}$ and $\mathcal{F}$ fulfil the statements i)-iv) of the proposition above, and let $w^*(z)$ be the conjugated harmonic function of $w(z)$ in G_1. Then, it is not difficult to see that the function

$$\varphi(z)=\exp\left(\frac{1}{c}\ (w(z)+iw^*(z))\right),\ z\in G_1,$$

is univalent, holomorphic and one to one on certain neighbourhood of each point of G_1, and besides $|\varphi(z)|=\phi(z)$, $z\in G_1$.

Let us denote $\tilde{\zeta}=\varphi(\zeta)$. Without loss of generality we can suppose that $\operatorname{Arg}\tilde{\zeta}\neq 0$. Let us define in $\tilde{G}=\varphi\{G_1\}$ the following compacts sets:

$$\tilde{\Delta}=\{z=re^{i\theta}:\ r_1\le r\le r_2;\ \theta_1\le\theta\le\theta_2\}$$

where

a) $\tilde{\Delta} \subseteq \tilde{G}$;

b) $1<r_1<|\tilde{\zeta}|<r_2<e^{1/\Gamma}$; $r_1,r_2\neq R$; $0<\theta_1<\operatorname{Arg}\tilde{\zeta}<\theta_2<2\pi$;

c) $\varphi|_\Delta$ is one to one where $\Delta=\varphi^{-1}\left\{\tilde{\Delta}\right\}$.

It is clear that every Δ is a compact neighbourhood of $\zeta \in G_1$. (the class of those Δ will be named $\mathcal{K}$), and it is sufficient to prove the lemma for some $\Delta \in \mathcal{K}$. So, we fix $\Delta \in \mathcal{K}$, $\Delta \subset D$, $\Delta\neq D$. For $t\geq 0$ we take

$$\tilde{\Delta}_t=\{z=re^{i\theta}:r_1-t\le r\le r_2+t;\ \theta_1-t\le\theta\le\theta_2+t\}$$

$$\Delta_t=\varphi^{-1}\{\tilde{\Delta}_t\}$$

Now we choose $T>0$ such that $\Delta_t\subset D$, $\Delta_t\in\mathcal{K}$, where $0\le t\le T$. (i.e. $\varphi:\Delta_t\longrightarrow\tilde{\Delta}_t$ is a conformal mapping for every $0\le t\le T$)

Since Δ_T is a compact subset of D, then

$$\inf_{z\in\Delta_t}\left|\left[\varphi^{-1}\right]'(z)\right|=L>0$$

Let us suppose that the fixed compact set Δ overlaps the level curve $[\varphi = R]$. We take $\delta=T\ r_1\ R^{-1}$ and $\varepsilon>0$ such that

$$\frac{4\varepsilon}{L}\sum_{k=1}^{\infty} k^{-2} = \delta/4$$

Let us denote $\eta_n = C(n+1) - C(n)$, $n=0,1,\ldots$; it follows that $\eta_n \searrow 0$, $\eta_n \geq 0$, as $n \longrightarrow \infty$, and moreover from the concavity of the function $C(x)$ we obtain

$$C(k) - C(n) - (k-n)\eta_n \leq 0, \quad k=1,\ldots \tag{10}$$

(cf. [8], p 119).

Let us analyze for every fixed $n \in \mathbb{N}$ the function

$$\psi_n(z) = (f - r_n)(z_{(n)})[\varphi(z)]^{-n} \exp(n\ \eta_n)\ d_n^{-1} n^p R^n \tag{11}$$

where $z_{(n)} = \varphi^{-1}\left[\varphi(z)\ \exp(-\eta_n)\right]$, $z_{(n)} \in \Delta_T \setminus P$, and $P = \bigcup_{n=1}^{\infty} P_n$.

It is clear that $z_{(n)} \longrightarrow z$ as $n \longrightarrow \infty$.

3. Let $z \in \Delta_T' = \Delta_T \cap E_R$

For each fixed natural number n, we denote by $\tilde{V}'_{n,k}(\varepsilon)$ the $\dfrac{\varepsilon}{mLk^2}$ neighbourhood of the set formed by the image through φ of the zeros of the polynomial

$$\gamma_{n,k}(z) = (Q_{n+k}\ Q_{n+k+1})(z_{(n)}), \quad k,n=1,2,\ldots$$

which belong to G_1, and let

$$V'_{n,k}(\varepsilon) = \varphi^{\ 1}\left\{\tilde{V}'_n(\varepsilon)\right\}$$

From Koebe's 1/4-theorem (cf. [9] p. 51 or [10] p. 93), if B is a disk whose center is the point $z=a$, and it has radius r, contained in the region $\tilde{G}$ where φ^{-1} has its domain, then the disk whose center is $\varphi^{-1}(a)$, and has radius $\left|\frac{r}{4}[\varphi^{-1}]'(a)\right|$ is contained in $\varphi^{-1}\{B\}$. Hence, bearing in mind (3) we obtain

$$|\gamma_{n,k}(z)| \geq c_1 \left(\frac{\varepsilon}{4mk^2}\right)^{2m}, \quad z \in \Delta_T \setminus V'_{n,k}(\varepsilon) \tag{12}$$

where c_1 neither depends on n nor k (henceforth the constants $c_2, c_3, \ldots$, will have the same meaning as c_1).

Let us take $\tilde{V}'_n(\varepsilon) = \bigcup_{k=1}^{\infty} \tilde{V}'_{n,k}(\varepsilon)$, $V'_n(\varepsilon) = \varphi^{-1}\left\{\tilde{V}'_n(\varepsilon)\right\}$. Then for $z \in \Delta_T' \setminus V'_n(\varepsilon)$

$$\psi_n(z)=\sum_{k=n}^{\infty} A_k \frac{\hat{a}_{k+m+1}(z_{(n)})}{b_{k-m+1}(z_{(n)})} \frac{[\varphi(z)]^{-n} \exp(n\,\eta_n)\, n^p R^n}{d_n(Q_k Q_{k+1})(z_{(n)})}$$

and by using (12) we obtain

$$|\psi_n(z)| \le c_2 \sum_{s=1}^{\infty} \left(\frac{\phi(z)}{R}\right)^s s^{4m},\ z\in\Delta'_T\setminus V'_n(\varepsilon) \qquad (13)$$

4. Let $z\in\Delta''_T = \Delta_T \cap \{z;\ \phi(z)>R\}$. Similar to section 3, we consider $\tilde{V}''_{n,k}(\varepsilon)$ as the $\frac{\varepsilon}{mLk^2}$-neighbourhood formed by the image through φ of the zeros of the polynomial

$$\tau_{n,k}(z)=\left(Q_{n-k}Q_{n-k+1}\right)(z_{(n)}),\ k,n=1,2,\dots$$

which belong to G_1, and $V''_{n,k}(\varepsilon)=\varphi^{-1}\left\{\tilde{V}''_{n,k}(\varepsilon)\right\}$, $\tilde{V}''_{n,k}(\varepsilon)=\bigcup_{k=1}^{\infty}\tilde{V}''_{n,k}(\varepsilon)$,

$V''_n(\varepsilon)=\varphi^{-1}\left\{\tilde{V}''_n(\varepsilon)\right\}$.

For $z\in\Delta''_T\setminus V''_n(\varepsilon)$ we have

$$\psi_n(z)=(f-r_1)(z_{(n)})[\varphi(z)]^{-n} \exp(n\,\eta_n)\, d_n^{-1} n^p R^n -$$

$$-\sum_{k=1}^{n-1} A_k \frac{\hat{a}_{k+m+1}(z_{(n)})}{b_{k-m+1}(z_{(n)})} \frac{[\varphi(z)]^{-n} \exp(n\,\eta_n)\, n^p R^n}{d_n(Q_k Q_{k+1})(z_{(n)})} = I_n(z)-\Sigma_n(z)$$

In the same way, taking into account the properties of the function $C(x)$, we obtain the bounds

$$|I_n(z)|\le c_3\left(\frac{R}{\phi(z)}\right)^n n^{4m+p},\ z\in\Delta''_T\setminus V''_n(\varepsilon)$$

and

$$|\Sigma_n(z)|\le c_4 \sum_{s=1}^{n-1}\left(\frac{R}{\phi(z)}\right)^s s^{4m+p},\ z\in\Delta''_T\setminus V''_n(\varepsilon)$$

so that, finally

$$|\psi_n(z)| \le c_5 \sum_{s=1}^{\infty}\left(\frac{R}{\phi(z)}\right)^s s^{4m+p},\ z\in\Delta''_T\setminus V''_n(\varepsilon). \qquad (14)$$

5. Let us consider the function $\Omega_n(z)=Q_n(z_n)\psi_n(z)$, analytic on Δ_T. Having in mind (13) and (14), we have the following estimate

$$|\Omega_n(z)|\le\begin{cases} c_6\sum_{s=1}^{\infty}\left(\frac{\phi(z)}{R}\right)^s s^{4m+p}, z\in\Delta_T'\setminus V_n'(\varepsilon); \\ c_6\sum_{s=1}^{\infty}\left(\frac{R}{\phi(z)}\right)^s s^{4m+p}, z\in\Delta_T''\setminus V_n''(\varepsilon) \end{cases}$$

Analyzing, as in [5], the function $\nu_k(x)=\sum_{s=1}^{\infty} x^s s^k$, $0\le x\le 1$, it easily follows that

$$\nu_k(x)=p_k(x)\ (1-x)^{-2^k},\ 0\le x\le 1,$$

where p_k is some polynomial in the variable x with degree $\le 2^{k+1}-1$. Therefore, taking $\mu=4m+p$, we have

$$|\Omega_n(z)|\le\begin{cases} c_6 p_\mu\left(\frac{\phi(z)}{R}\right)\left(1-\frac{\phi(z)}{R}\right)^{-2^\mu} & , z\in\Delta_T'\setminus V_n'(\varepsilon); \\ c_6 p_\mu\left(\frac{R}{\phi(z)}\right)\left(\frac{\phi(z)}{R}\right)^{2^\mu}\left(1-\frac{\phi(z)}{R}\right)^{-2^\mu} & , z\in\Delta_T''\setminus V_n''(\varepsilon) \end{cases}$$

Thus, we can easily define a continuous function $F(z)$ on G such that

$$|\Omega_n(z)| \le F(z)\ \left(1-\frac{\phi(z)}{R}\right)^{-2^\mu}, z\in\Delta_T\setminus V_n(\varepsilon),\ \phi(z)\ne R, \tag{15}$$

where $V_n(\varepsilon)= V_n'(\varepsilon) \cup V_n''(\varepsilon)$.

For an arbitrary set $e \le \mathbb{C}$ denote $\sigma(e)=\inf\left\{\sum_k |U_k|\right\}$, where the infimum is over all coverings $\left\{U_k\right\}$ of e by disks U_k, and $|U_k|$ is the diameter of U_k. From the definition of $V_n(\varepsilon)$ and the estimates above it is not difficult to show that $\sigma(\tilde V_n(\varepsilon))<\delta/2$, where $\tilde V_n(\varepsilon)=\varphi\left\{V_n(\varepsilon)\right\}$, (cf. [2]). From the choice $\delta=Tr_1/R$, it is not very hard to prove that for every $n\in\mathbb{N}$

$$\text{mes}\ \left\{t\in[0,T]:\ \partial\tilde\Delta_T \cap \tilde V_n(\varepsilon)\right\} < T/2$$

and therefore, for arbitrary n we can find $t_n\in (T/2,T)$ such that

$\partial\tilde{\Delta}_T \cap \tilde{V}_n(\varepsilon)\neq\emptyset$, or equivalently $\partial\Delta_{t_n} \cap V_n(\varepsilon)=\emptyset$.

Then from (15),

$$|\Omega_n(z)| \le F(z) \left(1-\frac{\phi(z)}{R}\right)^{-2^{\mu}}, z\in\partial\Delta_{t_n}, \quad \phi(z)\neq R.$$

(16) Let us take

$$\hat{\Omega}_n(z)=\Omega_n(z)\left(1-\frac{\varphi(z)}{R\exp\left(i(\theta_1-t_n)\right)}\right)^{2^{\mu}}\left(1-\frac{\varphi(z)}{R\exp\left(i(\theta_1+t_n)\right)}\right)^{2^{\mu}}.$$

The function $\hat{\Omega}_n$ is analytic on Δ_T, so from (16),

$$|\hat{\Omega}_n(z)| \le c_7 < \infty, \ z\in\Delta_{t_n}$$

Thus we have

$$|\Omega_n(z)| \le c_8, \ z\in\Delta_{T/4}$$

and the last inequality proves the lemma when $\Delta \cap \{\varphi(z)=R\}\neq\emptyset$. If this intersection is empty, the corresponding reasoning is simpler.

6. Let us now prove the theorem. If f is analytic on the point $\zeta\in G$, $\phi(\zeta)=R$, then from our lemma, for $z\in U$, U being the neighbourhood of ζ given by the lemma above, condition (9) takes place. Because of the existence of some infinite subset $\Lambda \subseteq \mathbb{N}$, such that

$$C(n)=\log\left(|A_n|\tau^n R^n n^p\right), \ n\in\Lambda,$$

we have

$$\left|\frac{(Q_n b_{n-m} f-P_n)(z)}{b_{n-m}(z)\ A_n\tau^n\phi^n(z)}\right| \le c_9, \ z\in U, \ n\in\Lambda \tag{17}$$

If $\operatorname{dist}(P_n,\zeta)\longrightarrow 0$, as $n\longrightarrow\infty$, and we denote by ζ_n a zero of $Q_n(z)$ such that $\operatorname{dist}(P_n,\zeta)=|\zeta-\zeta_n|$, it follows that $\zeta_n\longrightarrow\zeta$, so then, for large n, $\zeta_n\in U$. From (17) we obtain the estimate

$$\left|\frac{P_n(\zeta_n)}{b_{n-m}(\zeta_n)A_n\tau^n\phi^n(\zeta_n)}\right| \le c_9, \ n>n_0, n\in\Lambda \tag{18}$$

On the other hand, we are able to obtain from (4) that

$$\frac{P_n(\zeta_n)}{b_{n-m}(\zeta_n)A_n\tau^n\phi^n(\zeta_n)} = -\frac{1}{Q_{n+1}(\zeta_n)}\ \frac{\dfrac{\hat{a}_{n+m+1}(\zeta_n)}{b_{n+m+1}(\zeta_n)}}{\tau^n\phi^n(\zeta_n)}$$

and by using (18) we finally have

$$\left|\frac{1}{Q_{n+1}(\zeta_n)}\right| \le c_{10},\ n>n_o,\ n\in\Lambda,$$

but the last inequality is a contradiction, because it is obvious that $Q_{n+1}(\zeta_n) \longrightarrow 0$, as $n \longrightarrow \infty$. In conclusion $z=\zeta$, where $\phi(\zeta)=R$, is a singular point of the function $f(z)$, and the theorem posed in § 1. is so proved.

REFERENCES

1. Perron,O. Die Lehre von den Kettenbuchen.B. II.Stutgart: Teubner, 1957.

2. Gončar, A.A. The poles of the rows of Padé tables and the meromorphic extension of functions. Mat. Sb., 1981, 115, 4 (in russian). Eng. Transl. in Math. U.S.S.R., Sb.

3. Gončar, A.A. On the convergence of generalized Padé approximants of meromorphic functions. Mat. Sb., 1975, 98, 4 (in russian). Eng. Transl. in Math U.S.S.R. Sb.

4. Suetin, S.P., On the m-th row of a Padé tables. Mat Sb., 1983, 120,4 , (in russian); Eng. Transl. in Math.U.S.S.R., Sb.

5. Prokhorov, V.A., S.P. Suetin, V.V.Vavilov. Poles of the m-th row of a Padé table and singular points of functions. Mat Sb., 1983, 122, 4 (in russian); Eng. Transl. in Math. U.S.S.R. Sb.

6. López, G., V.A.Prokhorov, V.V.Vavilov, On an inverse problem for rows of a Padé table. Mat. Sb., 1979, 110, p 117-129 (in russian); Eng. Transl. in Math. U.S.S.R., Sb.

7. Agmon, M.S. Sur le series de Dirichlet; Ann. Ec. Norm. Sup., 66 (1949).

8. Mandelbroijt,S., Dirichlet Series. Principes and Methods, Moscow: Mir, 1966 (in russian).

9. Goluzin, G.M., Geometric theory of functions of a complex variable, Moscow: Nauka, 1966 (in russian); Eng. Transl. in A.M.S. Trans of Math Monographs, Vol. 26, 1969, Providence.

10. Fuchs W.H.J. Topics in the theory of functions of one complex variable; Van Nostrand Math. St., 1967.

11. Martínez Finkelshtein, A., Acerca de un problema inverso para filas de aproximantes multipuntuales de Padé de tipo Newtoniano; Ciencias Matemáticas, Vol VII, 1, 1986.

ON SIMULTANEOUS RATIONAL INTERPOLANTS OF TYPE (α,β)

René Piedra

Havana University, Cuba

Abstract. In this paper a convergence theorem for simultaneous rational interpolants of type (α,β) is established. This is a natural extension of the theorem of Montessus de Ballore for a row sequence of (scalar) Padé approximants. Results on differences of interpolants that tend to zero on some "large region", were first obtained by Walsh when he solved the case of polynomials interpolating on the roots of unity and at the origin. In the second part of this paper we also extend this study to differences of simultaneous rational interpolants of type (α,β).

Introduction.

Let D be a bounded region in $\mathbb{C}$. $E\subset D$ will denote a compact set whose complement is connected, also denote $G=D\setminus E$, $F=\bar{\mathbb{C}}\setminus D$ and assume that G is regular in the sense that the Dirichlet problem can be solved $(\infty\in F$, $0\in\Gamma)$. i.e. there exists $H(z)$, the harmonic function on G and continuous on $\bar{G}$ such that $H(z)/\partial E=0$ and $H(z)/\partial F=1$. Further, for every $\sigma\in(0,1)$, define $D_\sigma=\left\{z\in G \mid H(z)<\sigma\right\}\cup E$.

It's well known ([1] chap.VIII) that there exist tables of points $\alpha=\{\alpha_{n,k}\}$ and $\beta=\{\beta_{n,k}\}$ $k=1,2,\ldots,n+1$, $n\in\mathbb{N}$ that have no limit point exterior to E or F such that

$$\lim_n\left(\prod_{k=1}^{n+1}\left|\frac{z-\alpha_{n,k}}{z-\beta_{n,k}}\right|\right)^{1/n}=\exp\left[\frac{1}{c}\ (H(z)-1)\right]$$

uniformly on each compact subset of G, where c is the capacity of the condenser (E,F). (i.e. $c=\frac{1}{2\pi}\int_\Gamma \frac{\partial H}{\partial\eta}\,ds$, where Γ is a contour in G which "separates" E and F, $\frac{\partial}{\partial\eta}$ is the normal derivative in the direction from E to F, and ds is the arc length measure). We will substitute the factor $z-\beta_{n,k}$ by 1, in case $\beta_{n,k}=\infty$.

Let us denote $\omega_n^\alpha(z)=\prod_{k=1}^{n}(z-\alpha_{n-1,k})$, $\omega_n^\beta(z)=\prod_{k=1}^{n}(z-\beta_{n-1,k})$, and we assume that ∞ is not a limit point of β (in the other case, the normalization must be different: see [1] page 206).

We define the set P(E,D) as the set of elements (α,β) such that

$$\lim \left|\frac{w_n^{\alpha}(z)}{w_n^{\beta}(z)}\right|^{1/n} = \lambda \exp\left(\frac{H(z)}{c}\right)$$

uniformly on each compact subset of G, where λ is an absolute constant. It's easy to check that if $(\alpha,\beta)\in P(E,D)$ then $(\alpha,\beta)\in P(E,D_{\sigma})$ for every $\sigma\in(0,1)$. It's well known that $(\alpha,\beta)\in P(E,D)$ implies separate conditions on α and β (see [1],9.11). More precisely, we have

$$\lim|w_n^{\alpha}(z)|^{1/n}=\lambda_1\exp(\psi_1(z)) \ ; \ \lim|w_n^{\beta}(z)|^{1/n}=\lambda_2\exp(\psi_2(z)).$$

uniformly on each compact in E^c and F^c, respectively, where $\frac{\lambda_1}{\lambda_2}=\lambda$, $\psi_1(z)\geq 0$ and $\psi_2(z)\geq 0$ on G . $\psi_1(z)$ is a harmonic function on E^c and $\psi_2(z)$ is a harmonic function on F^c.

Let f_i , $i=1,2,\ldots,d$ be analytic functions on E : i.e. analytic functions on some open set U which contains E. Also we can suppose that $\bar{U}\subset D$. Let $m>0$ and a system of d-non-negative integers $\{\delta_i$, $i=1,2,\ldots,d\}$ be given such that $\sum_{i=1}^{d}\delta_i = m$.

Definition 1– We say that a system of rational functions

$$\left\{R_{n,i} = \frac{p_{n,i}}{q_n\, w_{n-m}^{\beta}},\ i=1,\ldots,d\right\}$$

is a (n,m) simultaneous rational interpolant of type (α,β) for $\{f_i$,$i=1,2,\ldots,d\}$ associated to $\{\delta_i$, $i=1,\ldots,d\}$ in the n–th row of α if for all sufficiently large n :

$$\deg(p_{n,i})\leq n-\delta_i \quad , \quad \deg(q_n)\leq m \quad , \quad \text{and}$$

$$\frac{q_n w_{n-m}^{\beta} f_i - p_{n,i}}{w_{n+1}^{\alpha}}$$

is analytic on U , for each $i=1,2,\ldots,d$.

We remark that this definition contains ,in particular , the case in which $\beta_{n,k}=\infty$ for every n,k, that we considered earlier (see [5]).

When $d=1$ Gonchar in [2] proved a Montessus de Ballore type convergence theorem for (α,β) tables . When $\beta_{n,k}=\infty$, for each n,k , and $d\in\mathbb{N}$ Graves–Morris and Saff in [3] proved an analogous theorem . Now , in §1 , using the same technique as in [3], this result is generalized to the more general case.

When $d=1$ and $\beta_{n,k}=\infty$. For all n ,k , the study of differences of interpolants to a fixed function had its origin in a paper of Walsh (see [11]) in which he considers a particular case of differences of polynomial interpolants and proved an overconvergence result: that is, that the differences of the said polynomial interpolants converge in a larger region than that of analyticity of the fixed analytic function to which they interpolate. Different extensions of Walsh's result can be found. In [4], using the arguments employed in [6], we studied the differences of rational interpolants type (α,β) (case d=1); in [5] we studied the case $d\in\mathbb{N}$ and $\beta_{n,k}=\infty$ for all n,k. Now, in §2, we obtain a new result for the differences of simultaneous rational interpolants of type (α,β).

§1 Convergence theorem.

The following definition was introduced in [3]:

Definition 2.- Let each of the functions $f_1(z),f_2(z),\ldots,f_d(z)$ be meromorphic on D_σ, $\sigma\in(0,1)$ and let non-negative integers δ_i, $i=1,2,\ldots,d$ be given for which $\sum_{i=1}^{d}\delta_i>0$, then the functions $f_i(z)$ are said to be polewise independent, with respect to the numbers δ_i, on D_σ if there do not exist polynomials $p_1(z),p_2(z),\ldots,p_d(z)$, at least one of which is non-null, satisfying

$$\deg(p_i(z))\le(\delta_i-1)\ ,\text{if } \delta_i\ge1\ ,\ p_i(z)\equiv0\ ,\ \text{if } \delta_i=0 \tag{1.1a}$$

and such that

$$\varphi(z)=\sum_{i=1}^{d}p_i(z)f_i(z) \tag{1.1b}$$

is analytic throughout D_σ (see [3] and [5]).

Theorem 1.- Suppose that each of the functions $f_1(z),f_2(z),\ldots,f_d(z)$ is analytic on D_σ, $\sigma\in(0,1)$, except for possible poles at the m (not necessarily distinct) points $z_1,z_2,\ldots,z_m$ in $D_\sigma\backslash E$ (if z_k is repeated exactly p times, then each $f_i(z)$ is permitted to have a pole of order at most p at z_k). Let $\delta_1,\delta_2,\ldots,\delta_d$ be non-negative integers such that $\sum_{i=1}^{d}\delta_i=m>0$ and such that the functions $f_i(z)$ are polewise independent on D_σ with respect to the δ_i's; suppose that $(\alpha,\beta)\in P(E,D)$. Let

$$\left\{R_{n,i}=\frac{p_{n,i}}{q_n\,w^{\beta}_{n-m}}\ ,\ i=1,2,\ldots,d\right\}$$

be a (n,m) simultaneous rational interpolant of type (α,β) for $f_1,f_2,\ldots,f_d$, associated to δ_i, i=1,2,..,d in the n-th row of α; then for each compact set K, where $K\subset D'_\sigma=D_\sigma\backslash\{z_1,z_2,\ldots,z_m\}$

$$\limsup_{n\to\infty}\left\{\max|f_i(z)-R_{n,i}(z)|;\ z\in K\right\}^{\frac{1}{n}}\le\exp\left(\frac{\max_{z\in K}H(z)-\delta}{c}\right) \tag{1.2}$$

where c is capacity of the condenser (E,F).

By the assumption of theorem 1, each $f_i(z)$ has poles in D_σ of total multiplicity at most m. Furthermore, if z_k is repeated exactly p times, then at least one $f_i(z)$ has a pole of order p at z_k. The latter assertion is a consequence of the assumption of polewise independence, as it is revealed in the following preliminary lemma, whose proof can also be found in [3].

Lemma 1.- Under the assumptions of theorem 1, write the list $z_1,z_2,\dots,z_m$ in the form $\zeta_1,\zeta_2,\dots,\zeta_s$ where the ζ_k's are distinct and each ζ_k is of multiplicity m_k, so that $Q(z)=\prod_{j=1}^{m}(z-z_j)=\prod_{k=1}^{s}(z-\zeta_k)^{m_k}$, $\sum_{k=1}^{s} m_k = m$. Then for every $k=1,2,\dots,s$ and every $t=1,2,\dots,m_k$ there exists a function $F_{k,t}(z)$ of the form $F_{k,t}(z) = \sum_{i=1}^{d} p_i(z)f_i(z)$ where the p_i's satisfy (1.1) and $F_{k,t}$ is analytic on D_σ, except for a pole of order t at the point ζ_k.

Proof of theorem 1.- It is well known that for each sufficiently large integer, there exist polynomials $Q_n(z)$ and $P_{n,i}(z)$, $i=1,2,\dots,d$ which satisfy $\deg Q_n(z)\le m$, $Q_n(z)\ne 0$; $\deg P_{n,i}(z)\le n-\delta_i$ and

$$\frac{Q_n w^\beta_{n-m} f_i - P_{n,i}}{w^\alpha_{n+1}}$$

is analytic on U for each $i=1,2,\dots,d$ (see U in definition 1).

We normalize $Q_n(z)$ by setting

$$Q_n(z)=\sum_{j=0}^{m} b_{n,j}z^j \; ; \; \sum_{j=0}^{m} b_{n,j}=1.$$

Then the polynomials $Q_n(z)$ are uniformly bounded on each compact subset of the plane.

We first show that, for $k=1,2,\dots,s$

$$\limsup |Q_n^{(j)}(\zeta_k)|^{1/n} \le \exp\left(\frac{H(\zeta_k)-\sigma}{c}\right), \quad j=1,2,\dots,m_k-1. \tag{1.3}$$

To establish (1.3), fix k and consider the function $F_{k,1}(z)$ in the lemma, which is analytic on D_σ except for a simple pole at ζ_k. Write $F_{k,1}(z)=\dfrac{g_{k,1}(z)}{z-\zeta_k}$, where $g_{k,1}(z)$ is analytic on D_σ and $g_{k,1}(\zeta_k)\ne 0$.

Using the polynomials $p_i(z)$ defined in the lemma when t=1, we obtain

$$\frac{Q_n w^\beta_{n-m} F_{k,1} - \tilde{P}_{n,1}}{w^\alpha_{n+1}}, \quad \text{where } \tilde{P}_{n,1}(z)=\sum_{i=1}^{d} p_i(z)P_{n,i}(z),$$

is analytic on U. Hence, $(z-\zeta_k)\tilde{P}_{n,1}(z)$ is the unique polynomial of degree at most n which interpolates $Q_n(z)g_{k,1}(z)w^\beta_{n-m}(z)$ at the n+1 points of α's n-th row. Then

$$Q_n(z)g_{k,1}(z)-(z-\zeta_k)\frac{\tilde{P}_{n,1}(z)}{w^{\beta}_{n-m}(z)}=\frac{1}{2\pi i}\int\limits_{[H(t)=\sigma']}\frac{w^{\alpha}_{n+1}(z)w^{\beta}_{n-m}(t)Q_n(t)g_{k,1}(t)}{w^{\alpha}_{n+1}(t)w^{\beta}_{n-m}(z)(t-z)}dt$$

where $H(\zeta_k)<\sigma'<\sigma$. Now, we can obtain

$$\limsup |Q_n(\zeta_k)|^{1/n}\leq\exp\left(\frac{H(\zeta_k)-\sigma}{c}\right).$$

Proceeding by induction , we take $1\leq k$, assume that

$$\limsup |Q^{(j)}_n(\zeta_k)|^{1/n}\leq\exp\left(\frac{H(\zeta_k)-\sigma}{c}\right), \tag{1.4}$$

$j=0,1,\ldots,l-2$; and we must show that (1.4) holds for $j=l-1$. Using the function $F_{k,1}(z)$ of the lemma, we obtain, as above, that

$$\frac{Q_n w^{\beta}_{n-m}F_{k,1}-\tilde{P}_{n,i}}{w^{\alpha}_{n+1}}$$

is analytic on U. Then, for any given compact set K, where $K\subset D_{\sigma}\setminus\{\zeta_k\}$, we may choose $H(\zeta_k)<\sigma'<\sigma$ and $\varepsilon>0$, so that, for every $z\in K$

$$\frac{Q_n(z)g_{k,1}(z)}{(z-\zeta_k)^{l-1}}-\frac{(z-\zeta_k)P_{n,1}(z)}{w^{\beta}_{n-m}(z)}=I_n(z)-J_n(z),$$

where

$$I_n(z)=\frac{1}{2\pi i}\int\limits_{[H(t)=\sigma']}\frac{w^{\alpha}_{n+1}(z)w^{\beta}_{n-m}(t)Q_n(t)g_{k,1}(t)}{w^{\alpha}_{n+1}(t)w^{\beta}_{n-m}(z)(t-\zeta_k)^{l-1}(t-z)}dt$$

and

$$J_n(z)=\frac{1}{2\pi i}\int\limits_{|t-\zeta_k|=\epsilon}\frac{w^{\alpha}_{n+1}(z)w^{\beta}_{n-m}(t)Q_n(t)g_{k,1}(t)}{w^{\alpha}_{n+1}(t)w^{\beta}_{n-m}(z)(t-\zeta_k)^{l-1}(t-z)}dt .$$

$$J_n(z)=\sum_{j=0}^{l-2}\frac{1}{2\pi i}\int\limits_{|t-\zeta_k|=\epsilon}\frac{w^{\alpha}_{n+1}(z)w^{\beta}_{n-m}(t)Q^{(j)}_n(t)g_{k,1}(t)}{w^{\alpha}_{n+1}(t)w^{\beta}_{n-m}(z)j!(t-\zeta_k)^{l-1-j}(t-z)}dt$$

hence

$$\lim_{\epsilon\to 0}\limsup_{n\to\infty} \| J_n(z) \|_K^{1/n} \le \exp\left[\frac{\max_{z\in K} H(z) - \sigma}{c}\right] ; \tag{1.5}$$

similarly

$$\lim_{\sigma'\to 0}\limsup_{n\to\infty} \| I_n(z) \|_K^{1/n} \le \exp\left[\frac{\max_{z\in K} H(z) - \sigma}{c}\right] . \tag{1.6}$$

Hence

$$\limsup_{n\to\infty} \left\| Q_n(z) g_{k,1}(z) - \frac{(z-\zeta_k)^1 \tilde{P}_{n,1}(z)}{w^{\beta}_{n-m}(z)} \right\|_K^{1/n} \le \exp\left[\frac{\max_{z\in K} - \sigma}{c}\right] . \tag{1.7}$$

Since the function $\varphi(z) = Q(z)g(z) - \dfrac{(z-\zeta_k)^1 \tilde{P}_{n,1}(z)}{w^{\beta}_{n-m}(z)}$ is analytic throughout D_σ , then (1.7) also holds for any compact set K, where $K\subset D_\sigma$, and we obtain

$$\limsup \left\{ \left[Q_n(z) g_{k,1}(z) \right]^{(1-1)} \Big|_{z=\zeta_k} \right\}^{1/n} \le \exp\left[\frac{H(\zeta_k)-\sigma}{c}\right] .$$

Now, (1.3) can be obtained easily.

Next, consider a basis of polynomials B

$$B=\left\{ B_{k,1}(z), k=1,2,\ldots,s; 1=0,1,\ldots,m_k-1 \right\}$$

such that: $\deg B_{k,1}(z) \le m-1$ for all k,1, and the polynomials interpolate at ζ_i according to

$$\left[B^{(j)}_{k,1}(z) \Big|_{z=\zeta_i} \right] = \delta_{ik}\delta_{j1} \; ; \; 1\le i\le s \; , \; 0\le j\le m_i -1 \; .$$

So we write

$$Q_n(z) = \sum_{k=1}^{s} \sum_{1=0}^{m_k-1} Q_n^{(1)}(\zeta_k) B_{k,1}(z) + b_{n,m} Q(z) .$$

From (1.3) we obtain inmediately that $\liminf_{n\to\infty} |b_{n,m}| > 0$. Thus, for n sufficiently large, we define

$$q_n(z) = \frac{Q_n(z)}{b_{n,m}} \; , \; p_{n,i}(z) = \frac{P_{n,i}(z)}{b_{n,m}} \; , \; i=1,2,\ldots,d.$$

Finally, to establish (1.2) , let K be a compact set of D'_σ . For $z\in K$ and $i=1,2,\ldots,d$

$$q_n(z) f_i(z) - \frac{p_{n,i}(z)}{w^{\beta}_{n-m}(z)} = I_{n,i}(z) - \sum_{k=1}^{s} J_{n,i,k}(z) .$$

where

$$I_{n,i}(z) = \frac{1}{2\pi i}\int_{[H(t)=\sigma']} \frac{w^{\alpha}_{n+1}(z)w^{\beta}_{n-m}(t)q_n(t)f_i(t)}{w^{\alpha}_{n+1}(t)w^{\beta}_{n-m}(z)(t-z)}dt$$

and

$$J_{n,i,k}(z) = \frac{1}{2\pi i}\int_{|t-\zeta_k|=\varepsilon} \frac{w^{\alpha}_{n+1}(z)w^{\beta}_{n-m}(t)q_n(t)f_i(t)}{w^{\alpha}_{n+1}(t)w^{\beta}_{n-m}(z)(t-z)}dt .$$

This gives us, using a similar technique as in (1.5) and (1.6), that

$$\limsup_{n\to\infty} \| q_n(z)f_i(z) - \frac{p_{n,i}(z)}{w^{\beta}_{n-m}(z)} \|^{1/n}_K \le \exp\left(\frac{\max_{z\in K} H(z) - \sigma}{c}\right) .$$

The inequality (1.2) now follows inmediately .

Corollary.- Under the assumtions of theorem 1, there exists, for each sufficiently large imteger n , a unique set

$$\left\{ R_{n,i}(z) = \frac{p_{n,i}(z)}{q_n(z)} , \ i=1,2,...d \right\}$$

of (n,m) simultaneous rational interpolant of type (α,β) for f_i, $i=1,2,...,d$ associated to δ_i , $i=1,2,...,d$ in the n-th row of α.

Proof.- Assume, on the contrary, that for some subsequence n of integers $\mathbb{N}$. another set $\left\{\tilde{R}_{n,i}(z)=\frac{\tilde{p}_{n,i}(z)}{\tilde{q}_n(z)}, i=1,2,...d\right\}$ of (n,m) simultaneous rational interpolant of type (α,β) such that $R_{n,j}\ne\tilde{R}_{n,j}$ for some j . Then , necessarely, $q_n(z)$ is not a constant multiple of $\tilde{q}_n(z)$. The proof of theorem 1 shows that for sufficiently large n , both $q_n(z)$ and $q_n(z)$, must be of precise degree m, so, we assume that both are monic and of degree m. We obtain that

$$\frac{(q_n(z)-\tilde{q}_n(z))w^{\beta}_{n-m}(z)f_i(z) - (p_{n,i}(z)-\tilde{p}_{n,i}(z))}{w^{\alpha}_{n+1}(z)}$$

is analytic on U for each i=1.2.....d , then theorem 1 implies that $\text{degree}(q_n - \tilde{q}_n)=m$, which is a contradiction.

§2. Differences of simultaneous rational interpolants of type (α,β).

Consider another scheme of interpolation $\alpha'=\{\alpha'_{n,k}\}$ k=1,2,...,n+1 in E and $w_n^{\alpha'}(z) = \prod_{k=1}^{n}(z-\alpha'_{n-1,k})$. A condition on the "nearness" between the tables α and α' is needed. Since the polynomials $w_j^{\alpha}(z)$ and $w_j^{\alpha'}(z)$ are monic polynomials of degree j, for each n there exist n unique constants $\mu_j(n)$ $0\le j\le n-1$ such that

$$w_n^{\alpha'}(z) = w_n^{\alpha}(z)+\sum_{j=0}^{n-1}\mu_j(n)w_j^{\alpha}(z) \quad \text{where} \quad w_0^{\alpha}(z)=:1 . \qquad (2.1)$$

We suppose that there exists b, $-\infty\le b<1$ such that

$$\limsup \left\{\sum_{j=0}^{n-1}|\mu_j(n)|(\lambda_1 e^{\sigma/c})^j\right\}^{1/n} \le \lambda_1 e^{b\sigma/c} < \lambda_1 e^{\sigma/c} . \qquad (2.2)$$

It was proven in [4] (see also [6]) that extremality of α and condition (2.2) yield the extremality of α', that is

$$\lim \left|\frac{w_n^{\alpha'}(z)}{w_n^{\beta}(z)}\right|^{1/n} = \lambda\exp\left(\frac{H(z)}{c}\right) .$$

In this case, as in [5], we assume that the functions $f_1,f_2,\ldots,f_d$ and the numbers $\delta_1,\delta_2,\ldots,\delta_d$ satisfy the following conditions in D_σ :

If $i\ne j$ then f_i has no common poles with f_j in D_σ . (2.3a)

Each f_i is analytic on E and meromorphic on D_σ with exactly δ_i poles in $D_\sigma\setminus E$, $i=1,2,\ldots,d$. (2.3b)

We note that conditions (2.3) are a particular case of polewise independence . It is clear also that polewise independence and (2.3a) (or (2.3b)) imply (2.3b) (or (2.3a)).

<u>Theorem 2.-</u> Under the assumptions of theorem 1 , (2.3a) and (2.2), if

$$R_{n,i}=\frac{p_{n,i}}{q_n w_{n-m}^{\beta}} \quad \text{and} \quad R'_{n,i}=\frac{p'_{n,i}}{q'_n w_{n-m}^{\beta}}$$

are the (n,m) simultaneous rational interpolants of type (α,β) and (α',β) respectively, then: for each compact subset K ,

$$K\subset\bar{D}_\tau\setminus\{z_1,z_2,\ldots,z_m\}, \quad \sigma\le\tau<1 ,$$

$$\limsup_{n\to\infty}\left\{\max_{z\in K}|R_{n,i}(z)-R'_{n,i}(z)|\right\}^{1/n} \le \exp\left(\frac{\tau-(2-a_i)\sigma}{c}\right) , \qquad (2.4)$$

where $$a_i=\max\left(b,\frac{\max\{H(u_j)\ ;\ u_j \text{ is pole of } f_k,\ k\ne i\}}{\sigma}\right) .$$

<u>Proof of theorem 2.-</u> The uniqueness of $R_{n,i}$ and $R'_{n,i}$, for every sufficiently large integer n, is a consequence of the corollary. By theorem 1, the polynomials of the denominators $q_n(z)$ and $q'_n(z)$ satisfy

$$\lim q_n(z) = \lim q'_n(z) = \prod_{j=1}^{m}(z-z_j) = q(z)$$

uniformly on every compact set of the plane . Let

$$Q_{(i)}(z) =: \prod_{j=\delta_i+1}^{m} (z-u_j) ,$$

where u_j , $j=\delta_i+1,\delta_i+2,\dots,m$ are the poles of f_k , $k\neq i$, if $\delta_i\neq m$ ($Q_{(i)}(z)=:1$ if $\delta_i=m$) and

$$q_n(z)=q_{n,\delta_i}(z)\cdot q_{n,\delta'_i}(z) \quad , \quad q'_n(z)=q'_{n,\delta_i}(z)\cdot q'_{n,\delta'_i}(z) ,$$

where if $\delta_i\neq m$ then, for any compact subset A of $\mathbb{C}$

$$\limsup \| q_{n,\delta'_i} - Q_{(i)} \|_A^{1/n} \leq \exp\left[\frac{\max \{ H(u_j) \ ; \ j=\delta_i+1,\dots,m \} - \sigma}{c}\right]$$

$$\limsup\|q'_{n,\delta'_i}-Q_{(i)}\|_A^{1/n}\leq\exp\left[\frac{\max \{ H(u_j) \ ; \ j=\delta_i+1,\dots,m \} - \sigma}{c}\right]$$

and if $\delta_i=m$ then,

$$q_{n,\delta_i}(z)=:1 \quad , \quad q'_{n,\delta'_i}(z)=:1 .$$

Now, define $J_{n,i}(z)=q_{n,\delta_i}(z)f_i(z)q'_{n,\delta_i}(z)$. Thus for each 1,2,....,d , $q'_{n,\delta_i}(z)p_{n,i}(z)$ is the unique polynomial in the set $\mathbb{P}_n$ of the polynomials of degree at most equal to n, which interpolates $q'_{n,\delta_i}(z)J_{n,i}(z)w^{\beta}_{n-m}(z)$ in $\alpha_{n,1},\alpha_{n,2},\dots,\alpha_{n,n+1}$ in the Hermite sense. Similarly, $q_{n,\delta_i}(z)p'_{n,i}(z)$ is the unique polynomial in $\mathbb{P}_n$ which interpolates $q'_{n,\delta'_i}(z)J_{n,i}(z)w^{\beta}_{n-m}(z)$ in $\alpha'_{n,1},\alpha'_{n,2},\dots,\alpha'_{n,n+1}$ .Since $J_{n,i}(z)$, are analytic on E, there exists a constant s>0 such that all $J_{n,i}(z)$, i=1,2,...,d , are analytic on $\left[H(z) \leq s \right] \cup E$. Then for each sufficiently large n, Hermite's formula gives

$$q'_n(z)p_{n,i}(z)=\frac{1}{2\pi i}\int_{[H(t)=s]}\frac{\left\{w^{\alpha}_{n+1}(t) - w^{\alpha}_{n+1}(z)\right\} q'_{n,\delta'_i}(z)q_{n,\delta'_i}(t)w^{\beta}_{n-m}(t)J_{n,i}(t)}{w^{\alpha}_{n+1}(t) \quad (t-z)}dt$$

and

$$q_n(z)p'_{n,i}(z)=\frac{1}{2\pi i}\int_{[H(t)=s]}\frac{\left\{w^{\alpha'}_{n+1}(t)-w^{\alpha'}_{n+1}(z)\right\}q_{n,\delta'_i}(z)q'_{n,\delta'_i}(t)w^{\beta}_{n-m}(t)J_{n,i}(t)}{w^{\alpha'}_{n+1,}(t)\quad(t-z)}dt$$

$$q'_n(z)p_{n,i}(z)-q_n(z)p'_{n,i}(z)=\frac{1}{2\pi i}\int_{[H(t)=s]}\frac{A_{n,i}(t,z)J_{n,i}(t)w^{\beta}_{n-m}(t)}{w^{\alpha}_{n+1}(t)w^{\alpha'}_{n+1}(t)\quad(t-z)}dt$$

$$A_{n,i}(t,z)=w^{\alpha}_{n+1}(t)w^{\alpha'}_{n+1}(t)\left[q_{n,\delta'_i}(t)q'_{n,\delta'_i}(z)-q'_{n,\delta'_i}(t)q_{n,\delta'_i}(z)\right]+$$

$$+w^{\alpha}_{n+1}(t)w^{\alpha'}_{n+1}(z)q'_{n,\delta'_i}(t)q_{n,\delta'_i}(z)-w^{\alpha'}_{n+1}(t)w^{\alpha}_{n+1}(z)\,q_{n,\delta'_i}(t)q'_{n,\delta'_i}(z)\,.$$

Next, let $\{u^*_j\}$, $j=1,2,...,s_i$, $s_i\le\delta_i$, denote the distinct poles of f_i in $D_\sigma\setminus E$. Let M be any constant such that $\max(0,b\sigma)<M<\sigma$ and all the poles of $f_i(z)$ lie in the interior of $[H(z)=M]$ (see (2.2) for b). By $c_j=\{t:|t-u^*_j|=\varepsilon\}$. $1\le j\le s_i$, we denote small circles that are mutually exterior and satisfy $c_j\subset D_M\setminus E$ for each $j=1,2,...,s_i$. Set $c_{s_i+1}=[H(t)=M]$. Thus

$$q'_n(z)p_{n,i}(z)-q_n(z)p'_{n,i}(z)=\sum_{j=1}^{s_i+1}I_{j,i,n}(z)\quad,\quad i=1,2,...,d\ ,\tag{2.5}$$

where

$$I_{j,i,n}(z)=\frac{1}{2\pi i}\int_{c_j}\frac{A_{n,i}(t,z)J_{n,i}(t)w^{\beta}_{n-m}(t)}{w^{\alpha}_{n+1}(t)w^{\alpha'}_{n+1}(t)\quad(t-z)}dt$$

and c_{s_i+1} is taken positively oriented, while the remaining contours c_j, $1\le j\le s_i$, are all negatively oriented.

Using (2.1) and noting that $\max_{[H(z)=s]}\psi_1(z)\le\max_{[H(z)=\tau]}\psi_2(z)$ if $0\le s\le\tau<1$, it is easy to prove according to (2.5), (2.2) and (1.3), that for $\sigma\le\tau<1$. (2.4) holds.

REFERENCES.-

1. J.L. Walsh.: Interpolation and Approximation by Rational Functions in the Complex Domain. A.M.S. Colloq. Publ. Vol.XX, Providence,R.I. 3-rd Ed 1960.

2. A.A. Gonchar.: On the convergence of generalized Padé approximants of meromorphic functions. MATH USSR Sbornik Vol. 27, No.4 , 1975.

3. Graves Morris and E.B. Saff.: A. de Montessus theorem for vector valued rational interpolants. Rational Approximation and inter polation. Lecture Notes in Math. proceedings, Tampa, Florida. 1983.

4. Piedra René.: On differences of approximants type (α,β). Rev. Ciencias Matemáticas, Vol. 6, No.3. 1985 (in spanish).

5. Piedra René.: On the differences of simultaneous rational interpo lants. Proceeding of XX Seminar on Banach Center, Poland , (1986). to appear.

6. E.B. Saff, A Sharma and R.S. Varga.:An extension to rational functions of a theorem of J.L.Walsh on differences of interpolating polynomials. R.A.I.R.O. Analyse Numerique/Numerical Analysis. Vol. 15. No-4, 1981 .

GÉNÉRALISATION DE FORMULES DE BORNAGE DE L-I ET APPLICATIONS AUX L^p

Celiar Silva Rehermann
Facultad de Matemática y Cibernética
Universidad de La Habana , Cuba

<u>Introduction</u> (Symboles et conventions)

Soit $f \in E = B_R(X)$ (application réelles bornées définies dans l'espace topologique X) ; $\mathcal{B}(z)$ est une base locale de voisinages du point z de X . Pour chaque $V \in \mathcal{B}(z)$, v_z est un membre de E positif tel que $v_z(z) = 0$ et que $x \notin V \Rightarrow v_z(x) \geq 1$. (Observez que cette v_z n'est pas continue et confronter avec les considérations de [3] où peut se faire que X manque de structure topologique). Si X est un space topologique complètement régulier, on peut choisir chaque v_z continue et avec l' image contenue dans [0,1]. Si (X,d) est un espace métrique et les membres de $\mathcal{B}$ sont les boules B(z,r), alors, pour chacune on peut prendre la fonction continue $v_{zr}(x) = \frac{1}{r} d(x,z)$, ou $(v_{zr})^{\alpha}$ avec $\alpha \in R_+^*$; en ce cas, on voit aussitot que si (L_n) est une suite d'applications de E dans E, alors la convergence uniforme sur une partie quelconque $Z \subset X$, avec un même r pour tout $z \in Z$, de la suite réelle $(L_n v_{zr})(z)$ est équivalente à la convergence uniforme sur Z de la suite $(L_n d(x,z))(z)$ des que les L_n soient homogènes.

En ce qui suit nous mettrons $L_z f=(Lf)(z)$ et représenterons avec I l' opérateur identique. En outre, dans cette section sera $\omega(f,V_z)=\sup\left\{|f(z)-f(x)| : x\in V_z\right\}$ et $\Omega(f,V_z)=\sup\left\{|f(x)-f(z)| : x\in X\setminus V_z\right\}$ si $V_z \neq X$, $\Omega(f,X) = 0$. Quand, selon le contexte, il n'a pas risque de confusion, nous mettrons simplement ω et Ω .

§ 1 Théorème 1.

Soit L une application de E dans E telle que est sous - linéaire (c'est à-dire sous-aditive et homogène) la restriction de la fonctionelle L_z au sous-espace engendré par v_z et la fonction constante 1, et monotone (par rapport à l'ordre de l'espace de Riesz E et de l'espace d'arrivée R) la restriction de L_z au sous-espace S engendré par $\{1, v_z, f\}$ (en particulier linéaire dans S et positive ou négative dans celui-ci).

Alors, on a la formule suivante:

$$|(L-I)_z f| \leq |f(z)(L-I)_z 1| + \Omega|L_z v_z| + \omega|L_z 1| \qquad \text{(I)}$$

Voici le détaiil de la démonstration pour L_z décroissante (cas que ne fut pas consideré en [2] ni en [3]).

Préfixe $z \in X$ et $V \in \mathcal{B}(z)$, prenons les fonctions auxiliaires h: $h(x)=f(x)-\Omega v_z-\omega$ et g: $g(x)=f(x)+\Omega v_z+\omega$. On a $h \leq f \leq g$; en effet, comme $\Omega \geq 0$, $\omega \geq o$, et $x \in V \Rightarrow \omega \geq |f(x)-f(z)|$, $x \notin V \Rightarrow (\Omega \geq |f(x)-f(z)|$ et $v_z(x) \geq 1)$ on a

$$h(x) \leq f(x)-|f(x)-f(z)| \leq f(z)+f(x)-f(z) \leq f(z)+|f(x)-f(z)| \leq g(x).$$

En résulte $L_z h \geq L_z f \geq L_z g$ car L_z est décroissante; par suite:

a) $L_z f-f(z) \leq f(z)(L_z 1-1)-\Omega L_z v_z-\omega L_z 1 \leq |f(z)(L_z 1-1)|+\Omega|L_z v_z|+\omega|L_z 1|$

b) $f(z)-L_z f \leq -f(z)(L_z 1-1)-\Omega L_z v_z-\omega L_z 1 \leq |f(z)(L_z 1-1)|+\Omega|L_z v_z|+\omega|L_z 1|$

d'où résulte inmédiatement l'inégalité (I).

Pour L_z croissante cette formule fut démontrée analoguement en [3].

§2. Conséquences sur la convergence d'une suite (L_n) d'operateurs qui verifient (I).

Tel comme en [3], mais référant maintenant au champ plus vaste de validité de (I), on peut démontrer propositions analogues aux de [3, II-A']

Théorème 2.

Soit E_z le sous ensemble de E constitué par les f continues en le point z, et $E_z^*=\{f: f(z)\neq 0\}$. Si on vérifie $\lim L_{nz}v_z=0$ ($\forall\ V_z\in\mathcal{B}(z)$), alors on a:

1°) ($f\in E_z\setminus E_z^*$ et ($L_{nz}1$) bornée) $\Rightarrow$ ($L_{nz}f \xrightarrow{n} f(z)$),

2°) ($\exists\ \varphi\in E_z^*$: $L_{nz}\varphi \xrightarrow{n} \varphi(z)$) $\Rightarrow$ ($\forall f\in E_z$) $L_{nz}f \longrightarrow f(z)$.

<u>Démonstration</u>

1°) Si $f(z)=0$, de (I) on en conclut inmédiatement.

2°) Soit $\varphi\in E_z^*$. Si $\varphi=1$, la conclusion surgit aussi sans peine de (I). Si $\varphi\neq 1$, il suffit donc de démontrer l'implication ($L_{nz}\varphi \xrightarrow{n} \varphi(z)$) $\Rightarrow$ ($L_{nz}1 \xrightarrow{n} 1$).

La proposition contre-réciproque de cette implication est:

(1 n'est pas la limite de $L_{nz}1$) $\Rightarrow$ ($\varphi(z)$ n'est pas la limite de $L_{nz}\varphi$), qui est équivalente à l'implication suivante:

(Il existe $\varepsilon>0$ tel que, quel que soit n, il existe $m>n$, qui vérifie $|L_{m,z}1-1|\geq\varepsilon$) $\Rightarrow$ (Il exite $\varepsilon'>0$ tel que, quel que soit n, il existe $m>n$, qui vérifie $|L_{nz}\varphi-\varphi(z)|\geq\varepsilon'$). Voici la démonstration de de cela:

Soit $\varepsilon'=\frac{1}{3}|\varphi(z)|\varepsilon$, et fixons n_2. Soit V_z tel que $\omega(\varphi,V_z)<\frac{1}{2}|\varphi(z)|$. (1)

Puisque $|\Omega(\varphi,V_z)\leq 2\sup\{|\varphi(z)|:x\in X\}$ et que $\lim L_{nz}v_z=0$, il existe n_0: $n\geq n_0 \rightarrow |\Omega(\varphi,V_z)L_{nz}v_z|<\varepsilon'$. (2)

Si $n_1=\max\{n_0,n_2\}$, par hypothése, il existe $m\geq n_1$ tel que $|L_{mz}1-1|\geq\varepsilon$, et par suite $|\varphi(z)||L_{mz}1-1|\geq 4\varepsilon'$. (3)

Nous verrons que ce même m vérifie à la fois $|L_{mz}\varphi-\varphi(z)|\geq\varepsilon'$.

Des inégalités a) et b) de §1 (appliquées a L_m et φ) on obtient:

a') $\varphi(z)-L_{mz}\varphi\geq -\varphi(z)(L_{mz}1-1)+\Omega L_{mz}v_z+\omega(L_{mz}1-1)+\omega$,

b') $L_{mz}\varphi-\varphi(z)\geq \varphi(z)(L_{mz}1-1)+\Omega L_{mz}v_z+\omega(L_{mz}1-1)+\omega$.

Si $\varphi(z)(L_{mz}1-1)<0$, de a') on infére:

$$\varphi(z)-L_{mz}\varphi\geq |\varphi(z)(L_{mz}1-1)|+\Omega L_{mz}v_z+\omega(L_{mz}1-1)+\omega.$$

Donc, on a toujours:

$$|\varphi(z)-L_{mz}\varphi| \geq |\varphi(z)|L_{mz}1-1| -|\Omega L_{mz}v_z| + \omega|L_{mz}1-1|.$$

Par suite, compte tenu de (1),(2) et (3), on a

$$|\varphi(z)-L_{mz}\varphi| \geq \epsilon'$$

ce qui achève la démonstration.

Rémarque. Quand les L_{nz} sont croissantes, la démonstration est analogue; ici nous avons détaillé le raisonnement concernant à L_{nz} décroissantes pour q'on puisse user les inégalités à la main a) et b). Or, quand les L_{nz} sont décroissantes il est clair que la thése de la proposition 2°) ne s'accomplit pas pour f=1. Donc, en ce cas on peut assurer qu'on a soit l'un, soit l'autre, des suivantes circonstances: La prémise $L_{nz}v_z \longrightarrow 0$ $(\forall v_z)$ est fausse; ou pour aucune $\varphi \in E_z^*$ s'accomplit l'hypothèse $L_{nz}\varphi \longrightarrow \varphi(z)$.

Convergence en norme. En supposant maintenant que les membres de E sont les fonctions continues bornées en $Z \subset X$ (par example des fonctions continues définies dans le sous-ensemble compact z de l'espace topologique X) on obtient ipso facto les résultats analogues sur la convergence dans l'espace normé E avec la norme $\|f\| = \sup\{|f(z): z \in Z\}$, quand on suppose que les L_{nz} sont croissantes (pour tout z et tout $n \in \mathbb{N}$). En particulier, si X est un sous-espace borné de $\mathbb{R}^m$ et $v_z(x)=(d(x,z))^2$, il arrive comm'on a vu en [3] que la convergence en E, $L_{nz}v_z \longrightarrow v_z$ $(\forall z \in X)$ on infére les convergences $L_n pr_i \longrightarrow pr_i$ et $L_n(pr_i)^2 \longrightarrow (pr_i)^2$, pour tout $i=\{1,\ldots,m\}$.

Observation. La conclusion subsiste quand au lieu de la croissance on suppose que chaque L_n vérifie l'implication $g \leq 1 \Rightarrow L_n g \leq L_n 1$. En effet, en référant à la descomposition $L = L^+ - L^-$ usée en [3,IV], puisque $L^+1 = \sup\{Lg: 0 \leq g \leq 1\}$ résulte $L^+1=L1$; par suite de $L_n 1 \longrightarrow 1$ on obtient $L_n^- 1 \longrightarrow 0$, donc $|L_n|(1)=L_n^+1+L_n^-1 \longrightarrow 1$. Alors, il suffit d'appliquer la conséquence sur la convergence uniforme de [3,IV,C] compte tenu de la observation inclue en [3,IV,B].

3. Cas d'un ensemble X quelconque (non nécessairement esp. top.).

Théorème 3. Si L est une application de E dans E q'est sous-linéaire la restriction de L_z au sous-espace engendré par 1 et la fonction

1_z: $1_z(z)=0$, $1_z(x)=1$ si $x\neq z$, ainsi que monotone la restriction au sous-espace engendré par $\{1,1_z,f\}$, alors, si $\omega(f,z)=\sup\{|f(x)-f(z)|:x\in X\}$, on a

$$|(L-I)_z f| \leq |f(z)(L-I)_z 1| + \omega(f,z)|L_z 1_z| \qquad \text{(II)}$$

Pour L_z décroissant, la démonstration est complètement analogue à la de (I), en considérant maintenant les fonctions auxiliaires h: $h(x)=f(x)-\omega(f,z)1_z$ et $g:g(x)=f(z)+\omega(f,z)$. Pour L_z croissante l'inégalité fut démontrée en [3].

<u>Conséquences.</u> En autre des conséquences sur la convergence consignées en [3], nous ajoutons ici les suivantes corollaires immédiats:

Si $L_z$est sous-linéaire et monotone on a:

1) Les zéros communes de f et la fonction ψ: $z\longrightarrow L_z 1_z$, sont de même zéros de $z\longrightarrow L_z v_z$. En effet, si $f(z)=0=L_z 1_z=0$ cela s'en suit évidement de (II).

2) On peut dire que lorsque $L_z 1_z=0$ la valeur $L_z f$ ne dépend pas de la f tout entière (c'est-à-dire, de l'ensemble de tous les pairs $(x,f(x))$ mais seulement de la valeur $f(z)$; autrement dit, si f_1 est une autre fonction telle que $f_1(z)=f(z)$, alors $L_z f_1=L_z f$.

En effet, il suffit d'appliquer 1) à la fonction $F=f-f(z)$, puisque en étant $F(z)=0$ on obtient $L_z f=(L_z 1)f(z)$.

4. <u>Autres</u> <u>consequences.</u>

<u>Théorème</u> <u>4.</u> Soit (L_n) la suite d'itérées d'une application L de l'espace topologique (non nécessairement vectoriel) E dans E, dont le graphe G est fermé dans ExE. Si on vérifie l'implication suivante

$$((\forall g\in K\subset E)\ L^n g\xrightarrow{n} g) \Longrightarrow ((\forall\in E)\ L^n f\xrightarrow{n} f),$$

alors est aussi vrai celle-ci:

$$((\forall g\in K\subset E)\ Lg = g) \Longrightarrow ((\forall\in E)\ Lf = f),$$

qui peut s'exprimer comme suit: Si K est "<u>Korovkin</u> <u>dense</u>" par rapport à la suite (L^n), alors on peut assurer que si la restriction à K de L est l'operateur identique sur K, il en est de même de L sur E.

Démonstration. Il est évident que pour toute $g \in K$, $(Lg=g) \Rightarrow (L^n g=g) \Rightarrow (L^n g \xrightarrow{n} g)$; par suite, selon l'hypothèse, $L^n f \xrightarrow{n} f$ pour toute $f \in E$, et de même $L^{n+1} f \xrightarrow{n} f$, c'est-à-dire $L(L^n f) \xrightarrow{n} f$. Or, $(L^n f, L(L^n f))$est une suite du graphe G (qui est un ensemble fermé de ExE); donc, sa limite (f,f)appartient a G, c'est-à-dire $f=Lf$.

Lemme 1. Si une application L de l'espace normé $E=B_{\mathbb{R}}(X)$ dans E est homogène et monotone, alors elle est "bornée" (c'est-à-dire, il existe $k \in \mathbb{R}_+$: $\|Lf\| \leq k\|f\|$ pour toutes les f de E).

En effet, on a $-\|f\|\cdot 1 \leq f \leq \|f\|\cdot 1$, et en vertu de la monotonie en résulte soit $-\|f\|\cdot L1 \leq Lf \leq \|f\|\cdot L1$, soit $\|f\|\cdot L1 \leq Lf \leq -\|f\|\cdot L1$. Par suite $\|Lf\| \leq \|f\|\cdot L1$, donc $\|Lf\| \leq \|f\|\cdot L1$. Il suffit,donc, de prendre $k=\|L1\|$.

Lemme 2. Si $L:E \longrightarrow E$ est "bornée" et si $L(0)=0$, alors L est continue en 0 (élément neutre aditif de l'espace normé E).

En effet, puisq'il existe k tel que $\|Lf\| \leq k\|f\|$, pour tout $\varepsilon>0$, on en déduit l'implication $\|f\|<\frac{\varepsilon}{k} \Longrightarrow \|Lf\|<\varepsilon$.

Lemme 3. Le graphe de toute application linéaire monotone L de $E=B_{\mathbb{R}}(X)$ dans soi-même, est fermé.

En effet, selon le Lemme 1, L est bornée, et comme (par la linéarité) $L0=0$, d'après le Lemma 2, L est continue au point 0 et par suite (en étant linéaire) L est continue. Alors, puisque l'espace d'arrivée E est normé, donc séparé, le graphe de L est fermé. (Voir, par exemple, [1,8 N°1]

Théorème 5. Soit L un opérateur linéaire et croissant sur l'espace E des fonctions continues et bornées définies sur l'espace topologique X. Si L a comme points fixes les éléments 1 et les v_z $(\forall z \in X,\ V \in \mathcal{B}(z))$, alors $L=I$.

En effet, en appliquant les conséquences sur les normes consignées en [2] à la suite (L^n) telle que $L_n=L^n$, de $L1=1$ et $Lv_z=v_z$, on en déduit $L^nf \xrightarrow{n} f$ ($\forall f\in E$); et, comme (d'après le Lemme 3) L a graphe fermé, il suffit d'appliquer le théorème 4 au cas $K=\{1\}\cup\{v_z : z\in X\}$.

Scolie. Si $X\triangleleft R^n$, il suffit que les fonctions pr_i et $(pr_i)^2$ soient des ponts fixes, pour qu'il soit de même des v_z.

Corollaire 1. Si 1 et les v_z sont des vecteurs propres de l'opérateur linéaire et monotone L, avec le même valeur propre non nul λ, alors L est l'homothétie de rapport $\lambda: f \longrightarrow \lambda f$.

En effet, si $L1=\lambda 1$ et $Lv_z=\lambda v_z$, est $\frac{1}{\lambda}L1=1$ et $\frac{1}{\lambda}Lv_z=v_z$, ce qui signifie que 1 et v_z sont des points fixes pour l'opérateur λL. Si L est croissant, ce qui exige $\lambda>0$, en résulte de même $\frac{1}{\lambda}L$ linéaire et croissant, et par suite, d'après le théorème 5, $(\frac{1}{\lambda}L)f=f$ d'où $Lf=\lambda f$. Si L est décroissant (ce qui exige $\lambda<0$), il suffit d'appliquer la conclusion antérieure à l'opérateur linéaire et croissant $-L$ (qui est une homothétie de rapport $-\lambda$).

Corollaire 2. Soient L et M opérateurs tels que chacun accomplit les hypothèses du théorème 5, et que prennent valeurs respectivement égales en les points 1 et v_z ($\forall z\in X$) de le espace E. Si en outre L est "bicroissant" (c'est-à-dire injectif et L^{-1} croissant) alors M=L.

Démonstration. Comme $L1=M1$ et $Lv_z=Mv_z$, en résulte $1=(L^{-1}\circ L)1= L^{-1}\circ(L1)= L^{-1}\circ(M1)= (L^{-1}\circ M)1$, et de même $v_z= (L^{-1}\circ M)v_z$. Par suite, parce que des hypothèses s'en suit que $L^{-1}\circ M$ est linéaire et croissant, d'après le théorème 5, $L^{-1}\circ M=I$, d'où M=L.

Rémarque. En vertu de l'unicité demontrée, on pourrait dire qu'un opérateur linéaire bicroissante L rests "déterminé" par ses valeurs L1 et v_z ($\forall z\in X$). Cependant est clair qu'il peut se faire que n'existe aucun opérateur linéaire bicroissant qui prend des valeurs préfixés en les points 1 et v_z. Un problème ouvert serait (en certains cas particulieres de l'espace E et du type d'opérateur L) chercher des conditions pour les valeurs L1 et Lv_z qui permettent assurer l'existence de tel opérateur , et, de plus, une fois donnés ces valeurs convenables, de calculer Lf pour une f quelconque.

5 Applications aux espaces L^p.

Soient $E=\mathcal{L}^p[a,b]$ ($p\geq 1$, réel); nous indiquerons la norme de f dans l'espace L^p avec la notation $\|f\|_p$, en réservant le symbole $\|f\|$ pour la norme L^∞ (donnée par ess sup f).

Si L est une application sous-linéaire de E dans E telle que pour z presque partout L_z est monotone, et pour chaque $r\in\mathbb{R}_+$ est F_r la fonction $z\longrightarrow L_z v_{zr}$ définie dans X (où v_{zr} est la fonction mentionnée dans l'introduction),et, enfin, $\omega_r:z\longrightarrow\omega(f,B(z,r))$, $\Omega_r:z\longrightarrow\Omega(f,B(z,r))$ alors , de la formule(I) de §1, découle

$$\|(L-I)f\|_p \leq A_p + B_p + C_p, \qquad \text{(III)}$$

ou A_p, B_p et C_p sont, respectivement, le plus petit composant des paires de nombres $(\|L1-1\|\|f\|_p, \|L1-1\|_p\|f\|_p)$, $(\|\Omega_r\|\|F_r\|_p, \|\Omega_r\|_p\|F_r\|)$, et $(\|\omega_r\|\|L1\|_p, \|\omega_r\|_p\|L1\|)$.

Si p=2, moyennant l'innégalité de Hölder, on obtient de même:

$$\|(L-I)f\|_2 \leq \|L1-1\|_1\|f\|_1 + \|\Omega_r\|_1\|F_r\|_1 + \|\omega_r\|_1\|L1\|_1 \qquad \text{(III')}$$

En applicant la formule (III) à chacun des opérateurs d'une suite (L^n) dont les termes accomplissent les hypothèses exigées a L, on infère aussitôt le théorème suivant (dont l'énnoncé inclu les fonctions F_{nz}: $z\longrightarrow L_{nz}v_{zr}$).

<u>Théorème</u> <u>6.</u> Si les suites $(L_n 1)$ et (F_{nr}) (pour tout r réel appartenant à quelque voisinage droit de 0) convergent dans l'espace normé L^p, respectivement, aux fonctions constantes 1 et 0, alors, si f est continue dans [a,b] (et par suite bornée et uniformément continue, ce qui entraîne $\lim_{r\to 0}\|\omega_r\|=0$) il résulte

aussi $\lim_n L_n f = f$ dans l'espace L^p.

<u>Scolie.</u> Comm'on sait, on se peut prendre v_{zr}^2 au lieu de v_{zr}; alors $F_{nr} \xrightarrow{n} 0$ découle de $L_n 1 \xrightarrow{n} 1$, $L_n x \xrightarrow{n} x$, $L_n x^2 \xrightarrow{n} x^2$.

D'une manière analogue, de la formule (II) de §3 (ou même de la (III) en prenant $r=b-a$, on déduit (compte tenu que $\omega(f,z) \le 2|f(z)| \le 2\|f\|$):

$$\|(L-I)f\|_p \le A_p + 2B_p \tag{IV}$$

et

$$\|(L-I)f\|_2 \le \|L1-1\|_1 \|f\|_1 + 2\|L1\|_1 \|f\|_1 . \tag{IV'}$$

<u>Bornage de la norme d'un opérateur linéaire positif L dans L^2, et de la norme $(\infty,2)$.</u>

1°) Soit L une application linéaire positive de $\mathcal{L}^2[a,b]$ dans lui même, et $f \ge 0$; alors le produit scalaire $(Lf,-f)$ est négatif, donc, en vertu d'une des formules que nous avons démontré en {4}, on a:

$$\|(L-I)f\|_2 \le \frac{\sqrt{2}}{2} \left| \|Lf\|_2 - \|f\|_2 \right| . \tag{V}$$

En combinant cette inégalité avec la (IV) pour $p=2$, on obtient:

$$\|Lf\|_2 \le [1+\sqrt{2}(\|L1-1\| + 2\|L1\|)]\|f\|_2 . \tag{VI}$$

Pour une f quelconque de L^2, on a

$$\|Lf\|_2 \le \|Lf^+ - Lf^-\|_2 \le \|Lf^+\|_2 + \|Lf^-\|_2 \le$$
$$[1+\sqrt{2}(\|L1-1\| + 2\|L1\|)](\|f^+\|_2 + \|f^-\|_2),$$

et comme selon la formule (1) de [4] est $\|f^+\|_2 + \|f^-\|_2 \le \sqrt{2}\|f^+ + f^-\|_2 = \sqrt{2}\| |f| \|_2 \le \sqrt{2}\|f\|_2$, on en conclut:

$$\|Lf\|_2 \le \sqrt{2}[1+\sqrt{2}(\|L1-1\| + 2\|L1\|)]\|f\|_2 . \tag{VII}$$

et par suite:

$$\|L\|_2 \le \sqrt{2} + 2(\|L1-1\| + 4\|L1\|) \tag{VIII}$$

2°) Puisque $\|f\|_2 \le (b-a)^{1/2}\|f\|$, de (VII) résulte:

$$\|L\|_{\infty,2} \le (b-a)^{1/2}(\sqrt{2} + 2(\|L1-1\| + 4\|L1\|) \tag{IX}$$

Par autre coté, comme $\|f\|_1 \le (b-a)\|f\|$, de (V) et (IV) découle

$$\|L\|_{\infty,2} \le (b-a)^{1/2} + \sqrt{2}(b-a)^2(\|L1-1\| + 2\|L1\|). \tag{IX'}$$

Rémarque: Selon la valeur de b-a, on peut que la borne donnée par (IX') soit moindre que celle-là fournie par (IX); par exemple, il est ainsi quand $b-a \leq 1$.

3°) En usant la décomposition $L=L^+-L^-$, on peut appliquer les resultats 1°) et 2°) à fin de trouver unes bornes de $\|L\|_2$ et de $L_{\infty,2}$, quel que soit l'opérateur linéaire continu sur L^2.

Références.

1. Bourbaki, N., Structures Topologiques, Chap . I, Paris, 1961.

2. Jiménez Pozo, M., Tesis de Candidatura a Doctor, La Habana, 1979.

3. Silva Rehermann, C., Variantes de una fórmula de acotación de Jiménez, Rev. Ciencias Mat., vol. I, no. 1, 1984.

4. Silva Rehermann, C., Una desigualdad y una igualdad entre normas, Memorias de la Fac. de Ciencias., vol. 1, no.1, La Habana, 1963.

ON C_0-SEMIGROUPS IN A SPACE OF BOUNDED CONTINUOUS FUNCTIONS IN THE CASE OF ENTRANCE OR NATURAL BOUNDARY POINTS

C.A.Timmermans

Dept. of Math., Un.of Techn.,

Delft, The Netherlands.

1. INTRODUCTION

In [1], Clément and the author investigated contraction semigroups $(T(t))$, $t \geq 0$, of class C_0 on the Banach space $C(\bar{I})$, where $\bar{I}$ is the two points compactification of an open interval (r_1, r_2), $-\infty \leq r_1 < r_2 \leq \infty$, (equipped with the supremum norm) generated by an operator A_0 of the form

$$(1.1) \qquad A_0 u = \alpha D^2 u + \beta D u.$$

Here α and β are continuous real valued functions on I, with $\alpha(x) > 0$ for $x \in I$. The domain of A_0, denoted by $D(A_0)$, is given by

$$(1.2) \qquad D(A_0) := \{u \in C(\bar{I}) \,|\, u \in C^2(I), \lim_{x \to r_1} A_0 u(x) = 0, \lim_{x \to r_2} A_0 u(x) = 0\}.$$

The boundary conditions in (1.1) are usually called Ventcel's boundary conditions, [7].

In order to state the conditions on α and β in the next theorems, we define the functions Wand Q as follows

$$W(x) := \exp\{-\int_{x_0}^{x} (\beta.\alpha^{-1})(t)dt\},$$

$$Q(x) := (\alpha W)^{-1}(x) \int_{x_0}^{x} W(t)dt.$$

$$R(x) := W(x) \int_{x_0}^{x} (\alpha W)^{-1}(t)dt.$$

Here x_0 denotes an arbitrary but fixed point in I.

In [1] we proved the following theorem, which gives *necessary* and sufficient conditions for A_0 to be the generator of a C_0-contraction semigroup on $C(\bar{I})$.

Theorem 1. Let α and β be real-valued continuous functions on I, and let A_0 be given by (1.1) and (1.2), then A_0 is the generator of a C_0-semigroup on $C(\bar{I})$ if and only if α and β satisfy

(H_1) $W \in L^1(r_1, x_0)$ or $Q \notin L^1(r_1, x_0)$ or both,

(H_2) $W \in L^1(x_0, r_2)$ or $Q \notin L^1(x_0, r_2)$ or both.

The reader familiar with the terminology of Feller [3], will recognize that (H_i) $(i = 1,2)$ is satisfied if and only if r_i is *not* an *entrance boundary point*, thus if r_i is a regular, an exit or a natural boundary point.
Let

$$(1.3) \qquad D(A) := \{u \in C(\bar{I}) \mid u \in C^2(I),\ \alpha D^2 u + \beta D u \in C(\bar{I})\},$$

and let $A : D(A) \to C(\bar{I})$ be defined by

$$(1.4) \qquad Au = \alpha D^2 u + \beta D u$$

In this paper we will give *necessary* and sufficient conditions for A to be the generator of a C_0-semigroup on $C(\bar{I})$.

2. MAIN THEOREMS

Let I denote a non-empty open interval of $\mathbb{R}$ (not necessarily bounded), $\bar{I}$ the two points compactification of I, and $\partial I := \bar{I} \backslash I$. $C(\bar{I})$ is the Banach space of real-valued continuous functions on $\bar{I}$ equipped with the supremum norm, denoted by $\|\cdot\|$.
The main theorems read :

THEOREM 2. Let α and β be real-valued continuous functions on I with $\alpha > 0$ on I. Let

$$(2.1) \qquad D(A) := \{u \in C(\bar{I}) \mid u \in C^2(I),\ \alpha D^2 u + \beta D u \in C(\bar{I})\}.$$

and let

$$Au = \alpha D^2 u + \beta D u$$

for $u \in D(A)$. Then $A : D(A) \to C(\bar{I})$, and the following conditions are satisfied :

(i) $D(A)$ is dense in $C(\bar{I})$

(ii) A is closed

(iii) $I - A$ is surjective in $C(\bar{I})$

THEOREM 3. Let A be defined as in Theorem 2, then A is the generator of a C_0-semigroup on $C(\bar{I})$ if and only if the following condition

(K) $\int_{r_1}^{x_0} R(x)dx = \infty$ and $\int_{x_0}^{r_2} R(x)dx = \infty$ is satisfied.

We also have:

THEOREM 4. Let α and β be real-valued continuous functions on I with $\alpha > 0$ on I. Let

$$D(A_1) := \{u \in C(\bar{I}) | u \in C^2(I),\ \alpha D^2 u + \beta Du \in C(\bar{I}),$$
$$\lim_{x \to r_1} (\alpha D^2 u) + \beta Du)(x) = 0\},$$

and let

$$A_1 u = \alpha D^2 u + \beta Du$$

for $u \in D(A_1)$. Then $A_1 : D(A_1) \to C(\bar{I})$ is the generator of a C_0-semigroup on $C(\bar{I})$ if and only if the following conditions are satisfied:

(i) $W \in L^1(r_1, x_0)$ or $Q \notin L^1(r_1, x_0)$ or both,

(ii) $\int_{x_0}^{r_2} R(x)dx = \infty$.

In the next section we will prove Theorem 2 and 3. The proof of Theorem 4 analogous.

3. PROOFS.

The proof of Theorem 2 is partly based on the following lemmata.

LEMMA 5. Let α and β be as in Theorem 2, and let $\lambda > 0$. Moreover, let A_1 (resp. A_2) be the set of positive increasing (resp. decreasing) solutions on I of

$$(3.1) \qquad u - \lambda(\alpha D^2 u + \beta Du) = 0$$

satisfying $u(x_0) = 1$. Then there exists a unique $u_1 \in A_1$ (resp. a unique $u_2 \in A_2$) which is minimal on $(r_1, x_0]$ (resp. $[x_0, r_2)$), i.e.

if $u \in A_1$ (resp. $u \in A_2$) then $u_1 \le u$ on $(r_1,x_0]$ (resp. $u_2 \le u$ on $[x_0,r_2)$).

The proof for u_2 can be found in [1], the proof for u_1 is similar. We define

$$M_i = \sup_{x\in I} u_i(x),\ i = 1,2.$$

Then we have

LEMMA 6. $M_1 < \infty$ (resp. $M_2 < \infty$) if and only if $R \in L^1(x_0,r_2)$ (resp. $R \in L^1(r_1,x_0)$).

PROOF. We only prove the lemma for M_1.

Necessity. Since u_1 satisfies the equation

$$(W^{-1}u_1')' = (\alpha W)^{-1}u_1 \tag{3.2}$$

we have for all $x \in (r_1,r_2)$

$$u_1(x)=\int_{x_0}^{x} W(t)\{\int_{x_0}^{t} (\alpha W)^{-1}(s).u(s)dx + u_1'(x_0)\}dt + u_1(x_0)\ . \tag{3.3}$$

Since u_1 is positive increasing we then have for $x > x_0$

$$u_1(x_0).\int_{x_0}^{x} R(t)dt \le \int_{x_0}^{x} W(t) \int_{x_0}^{t} (\alpha W)^{-1}(s)u(s)ds \le u_1(x) \le M_1.$$

So, if $M_1 < \infty$, then $R \in L^1(x_0,r_2)$.

Sufficiency. For $x \in (x_0,r_2)$ we have

$$0 \le u_1'(x) \le W(x).\int_{x_0}^{x} (\alpha W)^{-1}(s).u(s)ds \le u_1(x).R(x).$$

So $0 \le u_1'(x).(u_1(x))^{-1} \le R(x)$. Integration gives for $x \in (x_0,r_2)$

$$0 \le \log u_1(x) \le \int_{x_0}^{x} R(s)ds.$$

Thus, if $R \in L^1(x_0,r_2)$, then $M_1 < \infty$. □

Let V be the Wronskian $u_1'u_2 - u_2'\, u_1$, then clearly $V > 0$. We define the Green function $G : I \times I \to \mathbb{R}$ for (3.1) by

$$G(x,s) = u_1(x) \,.\, (\alpha V)^{-1}(s).u_2(s),\ x \le s,$$
$$= u_2(x) \,.\, (\alpha V)^{-1}(s).u_1(s),\ x > s,$$

and for each $g \in C(\bar{I})$ the function $u_g : I \to \mathbb{R}$ by

(3.4) $$u_g(x) = \int_{r_1}^{r_2} G(x,s)g(s)ds.$$

Then it follows by verification that $u_g \in D(A)$, and that u_g satisfies the equation

(3.5) $$u - \lambda Au = g.$$

(See also [3], Th. 13.1).

LEMMA 7. The mapping $I - \lambda A : D(A) \to R(I- A)$ is injective if and only if condition (K) holds. (I is the identity operator.)

PROOF. If condition (K) is satisfied it follows from Lemma 6 that $M_1 = M_2 = \infty$. Then the only bounded solution of (3.1) is $u = 0$. Thus $I - \lambda A$ is injective.
Conversely, if $I - \lambda A$ is injective, then u_g is the unique solution of (3.5). However, if $M_i < \infty$, $i = 1$ or 2, then for each constant C also $u_g + Cu_i$ should be an element of D(A) by the fact that $\lambda Au_i = u_i \in C(\bar{I})$. This is a contradiction, thus $M_i = \infty$, and, whith lemma 6, condition (K) holds. □

PROOF OF THEOREM 2.

(i) *D(A) is dense in $C(\bar{I})$.* In [1, prop.1] it is proved that $D_0(A)$ (see (1.2)) is dense in $C(\bar{I})$. Since $D_0(A) \subseteq D(A)$ for D(A) the same holds.

(ii) *A is closed.* Let $(u_n) \subset D(A)$, u and ν in $C(\bar{I})$ be such that $\lim_{n\to\infty} \| u_n - u \| = 0$ and $\lim_{n\to\infty} \| Au_n - \nu \| = 0$. We have to show that $u \in D(A)$ and $Au = \nu$. Let us denote Au_n by ν_n. For every $a,b \subset \mathbb{R}$ such that $[a,b] \subset I$ there is a constant $c > 0$ such that $\alpha(x) > c$ for all $x \in [a,b]$. Therefore the restriction on [a,b] of $\alpha D^2u + \beta Du$ is a regular Sturm-Liouville operator [2]. Since $u_n(a)$ (resp. $u_n(b)$) converges to u(a) (resp. u(b)), it follows from the classical theory [2] that u_n' and u_n'' are Cauchy sequences in C[a,b], and therefore that $u \in C^2[a,b]$, and $\alpha u'' + \beta u' = \nu$ on [a,b]. Since a and b are arbitrary, $u \in C^2(I)$ and $\alpha u'' + \beta u' = \nu$ on I. Since $\nu \in C(I)$, we see that $\lim_{x\to\partial I}(\alpha u'' + \beta u')(x)$ exists. Thus $u \in D(A)$ and $Au = \nu$.

(iii) *$I - A$ is surjective in $C(\bar{I})$.* This is a direct consequence of (3.4) - (3.5) with $\lambda = 1$. □

For the proof of Theorem 3 we need the following lemmata.

LEMMA 8. If $\lambda > 0$ and $u \in D(A)$, then $u - \lambda Au \geq 0$ implies $u \geq 0$ if and only if condition (K) is satisfied.

PROOF. If $R \notin L^1(r_1,x_0)$ and $R \notin L^1(x_0,r_2)$, then $I - \lambda A$ is injective and since $G > 0$ it follows from (3.4) that $g := u - \lambda Au \geq 0$ implies $u = u_g \geq 0$.

On the other hand, if $R \in L^1(r_1,x_0)$ or $R \in L^1(x_0,r_2)$ then $M_1 < \infty$ or $M_2 < \infty$. Say $M_1 < \infty$. Let $\bar{u} \in D(A)$ be such that $\bar{u} \geq 0$ and $\bar{u} - A\bar{u} \geq 0$. Let $C > (\lambda M_1)^{-1} . \lim_{x \to r_2} \bar{u}(x)$ and let $\underline{u} = \bar{u} - Cu_1$. Then $\underline{u} - \lambda A\underline{u} \geq 0$, however *not* $u \geq 0$.

LEMMA 9. A is dissipative if and only if condition (K) holds.

PROOF. From lemma 7 we know that condition (K) holds if and only if $I - \lambda A$ is injective. Assume $I - \lambda A$ is injective. We define $J_\lambda = (I-\lambda A)^{-1}$. Then J_λ is a positive linear operator ($u \geq 0 \Rightarrow J_\lambda u \geq 0$, lemma 8) on $R(I-\lambda A)$, and $J_\lambda e_0 = e_0 \in D(A)$, where e_0 is the constant function $e_0(x) = 1$. Then it follows that $\| J_\lambda g \| \leq \| g \|$ for all $g \in R(I-\lambda A)$, or

(3.6) $$\| u \| \leq \| u - Au \| \quad \text{for all } u \in D(A),$$

thus A is dissipative.
Conversely, assume A is dissipative, then it follows from (3.6) that $u - \lambda Au = 0$ implies $u = 0$, thus $I - \lambda A$ is injective. □

PROOF OF THEOREM 3.

It is known [4,Th. 3.1] that a necessary and sufficient condition for a closed operator A, with dense domain and for which $R(I-A) = C(\bar{I})$ to generate a strongly continuous semigroup of contraction operators, is that A is dissipative. Then Theorem 3 is a direct consequence of Theorem 2 (i), (ii), (iii) and lemma 9. □

EXAMPLES.

1. Let A be defined as in Theorem 2 with $I = \mathbb{R}$, $x_0 = 0$, $\alpha(x) = (1+x^2)^2$

and $\beta(x) = 0$. Then $W(x) = 1$, $(\alpha W)^{-1}(x) = (1+x^2)^{-2}$ and $R(x)= \frac{1}{2}$ arctan x $+ \frac{1}{2}\frac{x}{1+x^2}$. Thus condition (K) is satisfied and A is the infinitesimal generator of a C_0-semigroup on $C([-\infty,\infty])$.

2. Let A be defined as in Theorem 2 with $I = (-\frac{\pi}{2},\frac{\pi}{2})$, $x_0 = 0$, $\alpha(x) = 1$, $\beta(x) = -2\tan x$. Then $W(x) = (\cos x)^{-2}$, $(\alpha W)^{-1}(x) = \cos^2 x$, $R(x) = \frac{1}{2}\tan x + \frac{x}{2\cos^2 x}$. Thus condition (K) is satisfies and A is the infinitesimal generator of a C_0-semigroup on $C([-\frac{\pi}{2},\frac{\pi}{2}])$.
Remark. This Example can be obtained from example 1 by means of the diffeomorphism $\phi : (-\infty,\infty) \to (-\frac{\pi}{2},\frac{\pi}{2})$ with $\phi(x) =$ arctan x.

3. Let A be defined as in Theorem 2 with $I = (-1,1)$, $x_0 = 0$, $\alpha(x) = 1$ for $x \in I$,

$$\beta(x) = \frac{2p}{1-x^2}, \quad x \geq 0$$
$$= \frac{-2p}{1-x^2}, \quad x < 0$$

where $p > 1$. Then $W(x) = (\frac{1-x}{1+x})^p$ for $x \geq 0$, and $W(x) = (\frac{1+x}{1-x})^p$ for $x < 0$. If is easily verified that condition (K) holds. Thus A is the infinitesimal generator of a C_0-semigroup on $C([-1,1])$.

REMARK

Let α and β be such that condition (K) is satisfied. Then if moreover condition (H_i), $i = 0$ or 2 is satisfied, then one can prove that $\lim_{x\to r_i} Au(x) = 0$. In this case the boundary point r_i is called natural (see Feller [3]). In the case that both boundary points are natural Theorems 1 and 3 are both applicable.

Example.

$I = \mathbb{R}$, $x_0 = 0$, $\alpha(x) = 1$, $\beta(x) = 0$. Then $W(x) = 1$ $(\alpha W)^{-1}(x) = 1$, $Q(x) = R(x) = x$.

$D(A_0) = \{u \in C([-\infty,\infty]) \mid u \in C^2(-\infty,\infty),\ u'' \in C([-\infty,\infty])\}$,

$D(A_1) = \{u \in C([-\infty,\infty]) \mid u \in C^2(-\infty,\infty) \lim_{x\to\pm\infty} u''(x) = 0\}$.

Then $D(A_0) = D(A_1)$, and $A_0 = A_1$ is the infinitesimal generator of a C_0-semigroup in $C([-\infty,\infty])$, the so called Gauss-Weierstrass semigroup.

Acknowledgement.

The author would like to thank Ph.Clément for his valuable comments.

REFERENCES

1. Clément, Ph. and C.A. Timmermans, On C_0-semigroups generated by differential operators satisfying Ventcel's boundary conditions.

2. Coddington, E.A. and N.Levinson, Theory of Ordinary Differential Equations, Mc Graw Hill, N.Y.1955.

3. Feller, W., The parabolic differential equations and the associated semigroups of transformations, Ann. of Math.55,468-519 (1952).

4. Lumer, G. and R.S.Phillips, Dissipative operators in a Banach space, pacific. J.Math., 11, 679-698 (1961).

5. Martini, R., A relation between semigroups and sequences of approximation operators, Indag. Math.,35, 456-465 (1973).

6. Protter, M. and H.Weinberger, Maximum Principles in Partial Differential Equations, Prentice-Hall, Engelewood Cliffs, New-Jersey (1967).

7. Ventcel's, A.D., On boundary conditions for multidimensional diffusion processes, Th. of Prob.Appl., 4, 164-177 (1959).

ON THE APPROXIMATION OF RIEMANN INTEGRABLE FUNCTIONS BY BERNSTEIN POLYNOMIALS

E. van Wickeren

Aachen University of Technology

Abstract. This paper is concerned with the approximation of Riemann integrable functions by Bernstein polynomials. In connections with the (sequential) Riemann convergence, introduced recently, a direct theorem is established in terms of a weighted τ-modulus. Then the sharpness of this result as well as of further approximation theorems, well-known in this context, is discussed with the aid of our previous quantitative extensions of the uniform boundedness principle.

Let $C = C[0,1]$ and $R = R[0,1]$ be the space of functions, continuous and Riemann integrable on $[0,1]$, respectively. These spaces are Banach spaces under the sup-norm $\|\cdot\|_\infty$. The Riemann integral of $f \in R$ will be abbreviated by $\int f$. Denote the Bernstein polynomial of $f \in R$ by

$$B_n f := \sum_{k=o}^{n} f(k/n) p_{kn}, \qquad p_{kn}(x) := \binom{n}{k} x^k (1-x)^{n-k}.$$

In [1] a quantitative result was established to measure the approximation rate of $B_n f$ to $f \in R$ in the L^p-metric $\|f\|_p := (\int |f|^p)^{1/p}$, $1 \leqslant p < \infty$. Indeed,

$$\|B_n f - f\|_p \leqslant M\tau_2(f, n^{-1/2})_p \qquad (f \in R), \tag{1}$$

where the τ-modulus is defined by

$$\tau_2(f,\delta)_p := \|\omega_2(f,\cdot,\delta)\|_p, \qquad \omega_2(f,x,\delta) := \sup\{|\Delta_h^2 f(t)| : t \pm h \in U(x,\delta),\ h \geqslant 0\},$$

$$\Delta_h^2 f(t) := f(t-h) - 2f(t) + f(t+h), \ U(x,\delta) := [x-\delta, x+\delta] \cap [0,1].$$

An improvement of (1) was given in [11] in terms of the weighted modulus

$$\tau_2^*(f,\delta)_p := \|\omega_2^*(f,\cdot,\delta)\|_p, \qquad \varphi(x) := x(1-x), \qquad \rho(x,\delta) := \delta\varphi^{1/2}(x) + \delta^2/2,$$

$$\omega_2^*(f,x,\delta) := \sup\{|\Delta_h^2 f(t)| : t \pm h \in U(x,\rho(x,\delta)), h \geq 0\}$$

(note that $\omega_2(f,x,\delta)$ as well as $\omega_2^*(f,x,\delta)$ are Lebesgue measurable (cf. [14, Theorem 1.3]). In fact,

$$\|B_n f - f\|_p \leq M\tau_2^*(f,n^{-1/2})_p \qquad (f \in R) . \tag{2}$$

The sharpness of (1), thus in particular of (2), can be shown by an application of (the general) Theorem 5 in connection with the main result in [3,4]. To this end, let $\omega(\delta)$ be an abstract modulus of continuity, i.e., $\omega(\delta)$ is continuous and increasing in $\delta > 0$ satisfying $(0 < \delta_1 < \delta_2)$

$$\omega(\delta_2)/\delta_2 \leq \omega(\delta_1)/\delta_1, \qquad \omega(\delta) = o(1), \qquad \delta = o(\omega(\delta)) \tag{3}$$

(all Landau symbols are to be understood for $\delta \to 0+$ and $n \to \infty$, respectively).

<u>*Theorem 1.*</u> *For each* ω *subject to (3) there exists (even) a function* $f_\omega \in C$ *such that* $(1 \leq p < \infty)$

$$\tau_2(f_\omega,\delta)_p = O(\omega(\delta^2)) , \tag{4}$$

$$\|B_n f_\omega - f_\omega\|_p \neq o(\omega(1/n)) . \tag{5}$$

From a functional analytic point of view, however, the convergence in the L^p-metric (cf. (1,2))is somewhat unnatural for R since B_n is an unbounded operator on R (even on C) with respect to this metric. To avoid this problem, let us consider the so-called Riemann convergence, recently introduced in [7] (see also [16]) even on the larger space $B = B[0,1]$ of bounded functions. This sequential convergence is given by the following definition: A sequence $\{f_n\} \subset B$ is called Riemann convergent to $f \in B$ if (with upper Riemann integral $\overline{\int}$)

$$\text{(i)} \quad \|f_n\|_\infty = O(1), \qquad \text{(ii)} \quad \overline{\int} \sup_{k \geq n} |f_k - f| = o(1) .$$

Under this convergence $R \subset B$ is the completion of C or, in other words, R is not only complete, but C is dense in R. This is the main reason that also in R an appropriate Banach - Steinhaus theorem can be established, fundamental in approximation theory (see [7]).

Now the Bernstein polynomials form indeed a Riemann approximation process, i.e., $B_n f$ is Riemann convergent to $f \in R$, which is an immediate consequence of a Bohman - Korovkin theorem given in [13]. In fact,

Theorem 2. For $f \in R$ *and* $n \in \mathbb{N}$ *one has*

$$\overline{\int} \sup_{k \geq n} |B_k f - f| \leq M\tau_2^*(f, n^{-1/2})_1 . \tag{6}$$

That this is an improvement of (2) (for $p = 1$) may be seen by

Theorem 3. Let ω *subject to (3) be such that* $\delta^{1/2} = o(\omega(\delta))$. *Then there exists* $f_\omega \in C$ *satisfying (4,5) for* $p = 1$ *and simultaneously*

$$\overline{\int} \sup_{k \geq n} |B_k f_\omega - f_\omega| \neq O(\| B_n f_\omega - f_\omega \|_1) . \tag{7}$$

Concerning inverse theorems it was recently shown in [12] that for $f \in C$

$$\| B_n f - f \|_p = O(n^{-\theta}) \quad \Rightarrow \quad \tau_2^*(f, \delta)_p = O(\delta^{2\theta}) , \tag{8}$$

provided $1 < p < \infty$, $1/p < \theta < 1$. On the larger space R, however, this cannot hold true so that inverse results to (2) (even to (6)) are impossible.

Theorem 4. Let $\varepsilon(\delta)$ *be a positive function of* $\delta > 0$ *with* $\varepsilon(\delta) = o(1)$. *Then there exists* $f_\varepsilon \in R$ *such that*

$$\| B_n f_\varepsilon - f_\varepsilon \|_p = 0, \qquad \overline{\int} \sup_{k \geq n} |B_k f_\varepsilon - f_\varepsilon| = 0 \qquad (n \in \mathbb{N},\ p \geq 1) , \tag{9}$$

$$\tau_2^*(f, \delta)_p \neq o(\varepsilon(\delta)) \qquad (p \geq 1) . \tag{10}$$

PROOFS

To prove Theorem 1,3,4 let us recall the quantitative extension of the uniform boundedness principle to be applied here (see [6,15] and the literature cited there). Let X be a Banach space with norm $\| \cdot \|_X$ and X* be the space of sublinear, bounded functionals F on X, i.e. ($f, g \in X$, α scalar),

$$|F(f+g)| \leq |Ff| + |Fg|, \qquad |F(\alpha f)| = |\alpha| |Ff| ,$$

$\sup\{|Ff| : f \in X, \|f\|_X \leq 1\} < \infty$.

<u>*Theorem 5.*</u> *Let* $\{\psi_n\}$ *be a positive, decreasing nullsequence , let* $\sigma(\delta) > 0$, *and let* ω *be subject to (3). Suppose that for* $S_\delta, T_n, R_n, V_n \in X^*$ *there are elements* $h_n \in X$ *satisfying*

$$\|h_n\|_X = O(1) , \tag{11}$$

$$|S_\delta h_n| \leq M \min\{1, \sigma(\delta)/\psi_n\} \qquad (\delta > 0,\ n \in \mathbb{N}) , \tag{12}$$

$$T_n h_n \neq o(1) , \tag{13}$$

$$R_n h_n \neq o(1) , \tag{14}$$

$$V_n h_n = O(1) , \tag{15}$$

$$V_n h_j = o(\omega(\psi_n)) \qquad (j \in \mathbb{N}). \tag{16}$$

Then there exists a counterexample $f_\omega \in X$ *with*

$$S_\delta f_\omega = O(\omega(\sigma(\delta))) , \tag{17}$$

$$T_n f_\omega \neq o(\omega(\psi_n)) , \tag{18}$$

$$R_n f_\omega \neq o(V_n f_\omega) . \tag{19}$$

Returning to Bernstein polynomials, the proofs of Theorem 1,3 heavily depend on the result of [3,4] which states that for each $0 < \beta \leq 1$, $0 \leq \alpha \leq \beta$ there exists $f_{\alpha\beta} \in C$ satisfying

$$|\Delta_h^2 f_{\alpha\beta}(x)|/\varphi^\alpha(x) \leq M(h^2/\varphi(x))^\beta \qquad (0 < h \leq x \leq 1 - h) ,$$

but nevertheless

$$\limsup_{n\to\infty} n^\beta \ |B_n f_{\alpha\beta}(x) - f_{\alpha\beta}(x)| \geq \varphi^\alpha(x) ,$$

simultaneously for almost all $x \in (0,1)$. This result shows the sharpness of a direct theorem given in [2,8] and follows as an application of an extended version of

Theorem 5 (see [5]).

<u>Proof of Theorem 1.</u> To apply Theorem 5 let $X = C$ (with norm $\|\cdot\|_\infty$) and set $S_\delta f = \tau_2(f,\delta)_p$, $T_n f = \|B_n f - f\|_p$ (conditions (14-16) as well as (19) are irrelevant here, therefore set, e.g., $R_n = T_n$, $V_n = 0$). Let us recall that in [3,4] elements $g_n \in C$ were constructed such that $(0 \leq \alpha \leq 1, c_o, c_1 > 0)$

$$\|g_n\|_\infty = O(1), \tag{20}$$

$$|\Delta_h^2 g_n(t)| \leq M \min\{1, nh^2\varphi^{\alpha-1}(t)\} \qquad (h \leq t \leq 1-h), \tag{21}$$

$$|B_n g_n(x) - g_n(x)| \geq c_o \varphi^\alpha(x) \tag{22}$$

for x of some subset $\Lambda_n \subset [0,1]$ with (meas = Lebesgue measure)

$$\liminf_{n\to\infty} \operatorname{meas}(\Lambda_n \cap (a,b)) \geq c_1(b-a) \tag{23}$$

for each subinterval $(a,b) \subset [0,1]$. Now fix $\alpha \in (1-1/p, 1)$. Setting $h_n = g_n$, obviously (11) is satisfied. By (21), $\omega_2(g_n, x, \delta) \leq M$, thus $S_\delta g_n \leq M$. Moreover, if $\varphi(x) \geq 2\delta$ it follows that $\varphi(x) \leq \varphi(t) + |t-x| \leq \varphi(t) + \delta$, thus $\varphi(t) \geq \varphi(x)/2$ for $t \in U(x,\delta)$. Now $t \pm h \in U(x,\delta)$ implies $t \in U(x,\delta)$ and $h \leq \delta$ so that by (21)

$$|\Delta_h^2 g_n(t)| \leq Mn\delta^2\varphi^{\alpha-1}(x), \qquad \int_{\varphi(x)\geq 2\delta} \omega_2^p(g_n,x,\delta)\,dx \leq M(n\delta^2)^p$$

since $\|\varphi^{\alpha-1}\|_p < \infty$. On the other hand, if $\varphi(x) \leq 2\delta (\leq 1)$ and $t \pm h \in U(x,\delta)$, then $h \leq t \leq 1-h$, thus $\varphi(t) \geq \varphi(h) \geq h/2$. Hence

$$\int_{\varphi(x)\leq 2\delta} \omega_2^p(g_n,x,\delta)\,dx \leq M(n\delta^{\alpha+1})^p\delta \leq M(n\delta^2)^p$$

which yields (12) with $\sigma(\delta) = \delta^2$, $\psi_n = 1/n$. To establish (13) apply (22,23) to derive

$$T_n g_n \geq \int_{\Lambda_n\cap(1/4,3/4)} |B_n g_n(x) - g_n(x)|\,dx \geq c_o\varphi^\alpha(1/4)\operatorname{meas}(\Lambda_n \cap (1/4,3/4)) \neq o(1). \qquad \square$$

<u>Proof of Theorem 3.</u> Since $n^{-1/2} = o(\omega(1/n))$ there exists $m_n \in \mathbb{N}$, tending to infinity, such that

$$m_n = o(n^{1/2}), \qquad 1/m_n = o(\omega(1/n)) \tag{24}$$

(e.g., the integral part of $(n^{1/2}/\omega(1/n))^{1/2}$). In the setting of the first proof consider additionally

$$R_n f = \begin{cases} \overline{\int} \sup_{k \geq m_n} |B_k f - f| , & n \text{ even} \\ 0 , & n \text{ odd} \end{cases} , \qquad V_n f = \begin{cases} A_n \| B_{m_n} f - f \|_1 , & n \text{ even} \\ \omega^2(1/n) \|f\|_\infty , & n \text{ odd} \end{cases}$$

where A_n tends to infinity with

$$A_n = o(m_n \omega(1/n)), \qquad A_n = o(n/m_n^2) . \tag{25}$$

Let H be a function, infinitely differentiable on the real axis such that $0 \leq H(x) \leq 1$, $H(0) = 1$, and $H(x) = 0$ if $|x| \geq 1$. Setting

$$f_n(x) = \sum_{k=1}^{m_n} H\left(\frac{n}{m_n}\left(x - \frac{2k-1}{2m_n}\right)\right)$$

one has $\|f_n\|_\infty = 1$ and $\int f_n = (m_n^2/n) \int H = o(1)$ (note that the terms in the sum have disjoint support if n is large enough (see (24)). Since $f_n(j/2m_n) = 0$ if j even, and $= 1$ else, one has $B_{m_n} f_n = 0$, $\int B_{2m_n} f_n = m_n/(2m_n+1)$ so that for even n (cf. (25))

$$R_n f_n \geq \int B_{2m_n} f_n - \int f_n \neq o(1) , \qquad V_n f_n = A_n \int f_n = O(A_n m_n^2/n) = o(1) ,$$

$$S_\delta f_n \leq M \delta^2 \int |f_n''| = M \delta^2 n \int |H''| .$$

Altogether, it follows that for $h_n = f_n$ if n even, and $= g_n$ if n odd, conditions (11-15) are fulfilled. In view of (1.12,25) also (16) follows since

$$V_n h_j = \left. \begin{cases} O(A_n/m_n) , & n \text{ even} \\ O(\omega^2(1/n)) , & n \text{ odd} \end{cases} \right\} = o(\omega(1/n)) .$$

Hence Theorem 5 establishes (4,5) for some $f_\omega \in C$, in particular $f_\omega \neq 0$. But then (19) can only hold for even n which yields (7), too. □

Proof of Theorem 4. Consider the set $I_n := \{x \in [0,1] : x = \pi - k/n \text{ for } k \in \mathbb{N}\}$ of irrational numbers and set $h_n(x) = 1$ if $x \in I_n$, and $= 0$ else. Moreover, set

$$X = \{f \in R : \| B_n f - f \|_p = 0 \,, \quad \overline{\int} \sup_{k \geq n} |B_k f - f| = 0 \quad \text{for } n \in \mathbb{N},\ p \geq 1\}$$

which is a Banach space with respect to $\|\cdot\|_\infty$. Obviously, $B_k h_n = 0$ for all $k \in \mathbb{N}$, thus $h_n \in X$ satisfying (11). It is enough to consider $T_n f = \tau_2^*(f,\delta_n)_1$, $\delta_n := 2n^{-1/2}$ ($S_\delta = 0 = V_n$, $R_n = T_n$ are irrelevant here). Now if $x \in [0,1]$ there exists $k \in \mathbb{N}$ such that $\pi - k/n \leq x \leq \pi - (k-1)/n$. Without loss of generality assume $t := \pi - k/n \in (0,1)$. Then there exists $h \in (0,1/2n)$ with $t \pm h \in [0,1]$ which implies

$$|(t \pm h) - x| \leq 3/2n \leq \rho(x,\delta_n) \,, \qquad \omega_2^*(h_n,x,\delta_n) \geq |\Delta_h^2 h_n(t)| = 2,\ T_n h_n \geq 2 \,.$$

With $\psi_n = \varepsilon^2(\delta_n)$, $\omega(t) = t^{1/2}$ the assertion follows. □

Concerning the proof of Theorem 2 define for $f \in R$

$$\sigma(f,x,\delta) := \sup\{|f(y)| : y \in U(x,\rho(x,\delta))\} \,.$$

Lemma 6. *There exist kernels* K_i *such that for* $f \in R$

$$|f(u)| \leq \sigma(f,x,\delta) + M \int \sigma(f,t,\delta)[(u-x)^2\, K_1(x,t,\delta) + (u-x)^4\, K_2(x,t,\delta)]\, dt \,, \tag{26}$$

$$\int [\rho^2(x,\delta)\, K_1(x,t,\delta) + \rho^4(x,\delta)\, K_2(x,t,\delta)]\, dx \leq M \,, \tag{27}$$

uniformly for $\delta > 0$ *and* $u,x,t \in [0,1]$.

Proof. First of all, let us recall that there exists $a \geq 1$ such that for $x,y \in [0,1]$ with $|x-y| \leq \rho(x,\delta)$ (cf. [9])

$$\rho(x,\delta)/a \leq \rho(y,\delta) \leq a\, \rho(x,\delta) \,. \tag{28}$$

Obviously, (26) is trivial if $y \in U(x,\rho(x,\delta))$. Otherwise, i.e., if $|x-y| \geq \rho(x,\delta)$, consider the case $a^2\rho(x,\delta) \leq 2\rho(u,\delta)$. Then $|t-u| \leq \rho(x,\delta)/2$ implies $|t-x| \geq \rho(x,\delta)/2$ and, moreover, $|t-u| \leq \rho(u,\delta)/a^2 \leq \rho(u,\delta/a)$, thus $|t-u| \leq a\rho(t,\delta/a) \leq \rho(t,\delta)$ by (28). Setting

$$K_1(x,t,\delta) := \begin{cases} [\rho(x,\delta)(t-x)^2]^{-1} \,, & |t-x| \geq \rho(x,\delta)/2 \\ 0 \,, & \text{else} \end{cases}$$

(26) follows since

$$\int \sigma(f,t,\delta)\, K_1(x,t,\delta)\, dt \geq \int_{|t-u|\leq\rho(x,\delta)/2} \sigma(f,t,\delta)[\rho(x,\delta)(t-x)^2]^{-1}\, dt$$

$$\geq |f(u)|/2(u-x)^2 .$$

On the other hand, if $a^2\rho(x,\delta) > 2\rho(u,\delta)$ and $|t-u| \leq \rho(u,\delta)/a^2$, then $|t-x| \geq \rho(x,\delta)/2$. Again, $|t-u| \leq \rho(t,\delta) \leq a\rho(u,\delta)$ by (28) so that with

$$K_2(x,t,\delta) := \begin{cases} [\rho(t,\delta)(t-x)^4]^{-1} , & |t-x| \geq \rho(x,\delta)/2 \\ 0 , & \text{else} \end{cases}$$

one obtains (26) in view of

$$\int \sigma(f,t,\delta)\, K_2(x,t,\delta)\, dt \geq |f(u)| \int_{|t-u|\leq\rho(u,\delta)/a^2} [a\rho(u,\delta)(t-x)^4]^{-1}\, dt$$

$$\geq |f(u)|/a^3(u-x)^4 .$$

Concerning (27) substitute $v = |t-x|$. Then

$$|\rho(x,\delta) - \rho(t,\delta)| \leq \delta v^{1/2}, \qquad \rho^4(x,\delta) \leq M(\rho^4(t,\delta) + \delta^4 v^2) ,$$

which implies that for $v \geq \rho(x,\delta)/2$ (note that $\rho(t,\delta) \geq \delta^2/2$)

$$2v \geq \rho(t,\delta) - \delta v^{1/2} \geq \rho(t,\delta) - (2v\, \rho(t,\delta))^{1/2} ,$$

thus $v \geq \rho(t,\delta)/8$. Therefore

$$\int [\rho^2(x,\delta)\, K_1(x,t,\delta) + \rho^4(x,\delta)\, K_2(x,t,\delta)]\, dx$$

$$\leq 2 \int_{\rho(t,\delta)/8}^{1} [(\rho(t,\delta)+\delta v^{1/2})\, v^{-2} + M(\rho^4(t,\delta) + \delta^4 v^2)(\rho(t,\delta)v^4)^{-1}\, dv] \leq M . \qquad \square$$

<u>Proof of Theorem 2.</u> It is enough to show that ($\delta := n^{-1/2}/b, b := a+1$)

$$\bar{\int} \sup_{k\geq n} |B_k f| \leq M \int \sigma(f,x,\delta)\, dx \qquad (f \in R), \tag{29}$$

$$\overline{\int} \sup_{k \geq n} |B_k g - g| \leq n^{-1} \int \varphi(x)|g''(x)|\, dx , \tag{30}$$

where $g \in AC^2 = AC^2[0,1]$, the space of continuously differentiable functions g such that g' is absolutely continuous on [0,1]. In fact, in view of [10, Theorem 3.1] there exists $g_n \in AC^2$ with

$$|f(x) - g_n(x)| \leq M\omega_2^*(f,x,\delta) \qquad (x \in [0,1]),$$

$$n^{-1} \int \varphi(x)\ |g_n(x)|\, dx \leq M\tau_2^*(f,n^{-1/2})_1 .$$

Since $\omega_2^*(f,y,\delta) \leq \omega_2^*(f,x,n^{-1/2})$ for $y \in U(x,\rho(x,\delta))$, if follows by (29, 30) that

$$\overline{\int} \sup_{k\geq n} |B_k f-f| \leq \overline{\int} \sup_{k\geq n} |B_k(f-g_n)| + \int |f-g_n| + \overline{\int} \sup_{k\geq n} |B_k g_n - g_n| \leq M\tau_2^*(f,n^{-1/2})_1 ,$$

thus the assertion. To establish (29) it is well-known that $(\psi_x(u) := (u-x)^2, k \geq n)$

$$B_k\ \psi_x(x) \leq \varphi(x)/n \leq M\rho^2(x,\delta), \qquad B_k\ \psi_x^2(x) \leq 3\varphi^2(x)/n^2 + \varphi(x)/n^3 \leq M\rho^4(x,\delta) .$$

Then apply (26) to $B_k f$ which yields

$$\overline{\int} \sup_{k\geq n} |B_k f(x)|\, dx \leq \int \sigma(f,x,\delta)\, dx + M \iint \sigma(f,t,\delta)[\rho^2(x,\delta)K_1 + \rho^4(x,\delta)K_2]\, dt\, dx .$$

Interchanging the order of integration (29) follows in view of (27). Concerning (30) there exists a kernel $S_u(x)$, convex in x, such that (cf. [2, 11])

$$B_k g(x) - g(x) = \int (B_k\ S_u(x) - S_u(x)) g''(u)\, du \qquad (x \in [0,1]).$$

The convexity implies that $S_u \leq B_{k+1}\ S_u \leq B_k\ S_u$ so that (cf. [11])

$$\int \sup_{k\geq n} |B_k g - g| \leq \iint [B_n S_u(x) - S_u(x)]|g''(u)|\, du\, dx \leq n^{-1} \int \varphi(x)|g''(x)|\, dx ,$$

and the proof is finished. □

Acknowledgements. The author would like to express his sincere gratitude to Herbert Mevissen and Rolf J. Nessel for their critical reading of the manuscript and many valuable suggestions. He also thanks Kamen G. Ivanov for pointing out a basic argument in the proof of Theorem 2.

REFERENCES

[1] A.S. Andreev, V.A. Popov: Approximation of functions by means of linear summation operators in L_p. In: Functions, Series, Operators. Vol. I (Proc. Conf. Budapest 1980, Eds.: B. Sz.-Nagy, J. Szabados). North-Holland, Amsterdam 1983, 127 - 150.

[2] H. Berens, G.G. Lorentz: Inverse theorems for Bernstein polynomials. Indiana Univ. Math. J. 21 (1972), 693 - 708.

[3] W. Dickmeis: Ein quantitatives Resonanzprinzip und Schärfe von punktweisen Fehlerabschätzungen fast überall. Habilitationsschrift, RWTH Aachen 1983.

[4] W. Dickmeis: On quantitative condensation of singularities on sets of full measure. (to appear).

[5] W. Dickmeis, R.J. Nessel, E. van Wickeren: On nonlinear condensation principles with rates. Manuscripta Math. 52 (1985), 1 - 20.

[6] W. Dickmeis, R.J. Nessel, E. van Wickeren: Quantitative extensions of the uniform boundedness principle. Jahresber. Deutsch. Math.-Verein. 89 (1987) (in print).

[7] W. Dickmeis, H. Mevissen, R.J. Nessel, E. van Wickeren: Sequential convergence and approximation in the space of Riemann integrable functions. J. Approx. Theory (in print).

[8] Z. Ditzian: Interpolation theorems and the rate of convergence for Bernstein polynomials. In: Approximation Theory III (Proc. Conf. Austin 1980, Ed.: E.W. Cheney). Academic Press, New York 1980, 341 - 348.

[9] K.G. Ivanov: On a new characteristic of functions, I. Serdica 8 (1982), 262-279.

[10] K.G. Ivanov: A constructive characteristic of the best algebraic approximation in $L_p[-1,1]$ $(1 \leq p \leq \infty)$. In: Constructive Function Theory '81 (Proc. Conf. Varna 1981, Eds. Bl. Sendov et al.). Publ. House Bulg. Acad. Sci., Sofia 1983, 357 - 367.

[11] K.G. Ivanov: Approximation by Bernstein polynomials in L_p metric. In: Constructive Theory of Functions (Proc. Conf. Varna 1984, Eds. Bl. Sendov et al.). Publ. House Bulg. Acad. Sci., Sofia 1984, 421 - 429.

[12] K.G. Ivanov: Converse theorems for approximation by Bernstein polynomials in $L_p[0,1]$ $(1 < p < \infty)$. Constr. Approx. 2 (1986), 377 - 392.

[13] H. Mevissen, R.J. Nessel, E. van Wickeren: On the Riemann convergence of positive linear operators. In: Proc. Constructive Function Theory - 86, Edmonton (to appear).

[14] Bl. Sendov, V.A. Popov: Averaged Moduli of Smoothness (Bulgarian). Publ. House Bulg. Acad. Sci., Sofia 1983.

[15] E. van Wickeren: A Baire approach to quantitative resonance principles. Numer. Funct. Anal. Optim. 9 (1987), 147 - 180.

[16] E. van Wickeren: Sequential convergence in the space of functions Riemann integrable on arbitrary sets. (to appear).

OPTIMIZATION CRITERIA FOR MULTIVARIATE STRATA CONSTRUCTION

SIRA ALLENDE AND CARLOS BOUZA
DEPARTAMENTO DE MATEMATICA APLICADA
FACULTAD DE MATEMATICA CIBERNETICA
UNIVERSIDAD DE LA HABANA

1. INTRODUCTION

In practice a common problem is to estimate parameters when a finite population $I=\{1,...,M\}$ is studied. A subset s of I is named sample. A sample design is a probability measure that assigns, to any s, a probability of being selected.

Frequently an unbiased estimator of a parametric function θ is constructed and its accuracy is measured by its variance. Therefore an important task in sampler's work is to maintain the variance as low as possible for a fixed sample size m.

Stratified random sampling was studied by Neymann. This design fixes that I is divided into P nonoverlapping populations I_t (stratum t) of size M_t and θ is given by a linear function

$$\theta=\sum_{t=1}^{P} W_t\theta_t;\ W_t=M_t/M$$

where θ_t is the parametric function θ evaluated in I_t. This design is commonly used because it enables the sampler with the possibility of disminishing the variance in different ways. One question is "how should the strata be constructed in order to reduce the variance?" Dalenius (1950) tackled this problem and different statisticians have worked with it since then. Common assumptions are that only one variable Y is measured in each $i\in I$,say y_i, a density function $f(y)$ is known and y_i belongs to a fixed interval $[b_o,b_P]$. Then the problem of Optimum Stratification (OS) is to obtain a set of P-1 real numbers $b_1<b_2<...<b_{P-1}$ such that $b_1>b_o$ and $b_{P-1}<b_P$ and the variance

$$V(\hat{\theta})=\sum_{t=1}^{P} W_t^2 V(\hat{\theta}_t)$$

is minimum.

Take E as the mathematical expectation operator and V as the variance. The solution of a system of equations that depends of the expectation and variances of the P strata, permits to obtain the bounds b_t,if θ is the population mean. Say, the solution of the following equation system

$$\frac{[b_t-E(\hat{\theta}_t)]^2+V(\hat{\theta}_t)}{[V(\hat{\theta}_t)]^{1/2}}=\frac{(b_t-E(\hat{\theta}_{t+1})]^2+V(\hat{\theta}_{t+1})}{[V(\hat{\theta}_{t+1})]^{1/2}}\qquad t=1,...,P-1$$

fixes OS.

As the involved parameters are calculable if the strata are fixed only approximate solutions may be obtained by the use of numerical iterative procedures. See Cochran (1981) for further details.

Common survey practice deals with several variables $Z^1,\ldots,Z^k$. Therefore the population information is a set of k points of the M-dimensional Euclidean space. A similar analysis may be held with respect to the strata information. Now the variance of each $Z^h, h=1,\ldots,k$, should be minimized by the **OS**. Then a simultaneous minimization should be conducted. This is a multivariate extension of the usual **OS** problem. Its analytical characterization has not been reported but the optimization approach to it may be fixed as

$$\operatorname{Min} V(\hat{\theta}^h) = \operatorname{Min} \sum_{t=1}^{P} W_t^2 V(\hat{\theta}_t)$$

$$\text{Subject to: } E(\hat{\theta}_t)=\theta_t$$

where $z_i=(z_i,\ldots,z_i)$ is known for any $i\in I$.

This problem may be tackled by the use of optimization techniques. Mulvey (1983) proposed a model for the multivariate **OS**. An algorithm for solving it was developed and its capability for obtaining optimum strata is given.

The present paper also tackles the multivariate **OS**. A thermodynamical approach is used for stratifying and its behaviour is studied through different case-studies. This method is compared with other heuristic one. The gain in accuracy is used for evaluating the proposed approaches to multivariate **OS**.

2. **OS** in the multivariate case

The existence of information related with the interest variables permits to pose the **OS** as a deterministic optimization problem. Mulvey(1983) proposed a model structured in such a way that the use of the subgradient and Lagrangian relaxation methods permits to obtain a solution if $N\leq M$. A sequence of improving bounds of the objective function is generated ad an approximate solution is obtained. The model fixes that

$$d_{ij}=\sum_{h=1}^{k} P_h (z_i^h - z_j^h)^2$$

$$\min \sum_{i\in I} R_i \sum_{j\in J} d_{ij} x_{ij}$$

$$\sum_{j\in J} x_{ij}=1; \text{ for any } i$$

$$\sum_{j\in J} y_j=P$$

$$x_{ij}\leq y_j; \text{ for any } i,j$$

$$x_{ij}\in\{0,1\}, y_j\in\{0,1\}$$

wnere

P_h= weight assigned to Z^h

R_i= weight assigned to i

J= set of possible strata centroids with cardinality N

The structure of the problem permits to solve itby means of the subgradient method and Lagrangian relaxation methods if N≤M. A sequence of improving lower bounds of the objective function is generated and an approximate solution is obtained.

In constrast with this approach we assume that experts fix the set of N possible centroids,J. For each i∈I and j∈J the values of the classification variables are known. Take C_j^h as the value of Z^h in the centroid j. Then the stratification aims are related with the minimization of

$$(2.1)\quad \sum_{i\in I}\ \sum_{j\in J}\ \sum_{h=1}^{h} P_h(z_i^h-C_j^h)^2 x_{ij} = \sum_{i\in I}\ \sum_{j\in J} d'_{ij}x_{ij}$$

and the stratification is expressed by the constraints

$$(2.2)\quad \left|\begin{array}{l} \sum_{j\in J} x_{ij},\ \text{for any } i\in I \\ \sum_{i\in I} x_{ij}\le M y_j,\ \text{for any } j\in J \\ \sum_{j\in J} y_j\le p \\ x_{ij}\in\{0,1\} \\ y_j\in\{0,1\} \end{array}\right.$$

Note that this model has M+N+1 constraints while Mulvey's one has M+NM+1.

In addition we remark that this problem has a trivial solution if we want to obtain

$$\text{Min} \sum_{\substack{i\in I\\ j\in J}} d'_{ij}x_{ij} \quad \left|\begin{array}{l} \sum_{j\in J} x_{ij}=1,\ i\in I \\ x_{ij}\in\{0,1\} \end{array}\right.$$

The combinatorial structure of this model leads to the possibility of combining the solution of this problem and the heuristic selection of P centroids from J.

3. The heuristic methods

Heuristic methods are playing a very important role in the solution of different types of integer programming problems. The common features of this methods are:

(i) To find an adequate initial solution

(ii) To obtain bounds of the objective function or to solve a hard problem

An important class of heuristics is the probabilistic methods and the exchange algorithms. In the last years the thermodynamically motivated method has received special attention. It is a probabilistic method with a good perfomance in applications related with the Salesman Problem, the Quadratic Assignment Problem and others. See for example Burkhard-Rendll (1983) and Kovacs (1980).

To solve the OS model we propose a symbiosis of exact and probabilistic algorithms. Two versions of the exchange algorithm are considered:

(1) A thermodynamic procedure (TH)

(2) An exchange algorithm that exchanges only if it leads to a decrease of the objective funcion value (NT)

Using the notation of Burkhard-Rendll the stratification problem is formally described by a finite set $X=(X_1,\ldots,X_m)$ of feasible solutions and a mapping h that associates to each feasible solution x_j the objective function $h(x_j)$. The goal to be attained is to find the solution of

$$\min_{X_j \in X} h(X_j)$$

In order to describe our algorithm we define the set

$$P=\{\pi\in 2_J \mid \text{Card } \pi=P\}$$

where 2_J denotes the power set of J. The elements of P are the subsets of the possible centroids with cardinality P. It will be named "the configurational set".

The initial solution is constructed in the following way:

(1) An initial configuration π^1 is generated

(2) The relaxed problem

$$\text{Min}\{\sum_{\substack{i\in I\\ j\in J}} d_{ij}x_{ij} \mid \sum_{j\in\pi^1} x_{ij}, i\in I\}$$

is solved.

In general we define, for each $\pi^w \in P$, the problem

$$R_w: \text{Min}\{ \sum_{\substack{i\in I\\ j\in J}} d'_{ij}x_{ij} \mid \sum_{j\in\pi^w} x_{ij}=1, i\in I\}$$

Let X^w be its optimal solution. This procedure restricts us to the set of feasible solutions $x=\{x^1,x^2,\ldots,x^w,\ldots,x^\alpha\}$, where $\alpha=C^N{}_P$.

The exchange method is described by an operator T such that the calculation involved in the generation of the new feasible solution under it, should be simpler than producing a new feasible solution at random. In our case T is described in the following way:

$$T=T_1{}^\circ T_2{}^\circ T_3 \left| \begin{array}{l} T_3: X \text{---}> P,\ T_3(X^w)=\pi^w \\[1em] T_2: P \text{---}> P,\ T_2(\pi^w)=T_2(\pi^w) \\[1em] T_1: P \text{----} x,\ T_1(\pi^v)=X^v \end{array}\right.$$

Note that T_3 and T_1 are described by the old configuration π^v and the new one π^w.

The new configuration π^w is obtained by generating randomly two independent vectors with Bernouilly distributed components. The parameter is common and is equal to $(1-P)^{-1}$.

The direct construction of a new configuration requires of the perfomance of, at least, P random experiments. However T permits to solve the problem but generating only two variables. It is remarkable that the sets π^v and π^w differ exactly in only one element.

In the next step, a new feasible generated solution should be evaluated in order to determine if it is accepted or not.

Take, for $i\in I$

$$\beta(i)=\begin{cases} j \text{ if } x_{ij}(a)=1 \\ 0 \text{ otherwise}\end{cases}$$

then

$$\delta_h = \sum_{i\in I_v(c)} \text{Min}(d'_{i\beta(i)}-d'_{iq},0) - \sum_{i\in I_v} d'_{i\beta(i)+h*}$$

where

$$h^{*}=\mathrm{Min}\{\sum_{j\in\pi^{v}} d'_{ij}x_{ij} \mid \sum_{j\in\pi^{v}} x_{ij}=1, i\in I_{v}\}$$

Now is questioned if x^{v} is accepted. The answer depends on δ_h. Therefore, at first it should be examined how δ_h is computed. Let π^{*} be any configuration and x^{*} a corresponding optimal solution. We denote I_j as the set of the population elements covered by the centroid j of the configuration. Then $\beta(i)$ gives the centroid which covers the observation. As is seen, the calculation of the new solution, and its evaluation, is considerably simplified. It is remarkable that now the relaxed problem has a smaller number of constraints.

Two possible criteria determine a pair of versions of the heuristic. They are:

(1) The new solution is accepted if $\delta_h \leq 0$

(2) The new solution is accepted if is satisfied $\delta_h<0$ or $\delta_h \geq 0$ and $\exp(-\delta_h/t)>r$, where r is a random variable uniformly distributed in [0,1] and t is a control parameter.

The first criteria has a logical basis. The second defines the thermodynamic method. Note that

(1) The probability of accepting the new feasible solution disminishes if δ_h increases

(2) The probability of accepting a solution that increases the objective function decreases with falling t. Therefore if no better feasible solution can be obtained, after a sufficient number of iterations no further change will be accepted and the procedures ends.

These properties domine the procedure because

-A certain number of solutions is generated for a constant t and only some solutions are accepted

-If there is not change in the objective function for the accepted solution the procedure ends. Otherwise t is replaced by a lower value $t^{*}=at$, where $a<1$.

Though a certain set of strata may not be optimal in a rigorous sense, the gain in accuracy related with it is measurable. For stratified sampling for a single variable Z_h it is meassured by

$$G^{h}=[\sum_{t=1}^{P} W_t(\sigma_t^{h}-\bar{\sigma}^{h})^{2}+\sum_{t=1} W_t(\bar{z}_t^{h}-\bar{z}^{h})^{2}]/n$$

where

$$\sigma_t^{h}=[\sum_{i\in I_t}(z_i^{h}-\bar{z}_t^{h})^{2}/N_t]^{1/2};\ \bar{z}_t^{h}=\sum_{i\in I_t} z_i^{h}/N_t$$

$$\bar{\sigma}^{h}=\sum_{t=1} W_t\sigma_t^{h};\ \bar{z}^{h}=\sum_{t=1}^{P} W_t\bar{z}_t^{h}$$

see Cochran (1981) for details.

In the multivariate case the importance of this gain is also connected with the importance of the variable Z^{h}. It seems to be appropiate to use as an overall measure of the gain in accuracy

$$G=\sum_{h=1}^{k} P(h)G^{h}$$

The relative gain in accuracy may be computed by comparing with simple sampling overall variance given by

$$V=\sum_{h=1}^{P} P(h)\sum_{i\in I} (z^{h}-\bar{z}^{h})^{2}]/nN$$

Mulvey (1983) calculated,in an example with 15 336 observations and 9 attributes, that the gain was of the 41%.

The proposed TH version and the NT were evaluated in eleven populations. The description of the populations is given in Table I and the Percent of gain inaccuracy in Table II.

TABLE 1. POPULATIONS STUDIED

Population	**Number of**			
	Observations	**Atributes**	**Examined**	**Accepted**
Pastures:				
Pangola I	30	3	6	3
Pangola II	30	3	10	3
Bermuda I	116	4	8	3
Bermuda II	99	4	10	5
Pangola/bermuda	146	3	10	5
Credit and Service Cooperatives				
Province I	35	7	5	3
Province II	105	7	5	3
Educational Results	15	7	15	3
Meteorological Data	36	2	8	3
Anthropological Data				
Children I	186	2	9	3
Children II	173	2	9	3

The populations were rather small but they illustrate the behaviour of the proposed algorithms.

TABLE 2. PERCENT OF THE GAIN IN ACCURACY

Population	**Heuristic Method**	
	Thermodynamic	Nonthermodynamic
Pastures:		
Pangola I	0,82	0,42
Pangola II	1,30	0,51
Bermuda I	10,08	31,83
Bermuda II	29,36	10,74
Pangola/Bermuda	25,14	11,65
Credit and Service Cooperatives		
Province I	21,97	12,03
Province II	49,18	47,85
Educational Results	12,18	1,31
Meteorological Data	31,26	15,21
Anthropological Data		
Children I	2,36	1,82
Children II	8,49	2,07

We can say that the thermodynamical motivated algorithm gives, generally, better results than the other algorithm. Only in one case TH does not yield a larger gain in accuracy, see Table II.

It is necessary to develope more experiments with data from different branches of the Economy and other particular Sciences in order to give concluding remarks. Experimentation in the search of 'optimum weights' P(L) is also needed.

References

Burkhard,.E. and Rendl, F. (1983): A thermodynamically motivated simulation procedure for combinatorial optimization problems. Bericht 83-12. Technische Universität. Graz

Cochran, W.G.(1981): Tecnicas de Muestreo. ESPASA.Mexico

Dalenius, T.(1950):The problem of optimum stratification.Skand. Akt.33,133-148.

Kovacs, L.B.(1980):Combinatorial methods of discrete programming. Akademiai Kiadó.Budapest

Mulvey, J.M.(1983): Multivariate stratified sampling by optimization . Management Science.29,6.715-724.

Proof of two conjectures by G. Chen and D. L. Russell on structural damping for elastic systems†

Shuping Chen*
Department of Mathematics
Zhejiang University
Hangzhou, China

Roberto Triggiani
Department of Mathematics
University of Florida
Gainesville, Florida 32611

Department of Applied Mathematics
Thornton Hall, University of Virginia
Charlottesville, Virginia 22903

1. Introduction, preliminaries, statement of main results.

1.1 Introduction

In a recent paper [C-R.1] the authors propose a class of mathematical models "exhibiting the empirically observed damping rates in elastic systems." As they show by studying various properties of the models proposed, the crucial mathematical feature which justifies their claim is that such models generate s.c. (strongly continuous), holomorphic semigroups. While we refer to [C-R.1] for a discussion of elastic systems and their damping rates as analyzed in past engineering literature, we restrict our interest here to some mathematical questions which are raised in the paper. More specifically, in [C-R.1] Goong Chen and David L. Russell pose two conjectures which--if proven correct--would cover pricisely the cases that they would like to include in their proposed model for elastic systems.

Research partially supported by the Air Force Office of Scientific Research under Grant AFOSR-84-0365 and by National Science Foundation under Grant NSF DMS-8301668.

*Visiting the Department of Mathematics, University of Florida, Gainesville, Florida 32611, under the Pao Yu-kong and Pao Zao-long Scholarship

The main goal of the present paper is to prove the two conjectures in [C-R.1]. In another paper [C-T.1], we shall extend our results here and prove the counter part of the two conjectures in [C.R.1] for a more general class of elastic systems, i.e. with $A^{1/2}$ here replaced by A^{α}, $1/2 \leq \alpha \leq 1$, [C-T.1]).

As in [C-R.1], it is assumed throughout that:

(H.1): A (elastic operator) is a self-adjoint operator on an Hilbert space X, strictly positive(1), with dense domain $\mathcal{D}(A)$ and compact resolvent $R(\lambda,A)$, the case of interest in physical applications;

(H.2): B (dissipative operator) is, for the time being, a positive, self-adjoint operator on X likewise with dense domain $\mathcal{D}(B)$ in X.

The mathematical model proposed in [C-R.1] to describe elastic systems is then

$$\ddot{x} + B\dot{x} + Ax = 0 \quad \text{on } X \tag{1.1}$$

or equivalently,

$$\frac{d}{dt}\begin{vmatrix} x \\ \dot{x} \end{vmatrix} = \mathcal{A}_B \begin{vmatrix} x \\ \dot{x} \end{vmatrix} \quad \text{on } E \equiv \mathcal{D}(A^{1/2}) \times X \tag{1.2a}$$

$$\mathcal{A}_B = \begin{vmatrix} 0 & I \\ -A & -B \end{vmatrix}, \text{ with domain } \mathcal{D}(\mathcal{A}_B) \text{ containing } \mathcal{D}(A) \times \mathcal{D}(B) \tag{1.2b}$$

where the inner product on E is given by(2)

$$\left(\begin{vmatrix} x_1 \\ x_2 \end{vmatrix}, \begin{vmatrix} y_1 \\ y_2 \end{vmatrix}\right)_E = (A^{1/2}x_1, A^{1/2}y_1)_X + (x_2,y_2)_X \tag{1.2c}$$

and where, in addition, the operator B is thought of as being "related in various ways to the positive square root of A" [C-R.1, p. 433].

Thus, the prototype model taken in [C-R.1] is $B = 2\rho A^{1/2}$, $0 < \rho$, i.e. for future reference

$$\ddot{x} + 2\rho A^{1/2}\dot{x} + Ax = 0 \quad \text{on } X;\ 0 < \rho < \infty, \tag{1.3}$$

or equivalently

$$\frac{d}{dt}\begin{vmatrix} x \\ \dot{x} \end{vmatrix} = \mathcal{A}_\rho \begin{vmatrix} x \\ \dot{x} \end{vmatrix} \quad \text{on } E \equiv \mathcal{D}(A^{1/2}) \times X \tag{1.4a}$$

$$\mathcal{A}_\rho \equiv \mathcal{A}_{B=2\rho A^{1/2}} = \begin{vmatrix} 0 & I \\ -A & -2\rho A^{1/2} \end{vmatrix}, \mathcal{D}(\mathcal{A}_\rho) = \mathcal{D}(A) \times \mathcal{D}(A^{1/2}) \tag{1.4b}$$

1.2 Preliminaries

We shall collect here some results to be invoked in subsequent sections. We begin with a list of relevant results which are either well known or readily verifiable.

(i) In the dissipation-free case B = 0, the operator

$$\mathscr{A}_o = \begin{vmatrix} 0 & I \\ -A & 0 \end{vmatrix}, \quad \mathscr{D}(\mathscr{A}_o) = \mathscr{D}(A) \times \mathscr{D}(A^{1/2}) \tag{1.5}$$

is skew-adjoint on E: $\mathscr{A}_o = -\mathscr{A}_o^*$ and thus it generates a unitary s.c. group $\exp(\mathscr{A}_o t)$ on E (conservative elastic system).(3)

(ii) In the damped case, the operator $\mathscr{A}_B$ given by (1.2b) [resp. $\mathscr{L}_B$, given by footnote 2] is densely defined on E [resp. on W] and dissipative here; hence $\mathscr{A}_B$ [resp. $\mathscr{L}_B$] is closeable on E [resp. on W], see [P.1, p. 16], and we shall use the same symbol $\mathscr{A}_B$ [resp. $\mathscr{L}_B$] to denote its closure.

(iii) Since B is positive, then (i) implies that $\mathscr{A}_B$ is dissipative on E and Lumer-Phillips Theorem then gives that $\mathscr{A}_B$ generates a s.c. semigroup of contractions on E.

(iv) The resolvent operator $R(\lambda,\mathscr{A}_\rho) = [\lambda I - \mathscr{A}_\rho]^{-1}$ of the operator $\mathscr{A}_\rho$ in (1.4b) is given by

$$R(\lambda,\mathscr{A}_\rho) = \begin{vmatrix} [\lambda I + 2\rho A^{1/2}]V_\rho^{-1}(\lambda) & V_\rho^{-1}(\lambda) \\ -AV_\rho^{-1}(\lambda) & \lambda V_\rho^{-1}(\lambda) \end{vmatrix} \tag{1.6}$$

$$V(\lambda) \equiv V_\rho(\lambda) = \lambda^2 I + \lambda 2\rho A^{1/2} + A \tag{1.7}$$

at least for Re $\lambda > 0$. We note that A and $V_\rho(\lambda)$ commute. This commutativity property will be freely used below. Similarly, the resolvent $R(\lambda, \mathscr{A}_B) = [\lambda I - \mathscr{A}_B]^{-1}$ of the operator $\mathscr{A}_B$ in (1.2b) (and section 1.2(ii)) is

$$R(\lambda,\mathscr{A}_B) = \begin{vmatrix} V_B^{-1}(\lambda)(\lambda I + B) & V_B^{-1}(\lambda) \\ -V_B^{-1}(\lambda)A & \lambda V_B^{-1}(\lambda) \end{vmatrix} \tag{1.8}$$

$$V_B(\lambda) = \lambda^2 I + \lambda B + A = V_\rho(\lambda) + \lambda(B - 2\rho A^{1/2}); \quad \frac{I - V_B^{-1}(\lambda)A}{\lambda} = V_B^{-1}(\lambda)(\lambda I + B) \tag{1.9}$$

at least for Re $\lambda > 0$.

1.3 Statement of main results.

In [C-R.1] G. Chen and D. L. Russell formulated the following two conjectures which, if correct, would cover precisely the cases that the authors would like to include in their proposed model (1.3) for elastic systems.

Assume that the operators A and B satisfy the standing hypotheses (H.1) and (H.2) of section 1. Then, the s.c. semigroup generated by the operator $\mathcal{A}_B$ in (1.2b) [see section 1.2(ii)-(iii)] is also holomorphic on $E \equiv \mathcal{D}(A^{1/2}) \times X$, provided in addition:

Conjecture #1: $\rho_1^2 A \leq B^2 \leq \rho_2^2 A$, $0 < \rho_1 < \rho_2 < \infty$, i.e., explicitly

$$\rho_1^2(Ax,x) \leq (B^2x,x) \leq \rho_2^2(Ax,x), \quad x \in \mathcal{D}(B) = \mathcal{D}(A^{1/2}); \tag{1.10}$$

or else, provided in addition

Conjecture #2: $\rho_1 A^{1/2} \leq B \leq \rho_2 A^{1/2}$, $0 < \rho_1 < \rho_2 < \infty$, i. e., explicitly

$$\rho_1(A^{1/2}x,x) \leq (Bx,x) \leq \rho_2(A^{1/2}x,x), \quad x \in \mathcal{D}(B^{1/2}) = \mathcal{D}(A^{1/4}) \tag{1.11a}$$

Only partial results in the direction of these conjectures are offered in [C-R.1]. These are Corollary 3.2 in [C-R.1] - a "local" result concerning conjecture #2 and Theorem 4.1 in [C-R.1] regarding conjecture #1, which requires however several additional technical assumptions which appear to be difficult to verify.

We note that assumptions (1.10a) and (1.11a) are not equivalent (unless A and B commute) as mentioned in [C-R.1]; however, it is known that (1.10a) implies (1.11a), see [K.2, Corollary 7.1 p. 146] and [X.1, p 5] (Löwner's Theorem). Thus, to give an affirmative response to both conjectures raised in [C-R.1], it suffices to study conjecture #2, the more general of the two, and prove the following

Theorem 1.1 a) Under the standing hypotheses (H.1]-(H.2), let assumption (1.11a) also hold. Then the s.c. semigroup generated by the operator $\mathcal{A}_B$ in (1.2b) [see section 1.2(ii)-(iii)] is also holomorphic on $E \equiv \mathcal{D}(A^{1/2}) \times X$.

b) Hence, as the spectrum determined growth condition [T.2] is satisfied, there is $\delta \equiv \sup \operatorname{Re} \sigma(\mathcal{A}_B) > 0$, such that

$$||e^{\mathscr{A}_B t}||_{\mathscr{L}(E)} \leq e^{-\delta t}, \qquad t>0$$

Note that assumption (1.11) is equivalent to

$$0 < \rho_1(y,y) \leq (A^{-1/4} B A^{-1/4} y,y) \leq \rho_2(y,y), \qquad y \in X \tag{1.11b}$$

the form in which we shall use it below.

The proof of Theorem 1.1 is given in section 4. In addition, the paper presents the following other results. In section 2 we unveil the rather special spectral structure of the operator $\mathscr{A}_\rho$ in (1.4b), as a direct sum of two normal operators, thereby strengthening and refining results of [C-R.1]. In section 3 we show, in particular, that Theorem 1.1 is false if $A^{1/2}$ is replaced by A^α, $\alpha < 1/2$.

We conclude by observing that, as already noted, a natural extension of Theorem 1.1 when "B behaves like $A^{1/2}$" holds true also when B behaves like A^α, $1/2 < \alpha < 1$. see [C-T.1].

2. The case $B = 2\rho A^{1/2}$. The operator $\mathscr{A}_\rho$ is the direct sum of two normal operators on E and generates a s.c. holomorphic semigroup.

For the operator A subject to assumption (H-1), let $\{\mu_n\}_{n=1}^\infty$, $\mu_n > 0$, be its eigenvalues and let $\{e_n\}_{n=1}^\infty$ be the corresponding eigenvectors subject to the normalization condition (2.2b) below: $Ae_n = \mu_n e_n$. For simplicity of exposition we assume that the μ_n's are all simple.

The following spectral properties of $\mathscr{A}_\rho$, which for future use we need only in the case $0 < \rho < 1$ of 'light' damping, show, in particular, that $\mathscr{A}_\rho$ has a special structure: it is the direct sum of two normal operators on E. This property does not seem to have been observed before: see Remark 2.1 below. For the case $1 \leq \rho < \infty$, see [C.T.1]. For the case $B = 2\rho I$ of viscous damping see [T.2] and [L-T.1, Application 4.4].

Lemma 2.1 (spectral properties of $\mathscr{A}_\rho$). Let $0 < \rho < 1$.
(I) The eigenvalues $\{\lambda_n^{+,-}\}_{n=1}^\infty$ and the corresponding normalized eigenvectors $\{\Phi_n^{+,-}\}_{n=1}^\infty$ of the operator $\;_\rho$ in (1.4b) (which has compact resolvent) on

$E = \quad (A^{1/2}) \times X$ are given by:

$$\lambda_n^{+,-} = \mu_n^{1/2} e^{\pm i\Psi}; \qquad e^{\pm i\Psi} = -\rho \pm i\sqrt{1-\rho^2}, \qquad \frac{\pi}{2} < \Psi < \pi \tag{2.1a}$$

$$\lambda_n^{+} = \overline{\lambda_n^{-}} \text{ (complex conjugate of } \lambda_n^{-}) \tag{2.1b}$$

solutions of

$$\lambda^2 + 2\rho\, \mu_n^{1/2} \lambda + \mu_n = 0; \tag{2.1c}$$

$$\Phi_n^{+,-} = \begin{vmatrix} e_n \\ \lambda_n^{+,-} e_n \end{vmatrix} \tag{2.2a}$$

$$\|\Phi_n^{+,-}\|_E \equiv 1 \longleftrightarrow 2\mu_n \|e_n\|_X^2 \equiv 1 \tag{2.2b}$$

(II) They possess the following properties.

(i) $\{\Phi_n^+\}_{n=1}^{\infty}$ is an orthonormal family on E. (2.3a)

$\{\Phi_n^-\}_{n=1}^{\infty}$ is an orthonormal family on E. (2.3b)

(ii)
$$(\Phi_n^+, \Phi_n^-)_E = \begin{cases} 0 & m \neq n \\ \tfrac{1}{2}[1 + e^{2i\Psi}] & m = n \end{cases} \tag{2.4a}$$

$$(\Phi_m^-, \Phi_n^+)_E = \begin{cases} 0 & m \neq n \\ \tfrac{1}{2}[1 + e^{-2i\Psi}] & m = n \end{cases} \tag{2.4b}$$

(iii) (completeness of eigenvectors of $\mathscr{A}_\rho$ on E)

$$\overline{\text{span}}\, \{\Phi_n^{+,-}\}_{n=1}^{\infty} = E \tag{2.5}$$

(iv) Setting

$$E^+ = \overline{\text{span}}\, \{\Phi_n^+\}_{n=1}^{\infty}; \quad E^- \equiv \overline{\text{span}}\, \{\Phi_n^-\}_{n=1}^{\infty} \tag{2.6}$$

we have

$$E = E^+ \oplus E^-; \; E^+ \cap E^- = \{0\} \quad \text{(direct, \underline{not} orthogonal sum)} \tag{2.7}$$

Moreover, from (2.3a-b) it follows that [S.1, p 250]:

$$\mathscr{A}_\rho^+ \equiv \mathscr{A}_\rho\Big|_{E^+} = \text{restriction of } \mathscr{A}_\rho \text{ on } E^+ \text{ is a normal operator on } E^+ \tag{2.8a}$$

$$\mathscr{A}_\rho^- \equiv \mathscr{A}_\rho\Big|_{E^-} = \text{restriction of } \mathscr{A}_\rho \text{ on } E^- \text{ is a normal operator on } E^- \tag{2.8b}$$

(v) [consequence of (2.7)-(2.8)]. For every $x \in E$, we have

$$x = x^+ + x^-, \qquad x^+ \in E^+,\ x^- \in E^- \tag{2.9a}$$

$$x^+ = \sum_{n=1}^{\infty} (x^+, \Phi_n^+)_E\, \Phi_n^+; \qquad x^- = \sum_{n=1}^{\infty} (x^-, \Phi_n^-)_E\, \Phi_n^- \tag{2.9b}$$

(III). Let now $\mathscr{A}_\rho^*$ denote the adjoint operator of $_\rho$ in E:

$(\mathscr{A}_\rho x, y)_E = (x, \mathscr{A}_\rho^* y)_E$, $x \in \mathscr{D}(\mathscr{A}_\rho)$, $y \in \mathscr{D}(\mathscr{A}_\rho^*)$. Then

a):

$$\mathscr{A}_\rho^* = \begin{vmatrix} 0 & -I \\ A & -2\rho A^{1/2} \end{vmatrix}; \qquad \mathscr{A}_\rho^*\mathscr{A}_\rho - \mathscr{A}_\rho\mathscr{A}_\rho^* = 4\rho \begin{vmatrix} 0 & A^{1/2} \\ A^{3/2} & 0 \end{vmatrix} \tag{2.10}$$

$$\mathscr{D}(\mathscr{A}_\rho) = \mathscr{D}(\mathscr{A}_\rho^*) = \mathscr{D}(A) \times \mathscr{D}(A^{1/2})$$

[so that $\mathscr{A}_\rho^*$ is normal if and only if $\rho = 0$, the undamped case]

b): the eigenvalues of $\mathscr{A}_\rho^*$ are given by $\overline{\lambda_m^{+,-}} = \lambda_m^{-,+}$, with corresponding

$$\text{(normalized) eigenvectors} = \begin{vmatrix} e_m \\ -\lambda_m^{-,+}\, e_m \end{vmatrix} \tag{2.11}$$

c): if we set

$$\nu_m^+ = \frac{\lambda_m^- - \lambda_m^+}{2\lambda_m^-}; \qquad \nu_m^- = \overline{\nu_m^+} = \frac{\lambda_m^+ - \lambda_m^-}{2\lambda_m^+} \tag{2.12}$$

then the following non-normalized eigenvectors of $\mathscr{A}_\rho^*$,

$$\Phi_m^{*-} = \frac{1}{\nu_m^-} \begin{vmatrix} e_m \\ -\lambda_m^-\, e_m \end{vmatrix}$$

$$\Phi_m^{*+} = \frac{1}{\nu_m^+} \begin{vmatrix} e_m \\ -\nu_m^+\, e_m \end{vmatrix} \tag{2.13}$$

corresponding to the eigenvalues

$$\overline{\lambda_m^+} \quad \text{and} \quad \overline{\lambda_m^-}$$

respectively, form a bi-orthogonal system with respect ot the eigenvectors $\{\Phi_m^{+,-}\}$ of $\mathcal{A}_\rho$, corresponding to its eigenvalues $\lambda_m^{+,-}$:

$$(\Phi_m^{*-}, \Phi_n^+)_E = (\Phi_m^{*+}, \Phi_n^-)_E = \text{Kroneker } \delta_{mn} \tag{2.14a}$$

$$(\Phi_m^{*+}, \Phi_n^+)_E = (\Phi_m^{*-}, \Phi_n^-)_E = 0, \qquad \forall\, n,m \tag{2.14b}$$

d):

Taking the E-inner product of x given by (2.9) with $\Phi_m^{*+,-}$ and using properties (2.14), we obtain

$$(x, \Phi_m^{*-})_E = (x^+, \Phi_m^+)_E; \qquad (x, \Phi_m^{*+})_E = (x^-, \Phi_m^-)_E \tag{2.15}$$

whereby the expansion (2.9) for x E becomes

$$x = \sum_{n=1}^{\infty} (x, \Phi_n^{*-})_E\, \Phi_n^+ + \sum_{n=1}^{\infty} (x, \Phi_n^{*+})_E\, \Phi_n^-, \tag{2.16}$$

more convenient than (2.9).

Proof. For lack of space, direct verification of the above statements is left to the reader. □

From the space decomposition (2.7) and expansion (2.16) we obtain, in particular, an explicit representation for $\mathcal{A}_\rho$ and the corresponding s.c. semigroup $\exp(\mathcal{A}_\rho t)$ The latter shows, by inspection, that $\exp(\mathcal{A}_\rho t)$ is, in fact, holomorphic on E in a suitable triangular sector around the positive real axis R^+.

Theorem. 2.2 Let $0 < \rho < 1$. (For $1 \leq \rho < \infty$, see [C-T.1]).

(i) The operator $\mathcal{A}_\rho$ in (1.4b) is (a special case of a spectral operator of scalar type in the terminology of [D.1][D-S.1]; more precisely, $\mathcal{A}_\rho$ on E is) the direct sum of the two normal operators $\mathcal{A}_\rho^+$ and $\mathcal{A}_\rho^-$ defined in (2.8) on E^+, and, respectively, on E^-. [A fortiori, $\mathcal{A}\rho$ is similar[4] to a normal operator on E, by virtue of Wermer theorem [D-S.1, III p. 1947]).

(ii) The expansion for $\mathscr{A}_\rho$ is given by (from (2.16)):

$$\mathscr{A}_\rho x = \sum_{n=1}^{\infty} \lambda_n^+ (x, \Phi_n^{*-})_E \Phi_n^+ + \sum_{n=1}^{\infty} \lambda_n^- (x, \Phi_n^{*+})_E \Phi_n^-, \qquad x \in \mathscr{D}(\mathscr{A}_\rho) \tag{2.17}$$

(iii) $\mathscr{A}_\rho$ generates a s.c. semigroup exp($_\rho t$) on E given explicitly by

$$e^{\mathscr{A}_\rho t} x = \sum_{n=1}^{\infty} e^{\lambda_n^+ t} (x, \Phi_n^{*-})_E + \sum_{n=1}^{\infty} e^{\lambda_n^- t} (x, \Phi_n^{*+})_E \Phi_n^- \tag{2.18}$$

which is holomorphic (analytic) for $t > 0$ or, more generally, on $\sum_{\Psi-\pi/2} = \{\lambda: |\arg \lambda| < \Psi-\pi/2\}$, where Ψ is defined in (2.11a). Moreover, as observed in section 1.2 (iii), $\exp(\mathscr{A}_\rho t)$ is contraction on E:

$$\| e^{\mathscr{A}_\rho t} \|^2_{\mathscr{L}(E)} \leq 1, \quad t \geq 0 \tag{2.19}$$

Remark 2.1. The results in Lemma 2.1 and Theorem 2.2 strengthen and refine those in [C-R.1], by providing more precise information about the spectral structure of $\mathscr{A}_\rho$. In particular, [C-R.1] asserts only that $\mathscr{A}_\rho$ is similar to a normal operator, while Theorem 2.2 above specifies that, in fact, $\mathscr{A}_\rho$ is the direct sum of two normal operators, a plainly more precise conclusion. As a consequence, the eigenvectors of $\mathscr{A}_\rho$ are only asserted in [C-R.1] to form a Riesz basis, while Lemma 2.1 above specifies that, in fact they are the union of two families $\{\Phi_n^+\}_{n=1}^{\infty}$ and $\{\Phi_n^-\}_{n=1}^{\infty}$, each of which is orthonormal on E^+ and E^-, repectively, with $E = E^+ \oplus E^-$. Thus, the explicit, direct expansions (2.16)-(2.18)--from the latter of which one can read off holomorphicity of $\exp(\mathscr{A}_\rho t)$ --are not present in [C-R.1] and do not appear to have been observed before.

In the following Corollary we collect information which we shall use in Section 4.

Corollary 2.3 a) Let $0 < \rho < 1$ (for $1 \leq \rho < \infty$, see [C-T.1]). There exists a positive constant c_ρ such that

(i) $\quad \|\lambda^2 V_\rho^{-1}(\lambda)\|_{\mathscr{L}(X)} + \|\lambda A^{1/2} V_\rho^{-1}(\lambda)\|_{\mathscr{L}(X)} \leq c_\rho$, for all λ with $\mathrm{Re}\lambda > 0$ (2.20)

(ii) $\quad \|AV_\rho^{-1}(\lambda)\|_{\mathscr{L}(X)} \leq c_\rho$ (2.21)

b) Next, assume the right hand side of inequality (1.11a-b), i.e. let B be a (positive self-adjoint) operator satisfying

$(Bx,x) \leq \rho_2(A^{1/2}x,x)$, $x \in \mathcal{D}(B^{1/2}) = \mathcal{D}(A^{1/4})$, equivalently $\|A^{-1/4}BA^{-1/4}\| \leq \rho_2$ (2.22)

Then, there exists a positive constant $C_{\rho\rho_2}$ (depending also on ρ_2) such that

$$\|[S-2\rho I]AV_\rho^{-1}(\lambda)\|^2 \leq C_{\rho\rho_2} \quad \text{for all } \lambda \text{ with } \mathrm{Re}\lambda > 0 \tag{2.23}$$

$$S \equiv A^{-1/4}BA^{-1/4} \in \mathcal{L}(X) \text{ (self-adjoint).} \tag{2.24}$$

Proof First by Theorem 2.2 the resolvent $R(\lambda, \mathcal{A}_\rho)$ of the s.c., holomorphic semigroup generator $\mathcal{A}_\rho$ on E satisfies the usual estimate

$$\|R(\lambda, \mathcal{A}_\rho\|_{\mathcal{L}(E)} \leq \frac{M_\rho}{|\lambda|} \text{ for all } \lambda \in \textstyle\sum_\rho = \{\lambda: |\mathrm{ang}\ \lambda| > \frac{\pi}{2} + \theta_\rho\} \tag{2.25}$$

for some $\theta_\rho > 0$

(a)(i) For $x \in X$, we have from (1.6), (1.2c) and (2.25)

$$\left\|R(\lambda,\mathcal{A}_\rho)\begin{vmatrix} 0 \\ x \end{vmatrix}\right\|_E^2 = \|A^{1/2}V_\rho^{-1}(\lambda)x\|^2 + \|\lambda V_\rho^{-1}(\lambda)x\|^2 \leq \frac{M_\rho^2}{|\lambda|^2}\|x\|^2 \quad \text{for } \mathrm{Re}\lambda > 0$$

and part (i) follows.

(ii) We use the identity (easily verified via (1.7))

$$AV_\rho^{-1}(\lambda) = I - (\lambda^2 I + \lambda 2\rho A^{1/2})V_\rho^{-1}(\lambda) \tag{2.26}$$

Then part (i) readily yields part (ii).

b) Equations (2.21) and (2.22) plainly imply (2.23).□

Remark 2.2 Conclusions (2.20), (2.21) follow also by direct computation via, say, the well known self-adjoint formulas [K.1 formulas (4.9) p. 230]. In section 4 we shall also need the following

Lemma 2.4 For $\rho > 0$ and A as in (H.1) we have:

$$\mathrm{Re}(\lambda A^{1/4}V_\rho^{-1}(\lambda)x,x) \geq 0 \quad \text{for all } x \in X, \text{ for all } \lambda \text{ with } \mathrm{Re}\lambda \geq 0 \tag{2.27}$$

Proof Immediate from (1.7) using the new variable $\xi(\lambda) = V_\mu^{-1}(\lambda)x$ hence $x = [\lambda^2 I + 2\rho\lambda A^{1/2} + A]\xi(\lambda)$ in (2.27).□

3. The case: $k_1A^{2\alpha} \leq B^2 \leq k_2A^{2\alpha}$, $0 < k_1 < k_2$, with $\alpha < 1/2$. Lack of holomorphicity of the s.c. semigroup generated by $\mathscr{A}_B$.

In this section we shall see that the choice of the power "$A^{1/2}$" as a term of comparison for B is not accidental, in the sense that if the self-adjoint operator B satisfies instead

$$k_1\|A^\alpha x\| \leq \|Bx\| \leq k_2\|A^\alpha x\|, \quad 0 < k_1 < k_2, \quad x \in \mathscr{D}(A^\alpha) = \mathscr{D}(B) \tag{3.1}$$

for $\alpha < 1/2$

in place of (1.10a), then the operator $\mathscr{A}_B$ in (1.2b)[closed as in section 1.2(ii)] is the generator of a s.c. semi-group on E, [see section 1.2(iii)] which however is not holomorphic in general. Indeed, even more information of negative character is contained in the following construction.

Proposition 3.1. Let the positive self-adjoint operator A as in (H.1) have eigenvalues $\{\mu_n\}_{n=1}^\infty$, $\mu_n > 0$, and corresponding eigenvectors $\{e_n\}_{n=1}^\infty$ forming an orthonormal basis in X. Define the operator $B:\mathscr{D}(B) \to X$ by

$$Be_n = b_ne_n, \quad b_n > 0 \tag{3.2}$$

so that B is positive self-adjoint and commutes with A. If

$$\frac{\mu_n}{b_n^2} \to \infty \quad \text{as } n \uparrow \infty \tag{3.3}$$

then the corresponding operator $\mathscr{A}_B$ defined by (1.2b) (and section 1.2(ii)) generates a s.c. semigroup on E which, however, is not holomorphic here.

Proof. Generation by $_B$ of a s.c. semigroup on E was already asserted in Section 1.2(iii). The eigenvalue-vector problem for $\mathscr{A}_B$ is

$$\mathscr{A}_B\begin{vmatrix}\Psi_1\\ \Psi_2\end{vmatrix} = \lambda\begin{vmatrix}\Psi_1\\ \Psi_2\end{vmatrix} \quad \text{i.e.} \quad (-A)\Psi_1 = \lambda B\Psi_1 + \lambda^2\Psi_1 \tag{3.4}$$

whose solution is given precisely by the eigenvectors $\{e_n\}$ of A:

$$(-A)e_n = (\lambda b_n + \lambda^2)e_n = -\mu_ne_n \tag{3.5}$$

Thus, the corresponding eigenvalues $\lambda_n^{+,-}$ of $\mathscr{A}_B$ are the solutions of the quadratic equation

$$\lambda^2 + b_n\lambda + \mu_n = 0 \tag{3.6a}$$

and are given by

$$\lambda_n^{+,-} = \frac{-b_n}{2} \pm i\sqrt{\frac{4\mu_n - b_n^2}{2}} \tag{3.6b}$$

Thus, if (3.3) holds, then $4\mu_n - b_n^2 > 0$ for all n sufficiently large and

$$\left|\frac{I_m\ \lambda_n^{+,-}}{\text{Re}\ \lambda_n^{+,-}}\right| = \sqrt{\frac{4\mu_n}{b_n^2}} - 1 \to \infty \tag{3.7}$$

so that the eigenvalues $\{\lambda_n^{+,-}\}$ of $\mathcal{A}_B$ fail to be contained in a triangular sector of the type

$$\{\lambda : |\arg(\lambda - a)| \geq \frac{\pi}{2} + \theta\} \tag{3.8}$$

for some a real and some $\theta > 0$. Thus, as in well known [F.1; P. 1], holomorphicity of the semigroup generated by $\mathcal{A}_B$ is out of question. □

The case of interest is recaptured as a corollary.

Corollary 3.2. Let A be as in Proposition 3.1 and let B be defined by (3.2), where now

$$b_n \sim \mu_n^\alpha, \quad \alpha < 1/2, \text{ as } n \to \infty \tag{3.9}$$

(meaning: $c\mu_n^\alpha \leq b_n \leq C\,\mu_n^\alpha$, $0 < c < C$)

Then, B satisfies (3.1):

$$c^2\|A^\alpha x\|^2 \leq \|Bx\|^2 \leq C^2\|A^\alpha x\|^2, \ \alpha < 1/2, \ x \in \mathcal{D}(A^\alpha) = \mathcal{D}(B) \tag{3.10}$$

and $_B$ generates a s.c. semigroup on E, which however is not holomorhpic here. □

4. Proof of Theorem 1.1

The authors have two proofs, one based directly on the resolvent $R(\lambda, \mathcal{A}_B)$ in (1.8), and one based on the idea of comparing $R(\lambda, \mathcal{A}_B)$ with $R(\lambda, \mathcal{A}_\rho)$. We choose to present here the latter. The second proof is sketched in section 5. (The two proofs have some technical estimates in common.)

4.1 General Outline

The idea of the proof is based on the following three step-procedure (see [L-T.1]).

(i) First we show generation by $_B$ of a s.c. holomorphic semigroup on the (larger) Hilbert space $Z = \mathcal{D}(A^{1/4}) \times [\mathcal{D}(A^{1/4})]'$ - which turns out to be the 'natural' space for assumption (1.11), - with norm

$$\|z\|_Z^2 = \|A^{1/4} z_1\|_X^2 + \|A^{-1/4} z_2\|_X^2, \quad z = [z_1, z_2] \tag{4.0}$$

(ii) Part (i) will then imply (see section 4.3 below) that $\mathcal{A}_B$ generates a s.c. semigroup which is holomorphic also on its domain, the space $Y = \mathcal{D}(A^{3/4}) \times \mathcal{D}(A^{1/4})$, with norm

$$\|z\|_Y^2 = \|A^{3/4} z_1\|_X^2 + \|A^{1/4} z_2\|_X^2, \quad z = [z_1, z_2] \in Y.$$

(iii) We finally apply the interpolation theorem [L-M. 1, Thm 5.1, p 27] to, say, the Hille-Yosida's characterizations in Z and Y of $R(\lambda, \mathcal{A}_B)$: we then obtain that for all λ in a suitable triangular sector $\sum$, as required, the operator $R(\lambda, \mathcal{A}_B)$ is a continuous operator from the interpolation space $[Y, Z]_{\theta = 1/2}$, explicitly

$$[\mathcal{D}(A^{3/4}) \times \mathcal{D}(A^{1/4}), \mathcal{D}(A^{1/4}) \times [\mathcal{D}(A^{1/4})]']_{\theta = 1/2} = \mathcal{D}(A^{1/2}) \times X = E$$

into itself and satisfies here the analogous Hille-Yosida characterization

$$\|R(\lambda, \mathcal{A}_B)\|_{\mathcal{L}(E)} \leq \frac{\text{const}}{|\lambda|}, \quad \lambda \in \textstyle\sum . \tag{4.1}$$

Thus $\mathcal{A}_B$ generates a s.c. holomorphic semigroup on E, as soon as we prove that it generates a s.c. holomorphic semigroup on Z.

4.2 Proof that $\mathcal{A}_B$ generates a s.c., holomorphic semigroup on $Z = \mathcal{D}(A^{1/4}) \times [\mathcal{D}(A^{1/4})]'$.

4.2.1 Step 1. We compare $\mathcal{A}_B$ in (1.2b) with $\mathcal{A}_\rho$ in (1.4b) by writing

$$\mathcal{A}_B = \mathcal{A}_\rho + \mathcal{P}_\rho ; \quad \mathcal{P}_\rho = \begin{vmatrix} 0 & 0 \\ 0 & 2\rho A^{1/2} - B \end{vmatrix} \tag{4.2}$$

for some $0 < \rho < 1$, henceforth kept fixed. (Later in (4.21) and ff, ρ will be chosen $0 < 2\rho < \rho_1$).

Then, $\lambda I - \mathscr{A}_B = \lambda I - \mathscr{A}_\rho - \mathscr{P}_\rho = [I - \mathscr{P}_\rho R(\lambda,\mathscr{A}_\rho)][\lambda I - \mathscr{A}_\rho]$ and

$$R(\lambda,\mathscr{A}_B) = R(\lambda,\mathscr{A}_\rho)[I - \mathscr{P}_\rho R(\lambda,\mathscr{A}_\rho)]^{-1} \tag{4.3}$$

valid at least for all λ in Re $\lambda > 0$.

Step 2. By section 2, $\mathscr{A}_\rho$, $0 < \rho < 1$, is the generator of a s.c., holomorphic semigroup of contractions on E, hence on Z. Thus we have the usual estimate

$$\|R(\lambda,\mathscr{A}_\rho)\|_{\mathscr{L}(Z)} \leq \frac{M_\rho}{|\lambda|} \quad \text{for all } \lambda \in \textstyle\sum_\rho$$
$$\textstyle\sum_\rho = \{\lambda: |\arg\lambda| \geq \frac{\pi}{2} + \theta_\rho\}, \quad \text{for some } \theta_\rho > 0. \tag{4.4}$$

Now, in order to show that $\mathscr{A}_B$ is the generator of a s.c., homomorphic semigroup of contractions on E, it suffices to show that

$$\|R(\lambda,\mathscr{A}_B)\|_{\mathscr{L}(Z)} \leq \frac{c_{r_o}}{|\lambda|}, \quad \text{for all } \lambda \text{ with Re } \lambda \geq \text{some } r_o > 0. \tag{4.5}$$

[F.1, p. 185][P.1]. To this end, by virtue of (4.3), (4.4), it then suffices to establish the following estimate:

$$\|[I - \mathscr{P}_\rho R(\lambda,\mathscr{A}_\rho)]^{-1}\|_{\mathscr{L}(Z)} \leq c_{r_o,\rho}, \qquad \text{for all } \lambda \text{ with Re}\lambda \geq \text{some } r_o > 0. \tag{4.6}$$

where $c_{r_o,\rho}$ is a constant depending on r_o and on the fixed $\rho > 0$. Showing (4.6) is the crux of our proof. To this end we first show that: for all

$z = [z_1, z_2] \in Z = \mathscr{D}(A^{1/4}) \times [\mathscr{D}(A^{1/4})]'$, there exist positive constants r_o (in fact, $r_o = 0$) and $k_{r_o,\rho}$ such that

$$\|[I - \mathscr{P}_\rho R(\lambda,\mathscr{A}_\rho)]z\|_Z \geq k_{r_o,\rho}\|z\|_Z, \text{ for all } \lambda \text{ with Re}\lambda \geq r_o > 0, \tag{4.7a}$$

Next we shall verify that

$$\text{closure } \{[I - \mathscr{P}_\rho R(\lambda,\mathscr{A}_\rho)]Z\} = Z, \quad \text{for all } \lambda \text{ with Re}\lambda \geq r_o > 0 \tag{4.7b}$$

Eqts (4.7a-b) are equivalent to (4.6)[T-L.1].

4.2.2 Proof of (4.7a) (with $r_o = 0$). Step 4. From (1.6), we see that the second component $[R(\lambda,\mathscr{A}_\rho)z]_2$ of the vector $R(\lambda,\mathscr{A}_\rho)z$ in Z is given by

$$[R(\lambda,\mathscr{A}_\rho)z]_2 = -AV_\rho^{-1}(\lambda)z_1 + \lambda V_\rho^{-1}(\lambda)z_2 \tag{4.8}$$

Thus, by (4.2)

$$\mathscr{P}_\rho R(\lambda,\mathscr{A}_\rho)z = \left|\begin{matrix} 0 \\ [2\rho A^{1/2} - B][R(\lambda, \quad_\rho)z]_2 \end{matrix}\right|$$

$$[I - \mathscr{P}_\rho R(\lambda,\mathscr{A}_\rho)]z = z + \left|\begin{matrix} 0 \\ T_1(\lambda)z_1 + T_2(\lambda)z_2 \end{matrix}\right| \tag{4.9}$$

where by (4.8)

$$T_1(\lambda) = [B - 2\rho A^{1/2}][-AV_\rho^{-1}(\lambda)] \tag{4.10}$$

$$T_2(\lambda) = [B - 2\rho A^{1/2}]\lambda V_\rho^{-1}(\lambda) \tag{4.11}$$

For $z = [z_1, z_2] \in Z = \mathscr{D}(A^{1/4}) \times [\mathscr{D}(A^{1/4})]' = Z_1 \times Z_2$, after using (4.0), we can write from (4.9)-(4.11):

$$\begin{aligned}\|[I - \mathscr{P}_\rho R(\lambda,\mathscr{A}_\rho)]z\|_Z^2 &= \|z\|_Z^2 + \|T_1(\lambda)z_1 + T_2(\lambda)z_2\|_{Z_2}^2 \\ &\quad + 2\mathrm{Re}\left\{(T_1(\lambda)z_1 + T_2(\lambda)z_2, z_2)_{Z_2}\right\} \\ &= \|z\|_Z^2 + \|Q_1(\lambda)A^{1/4}z_1 + Q_2(\lambda)A^{-1/4}z_2\|^2 \\ &\quad + 2\mathrm{Re}\left\{(Q_1(\lambda)A^{1/4}z_1 + Q_2(\lambda)A^{-1/4}z_2, A^{-1/4}z_2)\right\}\end{aligned} \tag{4.12}$$

where as usual $\|\ \|$ and $(\ ,\ )$ are the norm and inner product on X and where now:

$$Q_1(\lambda) = A^{-1/4}T_1(\lambda)A^{-1/4} = [S - 2\rho I][-AV_\rho^{-1}(\lambda)] \tag{4.13}$$

$$Q_2(\lambda) = A^{-1/4}T_2(\lambda)A^{1/4} = [S - 2\rho I]\lambda A^{1/2}V_\rho^{-1}(\lambda) \tag{4.14}$$

$S = A^{-1/4}BA^{-1/4} \in \mathscr{L}(X)$ (self-adjoint), see (2.24).

Thus we obtain from (4.12) after setting

$$x_1 = A^{1/4}z_1 \in X, \quad x_2 = A^{-1/4}z_2 \in X \tag{4.15}$$

$$\|[I - \mathscr{P}_\rho R(\lambda, \mathscr{A}_\rho)]z\|_Z^2 = \|x_1\|^2 + \|Q_1(\lambda)x_1\|^2 + 2\mathrm{Re}\{(Q_1(\lambda)x_1, x_2 + Q_2(\lambda)x_2)\} + \|x_2 + Q_2(\lambda)x_2\|^2 \tag{4.16}$$

$$\geq \|x_1\|^2 + (1 - \tfrac{1}{\varepsilon})\|Q_1(\lambda)x_1\|^2 + (1-\varepsilon)\|x_2 + Q_2(\lambda)x_2\|^2 \tag{4.17}$$

for any $0 < \varepsilon < 1$. Recalling (4.13)-(4.14) we obtain explicitly from (4.17)

$$\|[I - \mathscr{P}_\rho(\lambda, \mathscr{A}_\rho)]z\|_Z^2 \geq \|x_1\|^2 + (1 - \tfrac{1}{\varepsilon})\|[S - 2\rho I]AV_\rho^{-1}(\lambda)x_1\|^2 + (1-\varepsilon)\|[I + (S - 2\rho I)\lambda A^{1/2}V_\rho^{-1}(\lambda)]x_2\|^2 \tag{4.18}$$

Thus, for B satisfying (2.22), we apply Corollary 2.3b, Eq. (2.23) in (4.18) and since $1 - 1/\varepsilon < 0$ for $0 < \varepsilon < 1$ we arrive at the following Lemma

Lemma 4.1 let B satisfy (2.22). Then, for all λ with $\mathrm{Re}\lambda > 0$ and $0 < \varepsilon < 1$ we have (recall (4.15)):

$$\|[I - \mathscr{P}_\rho R(\lambda, \mathscr{A}_\rho)]z\|_Z^2 \geq [1 + (1 - \tfrac{1}{\varepsilon})C_{\rho\rho_2}]\,\|x_1\|^2 + (1 - \varepsilon)\|[I + (S - 2\rho I)\lambda A^{1/2}V_\rho^{-1}(\lambda)]x_2\|^2 \tag{4.19}$$

where, we can always achieve $1 + (1 - \frac{1}{\varepsilon})C_{\rho\rho_2} > 0$ and $(1-\varepsilon) > 0$ as desired, by taking ε so that $\dfrac{C_{\rho\rho_2}}{1 + C_{\rho\rho_2}} < \varepsilon < 1$. □

Step 6 By Lemma 4.1, to achieve (4.7a) as desired, it ramains to show that the operator

$$I + (S - 2\rho I)\lambda A^{1/2}V_\rho^{-1}(\lambda) = (S - 2\rho I)[(S - 2\rho I)^{-1} + \lambda A^{1/2}V_\rho^{-1}(\lambda)] \tag{4.20}$$

is boundedly invertible on $\mathscr{L}(X)$ uniformly in $\mathrm{Re}\,\lambda > 0$. By assumption (1.11) we have for $S = A^{-1/4}BA^{-1/4}$ in (2.24), after choosing $0 < 2\rho < \rho_1$:

$$0 < [\rho_1 - 2\rho]I \leq S - 2\rho I \leq [\rho_2 - 2\rho]I \tag{4.21}$$

$$\frac{1}{\rho_2 - 2\rho}I \leq [S - 2\rho I]^{-1} \leq \frac{1}{\rho_1 - 2\rho}I \tag{4.22}$$

In view of (4.22), in order to achieve (4.7a), all it remains to show is that there exists $c > 0$ such that

$$\|[(S - 2\rho I)^{-1} + \lambda A^{1/2}V_\rho^{-1}(\lambda)]x\| \geq c\,\|x\| \quad \text{for all } x \in X \tag{4.23}$$

This is established in the next Lemma.

Lemma 4.2 Under assumptions (H.1)-(H.2) and (1.11), inequality (4.23) holds true with $c = 1/(\rho_2 - 2\rho)$ so that for all λ with $\text{Re}\lambda > 0$:

$$\|[(S - 2\rho I)^{-1} + \lambda A^{1/2} V_\rho^{-1}(\lambda)]^{-1}\| \leq \rho_2 - 2\rho \tag{4.24}$$

Proof For $x \in X$ we compute

$$\|[(S - 2\rho I)^{-1} + \lambda A^{1/2} V_\rho^{-1}(\lambda)]x\| \, \|x\| \geq$$

$$\geq \left|([S - 2\rho I]^{-1} x, x) + (\lambda A^{1/2} V_\rho^{-1}(\lambda)x, x)\right|$$

$$= \{[([S - 2\rho I]^{-1} x, x) + \text{Re}(\lambda A^{1/2} V_\rho^{-1}(\lambda)x, x)]^2$$

$$+ [\text{Im}(\lambda A^{1/2} V_\rho^{-1}(\lambda)x, x)]^2\}^{1/2}$$

(using (2.27) in Lemma 2.4 and (4.22))

$$\geq ([S - 2\rho I]^{-1} x, x) \geq \frac{1}{\rho_2 - 2\rho} \|x\|^2 \tag{4.25}$$

and (4.23) follows. □

Thus, the proof for inequality (4.7a) is complete.

4.2.3 Verification of (4.7b). Step 7. It remains to verify (4.7b). Equivalently, we shall prove that in Z, equivalently in fact, in E, we have:

null space of $[I - \mathscr{P}_\rho R(\lambda, \mathscr{A}_\rho)]^* =$

null space of $[I - R(\overline{\lambda}, \mathscr{A}_\rho)\mathscr{P}_\rho] = \{0\}$, λ with $\text{Re}\lambda > 0$ (4.26)

where we note that $\mathscr{P}_\rho^* = \mathscr{P}_\rho$ [from (4.2) and the self-adjontness of A and B]. To show (4.25) we fix λ, and let $[I - R(\overline{\lambda}, \mathscr{A}_\rho)\mathscr{P}_\rho]z = 0$ for $z \equiv z_\lambda$ E, depending on λ, (explicit dependence on λ will be suppressed henceforth). Recalling (4.2) and (1.6) and setting $z = [z_1, z_2]$ E, we re-write this identity as

$$z_1 - V_\rho^{-1}(\overline{\lambda})[2\rho A^{1/2} - B]z_2 = 0 \qquad \text{(a)}$$

$$z_2 - \overline{\lambda} V_\rho^{-1}(\overline{\lambda})[2\rho A^{1/2} - B]z_2 = 0 \qquad \text{(b)} \tag{4.27}$$

Application of $V_\rho(\overline{\lambda})$, see (1.7), on (4.27b) yields

$$V_\rho(\bar\lambda)z_2 - \bar\lambda[2\rho A^{1/2} - B]z_2 = 0, \text{ i.e. } [\bar\lambda^2 I + \bar\lambda B + A]z_2 = V_B(\bar\lambda)z_2 = 0$$

which for $\mathrm{Re}\bar\lambda > 0$ gives $z_2 = 0$ (see below). Then (4.27a) yields $z_1 = 0$. Thus $z = 0$, as desired. That $V_B(\bar\lambda)z_2 = 0$, $\mathrm{Re}\bar\lambda > 0$, implies $z_2 = 0$ can be seen in a few ways: either by recalling that $\mathscr{A}_B$ is dissipative, [section 1.2(iii)] and hence $R(\lambda, \mathscr{A}_B)$ in (1.8) is well defined for $\mathrm{Re}\bar\lambda > 0$ and thus $V_B(\bar\lambda)$ is boundedly invertible on X; or else directly via $0 = (V_B(\bar\lambda)z_2, z_2) = \bar\lambda^2(z_2, z_2) + \bar\lambda(Bz_2, z_2) + (Az_2, z_2)$. Separating this identity in real and imaginary part yields easily $z_2 = 0$ for $\mathrm{Re}\bar\lambda > 0$. Eq(4.7b) is proved. □

4.3 Proof that $\mathscr{A}_B$ generates a s.c., holomorphic semi-group on $Y \equiv \mathscr{D}(A^{3/4}) \times \mathscr{D}(A^{1/4})$.

We proceed as in [L.T.1]. By section 4.2, the operator

$$\mathscr{A}_B : Z \supset \mathscr{D}(\mathscr{A}_B) = \mathscr{D}(\mathscr{A}_\rho) = \mathscr{D}(A^{3/4}) \times \mathscr{D}(A^{1/4}) \to Z = [\mathscr{D}(A^{1/4})]' \tag{4.28}$$

is the generator of a s.c., holomorphic semigroup $\exp[\mathscr{A}_B t]$ on Z. It then follows that $\exp[\mathscr{A}_B t]$ is a s.c., holomorphic semigroup also on the space $\mathscr{D}(\mathscr{A}_B)$ equipped with its natural norm derived from its underlying space Z

$$\|y\|_{\mathscr{D}(\mathscr{A}_B)} \equiv \|\mathscr{A}_B y\|_Z, \quad y \in \mathscr{D}(\mathscr{A}_B) \tag{4.29}$$

(recall that $\lambda = 0$ is a point in the resolvent set of $\mathscr{A}_B$). But the set $\mathscr{D}(\mathscr{A}_B)$ coincides (set theoretically) with the set $\mathscr{D}(A^{3/4}) \times \mathscr{D}(A^{1/4})$, (4.28), whose original norm is given for $y = [y_1, y_2]$ by

$$\|y\|^2_{\mathscr{D}(A^{3/4}) \times \mathscr{D}(A^{1/4})} = \|A^{3/4} y_1\|^2_X + \|A^{1/4} y_2\|^2_X . \tag{4.30}$$

We prove that the two norms (4.29) and (4.30) are equivalent on the space $Y \equiv \mathscr{D}(A^{3/4}) \times \mathscr{D}(A^{1/4})$. In fact, by (4.28) and (4.0) the operator $\mathscr{A}_o$ (see (1.5)) yields

$$\|\mathscr{A}_o y\|^2_Z = \left\| \begin{matrix} y_2 \\ -Ay_1 \end{matrix} \right\|^2_Z = \|A^{1/4} y_2\|^2_X + \|A^{3/4} y_1\|^2_X = \|y\|^2_{\mathscr{D}(A^{3/4}) \times \mathscr{D}(A^{1/4})} \tag{4.31}$$

By the closed graph theorem, we have $\mathscr{A}_B \mathscr{A}_o^{-1}$ and $\mathscr{A}_o \mathscr{A}_B^{-1}$ both in $\mathscr{L}(Z)$. Thus, for $y \in \mathscr{D}(A^{3/4}) \times \mathscr{D}(A^{1/4})$, using (4.31)

$$\|\mathscr{A}_B y\|^2_Z \leq \|\mathscr{A}_B \mathscr{A}_o^{-1}\|^2_{\mathscr{L}(Z)} \|\mathscr{A}_o y\|^2_Z = \|\mathscr{A}_B \mathscr{A}_o^{-1}\|^2_{\mathscr{L}(Z)} \|y\|^2_{\mathscr{D}(A^{3/4}) \times \mathscr{D}(A^{1/4})} \tag{4.32}$$

$$\|y\|^2_{\mathcal{D}(A^{3/4})\times\mathcal{D}(A^{1/4})} \leq \|\mathcal{A}_o y\|^2_Z \leq \|\mathcal{A}_o\mathcal{A}_B^{-1}\|^2_{\mathcal{L}(Z)}\|\mathcal{A}_B y\|^2_Z \tag{4.33}$$

and norm-equivalence is established. Thus, $\exp[\mathcal{A}_B t]$ is a s.c., holomorphic semigroup on Y with norm (4.30).

5. Another proof of Theorem 1.1 based directly on $R(\lambda,\mathcal{A}_B)$ given by (1.8) (Sketch).

We shall reprove section 4.2 - that the s.c. semigroup generated by $\mathcal{A}_B$ is holomorphic on $Z \equiv \mathcal{D}(A^{1/4}) \times [\mathcal{D}(A^{1/4})]'$ - directly via $R(\lambda,\mathcal{A}_B)$, given by (1.8)-(1.9). We must establish that there exists $C > 0$ such that for all λ with $\mathrm{Re}\lambda > 0$ we have

$$\|\lambda R(\lambda,\mathcal{A}_B)\|_{\mathcal{L}(Z)} = \left\| \begin{vmatrix} A^{1/4} & 0 \\ 0 & A^{-1/4} \end{vmatrix} \begin{Vmatrix} I - V_B^{-1}(\lambda)A & \lambda V_B^{-1}(\lambda) \\ -\lambda V_B^{-1}(\lambda)A & \lambda^2 V_B^{-1}(\lambda) \end{Vmatrix} \begin{vmatrix} A^{1/4} & 0 \\ 0 & A^{1/4} \end{vmatrix} \right\|_{\mathcal{L}(W)} \leq C \tag{5.1}$$

$W = X \times X$; equivalently, that there exists $M > 0$ such that for all λ with $\mathrm{Re}\,\lambda > 0$ we have

$$\|A^{1/4}V_B^{-1}(\lambda)A^{3/4}\|_{\mathcal{L}(X)} \leq M \tag{5.2}$$

$$\|\lambda A^{1/4}V_B^{-1}(\lambda)A^{1/4}\|_{\mathcal{L}(X)} \leq M \tag{5.3}$$

$$\|\lambda^2 A^{-1/4}V_B^{-1}(\lambda)A^{1/4}\|_{\mathcal{L}(X)} \leq M \tag{5.4}$$

Proof of (5.2). From (1.9) we have

$$V_B^{-1}(\lambda) = [V_\rho(\lambda) + \lambda A^{1/4}(S - 2\rho I)A^{1/4}]^{-1} \tag{5.5}$$

with $S = A^{-1/4}BA^{-1/4} \in \mathcal{L}(X)$, as in (2.24), so that

$$\begin{aligned} A^{1/4}V_B^{-1}(\lambda)A^{3/4} &= \{[V_\rho(\lambda) + \lambda A^{1/4}(S - 2\rho I)A^{1/4}]A^{-1/4}\}^{-1}\, A^{3/4} \\ &= \{V_\rho(\lambda)A^{-1/4}[I + \lambda A^{1/2}V_\rho^{-1}(\lambda)(S - 2\rho I)]\}^{-1}\, A^{3/4} \\ &= \{(S - 2\rho I)[(S - 2\rho I)^{-1} + \lambda A^{1/2}V_\rho^{-1}(\lambda)]\}^{-1}\, AV_\rho^{-1}(\lambda) \\ &= [(S - 2\rho I)^{-1} + \lambda A^{1/2}V_\rho^{-1}(\lambda)]^{-1}\,(S - 2\rho I)^{-1}\, AV_\rho^{-1}(\lambda) \end{aligned} \tag{5.6}$$

Thus, recalling (2.21) in Corollary 2.3, (4.24) in Lemma 4.2, and (4.22), we obtain (5.2) with $M = C_\rho(\rho_2 - 2\rho)/(\rho_1 - 2\rho)$.

Proof of (5.3). We compute from (5.5)

$$\lambda A^{1/4} V_B^{-1}(\lambda) A^{1/4} = \lambda A^{1/4}\{A^{1/4}[I + (S - 2\rho I)\lambda A^{1/2} V_\rho^{-1}(\lambda)] V_\rho(\lambda) A^{-1/4}\}^{-1} A^{1/4}$$

$$= \lambda A^{1/2} V_\rho^{-1}(\lambda)[I + (S - 2\rho I)\lambda A^{1/2} V_\rho^{-1}(\lambda)]^{-1}$$

$$= \lambda A^{1/2} V_\rho^{-1}(\lambda)\{(S - 2\rho I)[(S - 2\rho I)^{-1} + \lambda A^{1/2} V_\rho^{-1}(\lambda)]\}^{-1}$$

$$= \lambda A^{1/2} V_\rho^{-1}(\lambda)[(S - 2\rho I)^{-1} + \lambda A^{1/2} V_\rho^{-1}(\lambda)]^{-1}(S - 2\rho I)^{-1} \tag{5.7}$$

Recalling (2.20) in Corollary 2.3, (4.24) in Lemma 4.2, and (4.22) we obtain (5.3) with $M = C_\rho(\rho_2 - 2\rho)/(\rho_1 - 2\rho)$

Proof of (5.4). From (5.5)

$$\lambda^2 A^{-1/4} V_B^{-1}(\lambda) A^{1/4} = \lambda^2\{A^{-1/4}[V_\rho(\lambda) + \lambda A^{1/4}(S - 2\rho I)A^{1/4}]A^{1/4}\}^{-1}$$

$$= \lambda^2\{[I + (S - 2\rho I)\lambda A^{1/2} V_\rho^{-1}(\lambda)] V_\rho(\lambda)\}^{-1}$$

$$= \lambda^2 V_\rho^{-1}(\lambda)\{(S - 2\rho I)[(S - 2\rho I)^{-1} + \lambda A^{1/2} V_\rho^{-1}(\lambda)]\}^{-1}$$

$$= \lambda^2 V_\rho^{-1}(\lambda)[(S - 2\rho I)^{-1} + \lambda A^{1/2} V_\rho^{-1}(\lambda)]^{-1}(S - 2\rho I)^{-1} \tag{5.8}$$

and (2.20), (4.24) and (4.22) yield (5.4) with $M = C_\rho(\rho_2 - 2\rho)/(\rho_1 - 2\rho)$. The proof of (5.1) is complete. □

Remark 5.1 Note that estimate (5.2), re-written as

$$\|A^{1/4} V_B^{-1}(\lambda) A^{3/4}\|_{\mathscr{L}(X)} = \|A^{1/2} V_B^{-1}(\lambda) A^{1/2}\|_{\mathscr{L}([\mathscr{D}(A^{1/4})]')} \le M \tag{5.9}$$

implies for all λ with $\operatorname{Re}\lambda > 0$

$$\|A^{1/2} V_B^{-1}(\lambda) A^{1/2}\|_{\mathscr{L}(\mathscr{D}(A^{1/4}))} = \|A^{3/4} V_B^{-1}(\lambda) A^{1/4}\|_{\mathscr{L}(X)} =$$

$$= \|[A^{3/4} V_B^{-1}(\lambda) A^{1/4}]^*\|_{\mathscr{L}(X)}$$

$$= \|A^{1/4} V_B^{-1}(\bar{\lambda}) A^{3/4}\|_{\mathscr{L}(X)} \le M \tag{5.10}$$

since plainly $[V_B^{-1}(\lambda)]^* = V_B^{-1}(\bar{\lambda})$ from (1.9).

Thus, by interpolation, (5.9) and (5.10) give for all λ with $\mathrm{Re}\,\lambda > 0$

$$\|A^{1/2} V_B^{-1}(\lambda) A^{1/2}\|_{\mathscr{L}(X)} \leq M \tag{5.11}$$

Similarly, estimate (5.4) rewritten as

$$\|\lambda^2 A^{-1/4} V_B^{-1}(\lambda) A^{1/4}\|_{\mathscr{L}(X)} = \|\lambda^2 V_B^{-1}(\lambda)\|_{\mathscr{L}([\mathscr{D}(A^{1/4})]')} \leq M \tag{5.12}$$

implies for all λ with $\mathrm{Re}\,\lambda > 0$

$$\begin{aligned} \|\lambda^2 V_B^{-1}(\lambda)\|_{\mathscr{L}(\mathscr{D}(A^{1/4}))} &= \|\lambda^2 A^{1/4} V_B^{-1}(\lambda) A^{-1/4}\|_{\mathscr{L}(X)} = \|[\lambda^2 A^{1/4} V_B^{-1}(\lambda) A^{-1/4}]^*\|_{\mathscr{L}(X)} \\ &= \|\bar{\lambda}^2 A^{-1/4} V_B^{-1}(\bar{\lambda}) A^{1/4}\|_{\mathscr{L}(X)} \leq M \end{aligned} \tag{5.13}$$

Thus, by interpolation, (5.12) and (5.13) give for all λ with $\mathrm{Re}\,\lambda > 0$

$$\|\lambda^2 V_B^{-1}(\lambda)\|_{\mathscr{L}(X)} \leq M \tag{5.14}$$

However, the situation is different for estimate (5.3), which we re-write now as

$$\|\lambda A^{1/4} V_B^{-1}(\lambda) A^{1/4}\|_{\mathscr{L}(X)} = \|\lambda A^{1/2} V_B^{-1}(\lambda)\|_{\mathscr{L}([\mathscr{D}(A^{1/4})]')} \leq M \tag{5.15}$$

since (5.15) does <u>not</u> imply now by the simple operation of adjointness (as for (5.9) and (5.12)) that

$$\|\lambda A^{1/2} V_B^{-1}(\lambda)\|_{\mathscr{L}(\mathscr{D}(A^{1/4}))} \leq M, \quad \mathrm{Re}\,\lambda > 0 \tag{5.16}$$

However, a direct proof - not by interpolation - can be given [C-T.1] to show, as desired that

$$\|\lambda A^{1/2} V_B^{-1}(\lambda)\|_{\mathscr{L}(X)} \leq M, \quad \mathrm{Re}\,\lambda > 0 \tag{5.17}$$

estimates (5.11), (5.14) and (5.17) <u>collectively are equivalent</u> to the sought after inequality:

$$\|\lambda\, R(\lambda, \mathscr{A}_B)\|_{\mathscr{L}(E)} \leq C, \quad \mathrm{Re}\,\lambda > 0.$$

FOOTNOTES

1. One may readily reduce the case of A being only non-negative to the case of A being strictly positive [C-R.1, p. 434].

2. Or, still equivalently, taking energy variables

$$\frac{d}{dt}\begin{vmatrix} A^{1/2}x \\ \dot{x} \end{vmatrix} = \mathscr{L}_B \begin{vmatrix} A^{1/2}x \\ \dot{x} \end{vmatrix} \quad \text{on } W = X \times X \qquad \mathscr{L}_B = \begin{vmatrix} 0 & A^{1/2} \\ -A^{1/2} & -B \end{vmatrix}$$

with domain containing $\mathscr{D}(A^{1/2}) \times [\mathscr{D}(A^{1/2}) \cap \mathscr{D}(B)]$ as in [C.R.1].

3. Equivalently,

$$\mathscr{L}_o = \begin{vmatrix} 0 & A^{1/2} \\ -A^{1/2} & 0 \end{vmatrix} \; ; \; \mathscr{D}(\mathscr{L}_o) = \mathscr{D}(A^{1/2}) \times \mathscr{D}(A^{1/2})$$

is skew-adjoint on W: $\mathscr{L}_o = -\mathscr{L}_o^*$ and generates a s.c. group on W.

4. Explicit construction of the corresponding similarity transformation can be obtained, along the lines of the construction given in [T.2], [L-T.1, Application 4.4] with I in these references replaced by $A^{1/2}$ now.

References

[C-R.1] G. Chen and D. L. Russell, A mathematical model for linear elastic systems with structural damping, Quarterly of Applied Mathematics, Jan. 1982, 433-454.

[C-T.1] S. Chen and R. Triggiani, Proof of extension of two conjectures of G. Chen and D. L. Russell on structural damping for general elastic systems, to appear.

[D.1] N. Dunford, A survey of the theory of spectral operators, Bulletin Amer. Math. Soc., 64, 5, (1958), 217-274.

[D-S.1] N. Dunford and J. Schwartz, Linear Operators, I (1958), II (1963), III (1971), Interscience Pubs, John Wiley, New York.

[F.1] H. O. Fattorini, The Cauchy Problem, Encyclopedia of Mathematics and its Applications Addison-Wesley, Benjamin/Cummings Inc., Reading, Massachusetts 01867, 1983.

[K.1] T. Kato, Perturbation of Linear Operators, Springer-Verlag, New York, 1966.

[K.2] S. G. Krein, Linear Differential Equations in Banach Space, Translations American Mathematical Society Vol 29, American Mathematical Society, Providence, Rhode Island 02904, 1971.

[L-M.1] J. L. Lions and E. Magenes, Nonhomogeneous Boundary Value Problems, Vol. I, Springer-Verlag, New York, 1972.

[L-T.1] I. Lasiecka and R. Triggiani, Finite rank, relatively bounded perturbations of C_o-semigroups, Part II: Spectrum allocation and Riesz basis in parabolic and hyperbolic feedback systems, Annali Matematica Pura e Applicata IV, Vol. CXLIII (1986), 47-100.

[L-T.2] I. Lasiecka and R. Triggiani, Feedback semigroups and cosine operators for boundary feedback parabolic and hyperbolic equations, J. Differ. Equats. 47(1983), 246-272.

[P.1] A. Pazy, Semigroups of Operators and Applications to Partial Differential Equations, Springer-Verlag, New York, 1983.

[S.1] M. Schechter, Principles of Functional Analysis, Academic Press, New York, 1971.

[T.1] A. E. Taylor and D. C. Lay, Introduction to Functional Analysis, Second Edition, John Wiley & Sons, 1980.

[T.2] R. Triggiani, Improving stability properties of hyperbolic damped equations by boundary feedback, Springer-Verlag Lecture Notes LNCIS (1985), 400-409.

[X.1] D. Xia, Spectral Theory of Hyponormal Operators, Birkhauser Verlag, Basel-Boston-Stuttgart, 1983.

AN ITERATIVE AGGREGATION ALGORITHM FOR LINEAR PROGRAMMING

L. García
Instituto de Cibernética, Matemática y Física
Academia de Ciencias de Cuba
Calle 0 #8, Habana 4, Cuba

1. INTRODUCTION

Aggregation has often been used to obtain an approximate but much more simpler model from a very large original one. In the case of linear programming, from the solution of a column aggregate problem we can obtain, by disaggregation, an approximate solution, x^o, of the original problem; furthermore, an optimal solution, say u^o, of the dual of the aggregate problem, give us some information about how good is this approximation x^o.

Attention can be focussed on the estimation of a bound for the error of that approximation. Zipkin [1980], for example, uses the dual evaluation $p^o := (c_j - u^o a_j)_{j=1,\ldots,n}$, to calculate an a posteriori bound for the error of the approximation cx^o of the optimal value of the original problem.

Another point of view has been to use aggregation to solve the original problem itself, by improving the aggregation operator in an iterative process. In this way, some authors have also used the information contained on p^o in order to correct the previous aggregation vector.

For the general L.P. problem Martínez [1973] has defined

$$y^{i+1} := y^i + \alpha_i \frac{p^i}{\| p^i \|}$$

where (α_i) is a sequence of positive scalars previously fixed, an such that $\alpha_i \xrightarrow[i]{} 0$, $\sum_i \alpha_i = +\infty$. If the vector y^{i+1} just defined is not a feasible aggregation vector - i.e. the aggregate problem is unfeasible - then it is redefined in an alternative way.

For Markov decision process, for example, Mendelssohn [1982] corrects the row-aggregation vector in a quite similar way: from a column point of view it would be

$$y^{i+1} := (x^i + p^i)_+ \quad .$$

Martínez's algorithm, although it applies to a general L.P. problem, it has some practical disadvantages since the sequence of points it produces may be not stable from the points of view of feasibility and evaluation.

Present paper shows an iterative aggregation algorithm for general L.P. problems which also uses the direction p^o to correct a previous aggregation vector, but which has not the disadvantages mentioned above.

2. NOTATIONS

We shall use the same notations and concepts as in García [1975] .
We consider

$$P : \max \{cx \mid Ax \leqslant b, x \geqslant 0\},$$

where A is an mxn-matrix, and b and c are respectively an m-column, and an n-row vectors. We suppose P to have a finite optimum, and we denote F (respectively F^*) the set of feasible (respectively optimal) solutions of P. Let Z^* be the optimal value of P.

We denote $Q := (J_k)_{k=1,\ldots,r}$, a partition of the column-index $J := \{1,\ldots,n\}$.
Given an n-vector $y \geqslant 0$, the <u>aggregate problem</u> $\tilde{P}(y,Q)$ is defined as follows:

$$\tilde{P}(y,Q) : \max \{\sum_{k=1}^{r} \tilde{c}_k \xi_k \mid \sum_{k=1}^{r} \tilde{a}^k \xi_k \leqslant b \ ; \ \xi_k \geqslant 0, \ k=1,\ldots,r\}$$

where

$$\tilde{c}_k := \sum_{j \in J_k} c_j y_j , \quad \tilde{a}^k := \sum_{j \in J_k} a^j y_j \quad k=1,\ldots,r ,$$

and a^j denotes the j-th column of matrix A.

Any feasible solution $t^o := (t_k^o)_{k=1,\ldots,r}$ of $\tilde{P}(y^o,Q)$ can be easily disaggregated as follows:

$$x_j^o := y_j \, t_k^o \quad \text{for all } j \in J_k, \ k=1,\ldots,r. \tag{1}$$

Moreover, the evaluations of t^o and x^o in their corresponding problems, coincide.
For a given partition Q, fixed, we shall say that a vector y^o is a <u>feasible</u> <u>aggregation vector</u> of P, if $\tilde{P}(y^o,Q)$ is a feasible problem; and y^o will be an <u>optimal</u> aggregation vector if, in addition, its <u>evaluation</u> $g(y^o,Q)$ - i.e. the optimal value of $\tilde{P}(y^o,Q)$ - is equal to Z^* . We denote $S(P,Q)$, respectivaly $S^*(P,Q)$, the set of feasible, respectively optimal, aggregation vectors.

For a given partition Q, the <u>aggregation problem</u> is defined as

$$\hat{P}(Q) : \max \{g(y,Q) \mid y \in S(P,Q)\}.$$

The set of optimal solutions of $\hat{P}(Q)$ is precisely the set $S^*(P,Q)$. Thus, solving P is equivalent, in a certain sense, to solving $\hat{P}(Q)$, and an iterative aggregation algorithm for P can be viewed as an iterative optimization algorithm for $\hat{P}(Q)$.

3. ALGORITHM

Let $y^i \in S(P,Q)$. Let x^i be the disaggregation of an optimal solution of $\tilde{P}(y^i,Q)$, and let u^i be an optimal solution of $\tilde{P}'(y^i,Q)$.

We define $p^i = (p^i_j)_{j=1,\dots,n} := (c_j - u^i a^j)_{j=1,\dots,n}$, and

$d^i := (d^i_j)_{j=1,\dots,n}$, where

$$d^i_j := \begin{cases} p^i_j & \text{if } x^i_j > 0 \text{ or } p^i_j \geqslant 0 \\ 0 & \text{if } x^i_j = 0 \text{ and } p^i_j < 0 . \end{cases}$$

From the point of view of $\hat{P}(Q)$, it would be desirable to improve the aggregation vector x^i, by searching in the direction d^i for the feasible aggregation vector with the best evaluation; that is by solving

$$\max \{g(y,Q) \mid y \in M(x^i, d^i)\} \tag{2}$$

where

$$M(x^i, d^i) := \{y \mid y \in S(P, Q) ; y = x^i + \alpha d^i \text{ for some } \alpha > 0\}.$$

In fact, instead of solving (2) we propose solving $\mathring{P}_Q(x^i, d^i)$ defined below.

We need the following notations:

$$K := \{1,\dots,r\}$$

$$K^i_- := \{k \mid 1 \leqslant k \leqslant r, \quad d^{ik}_- := (-\min(0, d^i_j))_{j \in J_k} \neq 0\} ;$$

$$K^i_+ := K \setminus K^i_- ,$$

$$\beta^{ik} := \min_{\substack{j \in J_k \\ d^i_j < 0}} \left\{ \frac{x^i_j}{-d^i_j} \right\} > 0 , \quad k \in K^i_- .$$

Then, problem $\mathring{P}_Q(x^i, d^i)$ is defined as follows:

$$\max \left[\sum_{k \in K^i_+} \left[(c^k x^{ik}) t_{1k} + (c^k d^{ik}) t_{2k} \right] + \sum_{k \in K^i_-} \left[c^k (x^{ik} - \beta^{ik} d^{ik}_-) t_{1k} \right. \right.$$
$$\left. \left. (c^k d^{ik}_+) t_{2k} + (c^k d^{ik}_-) t_{3k} \right] \right]$$

subject to:

$$\sum_{k \in K^i_+} \left[(A^k x^{ik}) t_{1k} + (A^k d^{ik}) t_{2k} \right] + \sum_{k \in K^i_-} \left[A^k (x^{ik} - \beta^{ik} d^{ik}_-) t_{1k} \right.$$
$$\left. + (A^k d^{ik}_+) t_{2k} + (A^k d^{ik}_-) t_{3k} \right] \leqslant b$$

$$t_{ik} \geqslant 0 \qquad k \in K^i_+ , \ i = 1,2 ; \ k \in K^i_- , \ i = 1,2,3$$

where supraindex k denotes that we are considering only the column indices which belong to J_k.

Note that in $\mathring{P}_Q(x^i, d^i)$, variables $\{x_j \mid j \in J_k\}$ are aggregated by means of two or three subvectors: x^{ik} and d^{ik}, if $k \in K_+^i$; or $(x^{ik} - \beta^{ik} d_-^{ik})$, d_+^{ik} and d_-^{ik}, if $k \in K_-^i$.

Consequently, we define the multiple disaggregation of an optimal solution t^i of $\mathring{P}(x^i, d^i)$ as y^{i+1}, where each subvector $y^{(i+1)k}$ is defined as follows:

$$y^{(i+1)k} := \begin{cases} x^{ik} t_{1k}^i + d^{ik} t_{2k}^i & \text{if } k \in K_+^i \\ (x^{ik} - \beta^{ik} d_-^{ik}) t_{1k}^i + d^{ik} t_{2k}^i + d_-^{ik} t_{3k}^i & \text{if } k \in K_-^i . \end{cases} \tag{3}$$

PROPOSITION

1. $\mathring{P}'_Q(x^i, d^i)$ has an optimal solution.

2. y^{i+1} satisfies

   a. $y^{i+1} \in F \subset S(P,Q)$.

   b. $$cy^{i+1} = g(y^{i+1},Q) \geqslant \max\{g(z,Q) \mid z \in M(x^i, d^i)\} \geqslant g(x^i,Q) \geqslant cx^i . \tag{4}$$

Proof:

1. It is easy to see that there exists a vector t such that its multiple disaggregation - see (3) - gives x^i, which is a feasible solution of P. Thus $\mathring{P}(x^i, d^i)$ has a feasible solution. Furthermore, disaggregation (3) of any feasible solution t of $\mathring{P}_Q(x^i, d^i)$ gives a feasible solution of P with the same evaluation. Therefore, since we suppose P to have a finite optimum, so do $\mathring{P}_Q(x^i, d^i)$.

2. As we mentioned above, disaggregation (3) maintains feasibility. In order to prove (b) it suffices to note that for any $z \in M(x^i, d^i)$ and any feasible solution $\xi \in R^r$ of $\tilde{P}(z,Q)$ the vector x, obtained as the disaggregation (1) of ξ, can also be obtained by the disaggregation (3) of some feasible solution, t, of $\mathring{P}(x^i, d^i)$. The same holds for $\tilde{P}(y^{i+1},Q)$ and $\tilde{P}(x^i,Q)$. ∎

ALGORITHM:

Let $Q = (J_k)_{k=1,\dots,r}$ and $\bar{Q} = (\bar{J}_l)_{l=1,\dots,s}$ be two partitions of J.

STEP 0. Choose $y^o \in S(P,\bar{Q})$. Let $i = 0$.

STEP 1. Solve $\tilde{P}(y^i,\bar{Q})$. Let ξ^i, λ^i be optimal solutions of this problem and its dual. Let x^i be the disaggregation (1) of ξ^i.

STEP 2. Calculate $p^i := (c_j - \lambda^i a^j)_{j=1,\dots,n}$:

- If $p^i \leqslant 0$: STOP, x^i is an optimal solutions of P.
- If $p^i \not\leqslant 0$, go to Step 3.

STEP 3. Solve problem $\mathring{P}(x^i, d^i)$: Let t^i, w^i be solutions of this problem and its dual, respectively.
Let y^{i+1} be the multiple disaggregation (3) of t^i.
Set $i = i$ 1, and go to Step 1.

Remarks: Notice that from Proposition 1 this algorithm is well defined. Moreover, $cx^{i+1} \geqslant cy^{i+1} \geqslant cx^i$.

If we take $Q = \bar{Q}$ it would have no sense, according to (4), to return to Step 1 after having solved $\mathring{P}(x^i, d^i)$.
Therefore, the last sentece of Step 3 will modified as follows:

"Set $\lambda^{i+1} := w^i$, $i := i+1$ and go to Step 2." (5)

That is, problem $\mathring{P}(x^i, d^i)$ allows improving the previous aggregation vector y^i and evaluates the new one - that is, it gives solutions of $\tilde{P}(y^{i+1}, Q)$ and its dual - ; thus, solving $\mathring{P}(x^i, d^i)$ gives all necessary information to restart the process.

4. CONVERGENCE

In its general version, this algorithm is similar to those considered in García [1986 , 1987] , since $\mathring{P}(x^i, d^i)$ plays the role of the auxiliar problem $\bar{P}(x^i, d^i)$ by which the correction of the aggregation vector is performed there. Function $f_{(x,u)}(\cdot)$ considered in that case will be here the multiple disaggregation defined in (3).

In order to apply the theoretical results of García [1986] , it is just necessary to slightly modificate the original version by taking ${}^{\varepsilon_o}d^i$, where

$$ {}^{\varepsilon_o}d_j^i := \begin{cases} P_j^i & \text{if} \quad x_j^i > \varepsilon_o \quad \text{and} \quad P_j^i < -\varepsilon_o \quad \text{or} \quad P_j^i > 0 \\ 0 & \text{in other case.} \end{cases} $$

Then for ε_o sufficiently small, it would be proved, as in García [1987] , that any accumulation point, x^* , will be a stationary point of the process; that is, it couldn t be improved by solving neither $\mathring{P}(x^*, d^*)$ nor $\tilde{P}(x^*, Q)$. But, in general, these conditions are only necessary conditions of optimality for fixed partitions.

5. IMPLEMENTATION

An experimental version has been made for an IBM PC, by using Turbo Pascal. In practice, it has proved to be necessary to consider the possibility for the user, to change the current partition at any iteration, in order not only to pass over non-optimal stationary points, but also to accelerate the convergence, or even to recognize optimal points. Therefore, in our implementation, the current partition can be changed interactively.

A particular class of partitions that have proved to be interesting are those in which some subsets J_k have an only one element: in these cases we can take the corresponding aggregate columns directly from the original problem, that is, the corresponding original variables remain non-aggregated (liberated) in the aggregate problem. Our implementation allows then, the user, to select the variables to liberate, if he wants, according to the information available: $(x_j)_{j=1,\dots,n}$ and $(d_j)_{j=1,\dots,n}$.

User can also decide to apply the algorithm using Step 3 as it was described in the algorithm or as in (5), that is solving only multiaggregate problems, except for the beginning (Step 1). Also any mixed strategy is possible.

The algorithm can also be used in two phases, looking for feasibility in the first phase, if needed. It is necessary, however, to remark that it is quite easy to find a feasible aggregation vector: for example Tables 1 and 2 corresponds to a minimization problem with 10 constraints and 100 variables, from which we had not a priori information, and we took the initial aggregation vector as $y^0 := (1,1,\dots,1) \in \mathbb{R}^{100}$. In each of the showed runnings a different strategy for changing the partition was followed; and in both of them, Step 3 was taken as in (5). Iteration 0 refers to the results of Step 1 of the algorithm.

In Table 1 we have continued the iterations until we have reached the optimal value $Z^* = 855.349347$ (stop rule of Step 2). Note that feasible solutions with evaluation $\bar{Z}$, such that $|\bar{Z} - Z^*| \leqslant (0.05)Z^*$ were found quite soon.

Experimentation is just at the beginning, and our implementation has to be improved in many aspects, so we cannot state any definitive conclusions; however, the algorithm seems to be very useful in order to find a good approximated solution.

Iteration	Number of Variables	Evaluation
0	25	1 750.5372
1	25	923.9446
2	25	880.3585
3	26	869.6935
4	26	863.0116
5	23	859.1675
6	17	856.4360
7	20	855.8379
8	20	855.3588
9	20	$Z^* = 855.3493$

TABLE 1

Iteration	Number of Variables	Evaluation
0	25	1 750.5372
1	25	945.0977
2	21	898.4007
3	26	883.4057
4	26	868.7281

TABLE 2

REFERENCES

[1] Garcìa,L.: Sobre la agregación de modelos lineales. Investigacion Operacional, 15,1975,30-46.

[2] Garcia,L.: A global convergence theorem for aggregation algorithms. (Submitted to Optimization,1986).

[3] Garcia,L.: Estudio de una clase de algoritmos de agregacion. Ph.D. Dissertation. Academia de Ciencias, 1987, Cuba.

[4] Martinez,F.: Método iterativo de agregacion en Programacion Lineal. Economia y Desarrollo, 19(1973).

[5] Mendelssohn,R.: An iterative aggregation procedure for Markov decision processes. Opns.Res.,30(1982)1,62-72.

[6] Zipkin,P.: Bounds on the effect of aggregating variables in linear programs. Opns.Res.,28(1980)403-418.

OPTIMAL CONTROL OF NON LINEAR RETARDED SYSTEMS WITH PHASE CONSTRAINTS

J.A. Gómez
Institute of Cybernetics, Mathematics and Physics. Cuban Academy of Sciences.

ABSTRACT

The paper deals with necessary conditions for optimal control problems of non-linear retarded systems with phase constraints. Here necessary conditions are obtained by an application of the Dubovitskii-Milyutin approach, with the direct calculation of tangent directions of the differential constraints and the introduction of a non standard corresponding tangent cone (in L_∞). We combine also the classical "peak variation" approach, in the context of piecewise continuous control functions, and we don't need the Milyutin time transformation, making the proof, in some sense, more "elementary". On the other hand we need a convex assumption in order to get the global maximum principle. The regularization of the Lagrange multiplier corresponding to the phase constraint is done following Ledzewicz- Kowalewska(1985).

1. INTRODUCTION

We consider the following problem,

$$\text{Minimize} \quad \varphi\,[x(t_1)] \tag{1.1}$$

subject to

$$x(t) = f[x(t),x(t-h(t)),u(t),t]\ , \quad t \in [t_0,t_1] \tag{1.2}$$

$$x(t) = g(t)\ , \quad t < t_0 \quad , \quad x(t_0) = x_0 \tag{1.3}$$

$$x(t) \in \mathbf{X}(t) = \left\{ x \in \mathbb{R}^n :\ \mathbf{Q}(x,t) \le 0 \right\},\ t \in [t_0,t_1] \tag{1.4'}$$

$$u(t) \in \mathbf{U}\ , \quad t \in [t_0,t_1], \tag{1.5}$$

where $x(t) \in \mathbb{R}^n$, $u(t) \in \mathbb{R}^r$, $f : \mathbb{R}^n \times \mathbb{R}^n \times \mathbb{R}^r \times [t_0,t_1] \text{------>} \mathbb{R}^n$,

$\varphi : \mathbb{R}^n \text{---->} \mathbb{R}$, $g : [t_0 - \tilde{h}, t_0] \text{---->} \mathbb{R}^n$, $h(t) \le \tilde{h}$ for all t,

$\mathbf{Q} : \mathbb{R}^n \times [t_0,t_1] \text{---->} \mathbb{R}$, $\mathbf{U} \subset \mathbb{R}^r$, and we make the following assumptions:

(A.1) φ is a continuously differentiable function with $\|D_x\varphi\|$ bounded in bounded sets.

(A.2) f is continuous respect to (x,y,u,t) and continuously differentiable respect to (x,y), with $\|D_x f\|$ and $\|D_y f\|$ bounded in bounded sets.

(A.3) h is positive and continuously differentiable, with derivative $\dot{h}(t) < 1$, for all t. In other words, $\nu(t) = t-h(t)$ is an increasing function on $[t_0,t_1]$.

(A.4) g is continuous, and $x_0 \in \mathbf{X}(t_0)$.

(A.5) $\mathbf{Q}$ is continuous in (x,t) and continuously differentiable respect to x, with $\|D_x\mathbf{Q}\|$ bounded in bounded sets.

(A.6′) The control functions are piecewise continuous, and $\mathbf{U}$ is a bounded set.

The problem, techniques and results that we will handle are not news, see for example Colonius(1982) for references, but the author wanted to answer himself the following questions:

a) Is it possible to obtain the <u>global</u> Maximum Principle for problem (1.1)-(1.5) with the Dubovitskii-Milyutin approach (see Girsanov(1972)), without the introduction of time transformations, which reduce it to apply the <u>local</u> one ?.

b) In order to get a Maximum Principle, is it possible to combine the classical <u>non-linear</u> techniques of "peak variations" with the beautiful and more recent (but <u>linear</u> !) functional analytical approach ?. Do this combination, if it exists, also works for retarded systems?.

c) How long we can go with <u>piecewise</u> <u>continuous</u> functions for controls ?.

In the solution of these questions, it appears convenient to rewrite the phase constraint (1.4′) in the following form:

$$x_{t_1} \in \mathbf{X} = \left\{ \Phi \in \mathcal{C}^n [-H,0] : \mathbf{Q}[\Phi(s),s] \le 0 , s \in [-H,0] \right\} \qquad (1.4)$$

where $x_t(s) = x(t+s)$, $s \in [-H,0]$, and $H = t_1 - t_0$.

It's clear that both conditions are equivalent. Introducing the operator:

$\bar{Q} : L^n_\infty [-H,0] \longrightarrow L_\infty [-H,0]$, defined by

$$\bar{Q}(\Phi)(s) = Q(\Phi(s), t_1 + s) \quad , \quad \text{a.e.} \quad s \in [-H,0] \,, \tag{1.6}$$

then $X = \left\{ \Phi \in \mathscr{C}^n([-H,0]) : \ \bar{Q}(\Phi)(s) \leq 0 \right\}$. Furthermore, $\bar{Q}$ is Frechét differentiable respect to essup norm, and

$$D_\Phi \bar{Q}(\Phi_o)(\Phi)(s) = D_x Q(\Phi_o(s), t_1 + s)\,\Phi(s) \,, \quad \text{a.e.} \quad s \in [-H,0],$$
$$\Phi_o \in \mathscr{C}^n[-H,0], \Phi \in L^n_\infty[-H,0]. \tag{1.7}$$

2. PRELIMINARIES

Let be u^o, x^o the optimal control and corresponding trajectory. Following classical methods (see for example Gabasov-Kirillova (1974)) we can write the increment $\Delta x(t) = x(t) - x^o(t)$ corresponding to another feasible pair x, u, in the following way,

$$x(t) = \mathscr{F}(t,\tau)\,\Delta x(\tau) + \int_{\nu(\tau)}^{\tau} \left\{ 1_{[\nu(\tau),\nu(t)]}(s) \quad \mathscr{F}(t,\lambda(s))\, D_y f^o(u^o(\lambda(s)),\right.$$
$$\left.\lambda(s))\ \lambda'(s) \right\} \Delta x(s)\, ds \ + \int_{\tau}^{t} \mathscr{F}(t,s)\ \Delta_{u(s)} f^o(u^o(s),s) ds + \eta \tag{2.1}$$

where:

$$f^o(u,s) := f(x^o(s), x^o(\nu(s)), u, s), \qquad 1_{[a,b]}(s) := \begin{cases} 1 & \text{for} \quad s \in [a,b] \\ 0 & \text{for} \quad s \notin [a,b] \end{cases}$$

$\Delta_u f^o(v,s) := f^o(u,s) - f^o(v,s)$, $\lambda(s) := \nu^{-1}(s)$, and

$\mathscr{F}(t,s,\tau)$ is the solution of the matrix differential equation:

$$D_s F(t,s) = -F(t,s)\, D_x f^o(u^o(s),s) - 1_{[\tau,\nu(t))}(s)\ F(t,\lambda(s))\ D_y f^o($$
$$u^o(\lambda(s)), \lambda(s))\ \lambda'(s) \quad , \qquad s \in [\tau,t]. \tag{2.2}$$

with initial conditions:

$$F(t,s) \equiv 0 \quad \text{for} \quad s > t > \tau \quad , \quad F(t,t) = E_n \begin{pmatrix}\text{identity} \\ \text{matrix}\end{pmatrix}. \tag{2.3}$$

We denote $\mathcal{F}(t,s):=\mathcal{F}(t,s,t_o)$, and by η the rest of non-linear terms respect to Δx. For a multiple peak variation:

$$u_p(t) := \begin{cases} u^o(t) & \text{for } t \notin \bigcup_{i=1}^{p} [\theta_i,\theta_i+\lambda_i\varepsilon) , \quad \theta_i \in [t_o,t_1], \\ v_i & \text{for } t \in [\theta_i,\theta_i+\lambda_i\varepsilon) \quad v_i \in \mathbf{U} , \lambda_i > 0 \end{cases} \tag{2.4}$$

$\varepsilon > 0$, we have the following:

Lemma 1: The trajectory corresponding to (2.4) can be written, for $\varepsilon > 0$ small enough, in the following way:

$$x_p(t) = x(t,u_p) = x^o(t) + \varepsilon \sum_{i=1}^{p} \lambda_i \ell_{\theta_i v_i}(t) + o(\varepsilon) \tag{2.5}$$

where,

$$\ell_{\theta v}(t) := \mathcal{F}(t,\theta)\ \Delta_v f^o(u^o(\theta),\theta) \quad , \quad t \in [t_o,t_1]. \tag{2.6}$$

Proof: Let be $p = 1$, $\tau = \theta + \lambda\varepsilon$, $t \in (\theta+\lambda\varepsilon,t_1]$.
For ε small enough $\nu(\theta+\lambda\varepsilon) < \theta$, and by (2.1) and uniqueness

$$\Delta x(t) = \mathcal{F}(t,\theta+\lambda\varepsilon)\ \Delta x(\theta+\lambda\varepsilon) + \int_{\theta}^{\theta+\lambda\varepsilon} \Big\{ 1_{[\theta,\nu(t))}(s)\ \mathcal{F}(t,r(s))\ D_y f^o(u^o(r(s)), r(s))\ r'(s) \Big\} \Delta x(s)\ ds + \eta \tag{2.7}$$

It is a well known result (see Gabasov-Kirillova (1974)) that under conditions (A.1)-(A.6) we have $\|\Delta x(t)\| \le K\varepsilon$, and then $\eta = o(\varepsilon)$, uniformly in $t \in [t_o,t_1]$. We get,

$$\Delta x(t) = \mathcal{F}(t,\theta+\lambda\varepsilon)\ \Delta x(\theta+\lambda\varepsilon) + o(\varepsilon) =$$
$$= \lambda\ \varepsilon\ \mathcal{F}(t,\theta)\ \Delta_v f^o(u^o(\theta),\theta) + o(\varepsilon), \quad t \in (\theta+\lambda\varepsilon,t_1]. \tag{2.8}$$

Taking now $\tau = \theta$, and $t \in (\theta,\theta+\lambda\varepsilon]$ in (2.1), similar arguments show that (2.8) remains true for $t \in (\theta,t_1]$, and then, by definition (2.4) and uniqueness, for all $t \in [t_o,t_1]$.

Assume, by induction, that for any peak variation of (p-1) points we have:

$$\Delta x(t,u_{p-1}) = x(t,u_{p-1}) - x^o(t) = \varepsilon \sum_{i=1}^{p-1} \lambda_i\ \ell_{\theta_i v_i}(t) + o(\varepsilon) , \tag{2.9}$$

We consider now the peak variation of p points (2.4), with $\theta_1 < \theta_2 < \ldots < \theta_p < t_1$. Take $\tau = \theta_p$, $u = u_{p-1}$, $t \in (\theta_p,t_1]$ in (2.1); then for ε small enough we have $\theta_{p-1} + \lambda_{p-1}\varepsilon < \theta_p$ and

$$\Delta x(t,u_{p-1}) = \mathcal{F}(t,\theta_p)\ \Delta x(\theta_p,u_{p-1}) + \int_{\nu(\theta_p)}^{\theta_p} \Big\{ 1_{[\nu(\theta_p),\nu(t))}(s)\ \ \mathcal{F}(t,\imath(s))\ D_y f^\circ(u^\circ(\imath(s)),\imath(s))\ \imath'(s) \Big\}\ \Delta x(s,u_{p-1})\ ds\ +\ o(\varepsilon) \qquad (2.10)$$

On the other hand for $\tau = \theta_p$, $u = u_p$, $t \in (\theta_p,t_1]$, we have in (2.1):

$$\Delta x(t,u_p) = \mathcal{F}(t,\theta_p)\ \Delta x(\theta_p,u_p) + \int_{\nu(\theta_p)}^{\theta_p} \Big\{ 1_{[\nu(\theta_p),\nu(t))}(s)\ \ \mathcal{F}(t,\imath(s))\ D_y f^\circ(u^\circ(\imath(s)),\imath(s))\ \imath'(s) \Big\}\ \Delta x(s,u_p)\ ds\ +\ \int_{\theta_p}^{t} \mathcal{F}(t,s)\ \Delta_{u_p(s)} f^\circ(u^\circ(s),s)\ ds\ +\ o(\varepsilon), \qquad t \in (\theta_p,t_1].$$

Using (2.10) and $u_p = u_{p-1}$ for $t \in [t_o,\theta_p)$ we have $\Delta x(t,u_p) =$ $= \Delta x(t,u_{p-1}) + \varepsilon\,\lambda_p \ell_{\theta_p v_p}(t) + o(\varepsilon)$, $t \in [t_o,t_1]$,

and using now (2.9) we obtain (2.6).∎

3. *CONES AND DUAL CONES*

We define $\mathcal{X} = \mathbb{R}^n \times L^n_\infty[-H,0]$, with the usual product norm. We consider also, $\{[-H,0],\Sigma,m\}$ as the measure space $[-H,0]$ with the σ-algebra Σ and the Lebesgue measure m; *ba*$[-H,0]$, denotes the family of additive set-functions $d\omega : \Sigma \longrightarrow \mathbb{R}$ satisfying the following conditions:

i) $A \in \Sigma$, $m(A)=0$ then $d\omega(A)=0$

ii) $d\omega$ is of bounded variation, $|d\omega|_{[-H,0]} < +\infty$.

Let be the cones:

$K_o = K^d_o \times L^n_\infty[-H,0]$, $K_1 = \mathbb{R}^n \times K^l_1$, $K_2 = K^d_2 \times K^l_2$, where

$$K^d_o := \Big\{ d \in \mathbb{R}^n :\ D_x\varphi\,(x^\circ(t_1))\ d\ <\ 0 \Big\},$$

$$K^l_1 := \Big\{ \ell \in L^n_\infty[-H,0] :\ \text{essup}\ [\ D_\Phi \bar{Q}(x^\circ_{t_1})(\ell);\ R\ ] < 0 \Big\},$$

$$K^d_2 := \Big\{ \lambda d \in \mathbb{R}^n :\ \lambda > 0\ ,\ d = \ell_{\theta v}(t_1)\ ,\ \theta \in [t_o,t_1],\ v \in U \Big\},$$

$$K^l_2 := \Big\{ \lambda\ell \in L^n_\infty[-H,0] : \lambda > 0,\ \ell = (\ell_{\theta v})_{t_1},\ \theta \in [t_o,t_1],\ v \in U \Big\},$$

and where:

$$\text{essup}\,[\Phi\,;\,R\,] := \sup\left\{\mu\in\mathbb{R} : m\left[s\in R : \Phi(s) > \mu\right] > 0\right\},$$

$$R := \left\{ s \in [-H,0] : \ \mathbf{Q}(x^{o}(t_1+s), t_1+s) = 0 \right\}.$$

K_o and K_1 are open and convex cones in $\mathscr{E}$. We have also

Lemma 2:

a) If $K_o \neq \emptyset$ then the dual cone of K_o has the form

$$K_o^* = (K_o^d)^* \times \{0\} \quad , \text{ where } (K_o^d)^* = \left\{ \lambda\, D_x\varphi\,(x^o(t_1))\ ,\ \lambda \le 0 \right\},$$

b) If the following assumption holds:

(A.7) $D_x\mathbf{Q}[x^o(t_1+s),\ t_1+s\,] \neq 0$, for almost all $s \in R$,

then $K_1^* = \{0\} \times (K_1^l)^*$, with $(K_1^l)^* = \left\{ \omega \circ D_\Phi\bar{\mathbf{Q}}\,(x^o_{t_1})\ ;\ \omega \in N_1^* \right\}$

and $N_1 = \left\{ \ell \in L^n_\infty\,[-H,0] : \ \ell\,(s) \le 0\ ,\ \text{a.e. } s \in R \right\}.$

Furthermore, if $\omega \in N_1^*$ then there exists an additive set-function $d\omega \in ba[-H,0]$ such that:

i) $d\omega$ is concentrated on the set R (i.e.,if $A \in \Sigma$, $A \cap R = \emptyset$, then $d\omega(A)=0$).

ii) $\omega\,[\ell\,] = \int_{-H}^{0} \ell(s)\, d\omega(s)$, for all $\ell \in L^n_\infty\,[-H,0]$.

c) The dual cone of K_2 has the form:

$$K_2^* = \left\{ (z,\omega) \in \mathscr{E}^* : z^T \ell_{\theta v}(t_1) + \omega\left[(\ell_{\theta v})_{t_1} \right] \ge 0\ ;\ \begin{matrix} \theta \in [t_o,t_1) \\ v \in \mathbf{U} \end{matrix} \right\}$$

Proof: a) is a well known result (Girsanov(1972)), and c) follows by definition of dual cone and the form of K_2 .

For b) we consider the open cone:

$N = \left\{ \ell \in L_\infty[-H,0] : \ \text{essup}\,[\ell\,;\,R\,] < 0 \right\}$. It's easy to see that N_1 is the norm clousure of N , and that we have

$K_1^l = \left\{ \ell \in L^n_\infty[-H,0] : D_\Phi\mathbf{Q}(x^o_{t_1})\,(\ell) \in N \right\}$. Then, by (A.7) the cone K_1^l is not empty and the first result follows from Farkas-Minkowsky Lemma.The form of the functionals belonging to N_1^* is a classical result (see for example Kantorovich et al.(1977)).

4. NECESSARY CONDITIONS

Lemma 3:

a) We have the relation $K_o \cap K_1 \cap K_2 = \emptyset$.

b) If in addition, the following assumption holds:

(A.8) For all $\theta \in [t_o, t_1)$, the set $f(x^o(\theta), x^o(\nu(\theta)), \mathbf{U}, \theta)$ is a convex set in $\mathbb{R}^n$,

then, we also have $K_o \cap K_1 \cap \mathrm{co}\, K_2 = \emptyset$, where co K denote the convex hull of the cone K.

Proof: If $(d, \ell) \in K_o \cap K_1 \cap K_2$, then

$d = \ell_{\theta v}(t_1)$, $\ell = (\ell_{\theta v})_{t_1}$, for some $\theta \in [t_o, t_1)$ and $v \in \mathbf{U}$.

In addition,

$$D_x\varphi\,(x^o(t_1))\,\ell_{\theta v}(t_1) < 0 \tag{4.1}$$

and there exists a negative number α such that

$$D_x\mathbf{Q}^T(x^o(t_1+s), t_1+s)\,\ell_{\theta v}(t_1+s) \le \alpha < 0, \text{ a.e. } s \in R. \tag{4.2}$$

By definition of $\ell_{\theta v}$ and $\mathcal{F}$, we have $\ell_{\theta v}(t_1+s) = 0$, for $s \in [-H, \theta - t_1)$, and $\ell_{\theta v}$ is continuous for $s \in [\theta - t_1, 0]$.

Then (4.2) is true for <u>all</u> $s \in R_\theta := R \cap [\theta - t_1, 0]$. Furthermore, there exists a number $\delta > 0$, such that (4.2) holds for all

$$s \in R_\delta(\theta) := \left\{ s \in [\theta - t_1, 0] : \inf_{\sigma \in R_\theta} |s - \sigma| \le \delta \right\}.$$

Consider now the peak variation $\tilde{u}$ with parameters θ, v, and the corresponding trajectory $\tilde{x}$. For $s \in [-H, \theta - t_1]$ we have $x^o(s) = \tilde{x}(s)$, and $\mathbf{Q}\,(\tilde{x}(t_1+s), t_1+s) \le 0$.

For $s \in R^c_\delta(\theta) := [\theta - t_1, 0] \setminus R_\delta(\theta)$, we have by definition and compactness,

$$\max_{s \in R^c_\delta(\theta)} \mathbf{Q}\,(x^o(t_1+s), t_1+s) = \rho < 0.$$

Using (A.5) and Lemma 1 for p = 1, we obtain,

$$\mathbf{Q}\,[\tilde{x}(t_1+s), t_1+s] = \mathbf{Q}\,[x^o(t_1+s), t_1+s] + \varepsilon\, D_x\mathbf{Q}^T[x^o(t_1+s), t_1+s]\,\ell_{\theta v}(t_1+s) + o(\varepsilon) < \rho/2,$$

for $\varepsilon \in (0, \varepsilon_o)$, and $s \in R^c_\delta(\theta)$.

Finally, for $s \in \bar{R}_\delta(\theta)$, we use again similar arguments and ,

$$\mathbf{Q}\,[\tilde{x}(t_1+s),t_1+s] \le \mathbf{Q}\,[x^o(t_1+s) + \varepsilon\,\ell_{\theta v}(t_1+s),t_1+s] +$$
$$+ M\,\|o(\varepsilon)\| = \mathbf{Q}\,[x^o(t_1+s),t_1+s] + \varepsilon\,D_x^T\mathbf{Q}\,[x^o(t_1+s),t_1+s]\,\ell_{\theta v}(t_1+s) + o_1(\varepsilon) \le \varepsilon\,\rho_1 + o(\varepsilon) < 0 ,$$

for $\varepsilon>0$, small enough.

The preceding inequalities show that $\tilde{x}(t)$ is a feasible trajectory. This is a contradiction, considering that $\varphi\,[\tilde{x}(t_1)] < \varphi[x^o(t_1)]$, for $\varepsilon > 0$, small enough. This proved a).

For b) we consider a convex combination of elements of cone K_2, which belongs to the intersection . This combination depends on points $\theta_1,\dots,\theta_p \in [t_o,t_1)$, and vectors $v_1,\dots,v_p \in \mathbf{U}$. By (A.8) we can consider only the case $\theta_1 < \theta_2 < \dots < \theta_p$, and we get a contradiction by the same arguments that before, but considering now the sets $R(\theta_i) := R \cap [\theta_i - t_1, \theta_{i+1} - t_1]$, $i = 1,\dots p$. ■

Lemmas 2 ,3 and the Dubovitskii-Milyutin Lemma (see Girsanov(1972), Lemma 5.11) give the existence of multipliers, not all zeros, $\bar{z} \in \mathbb{R}^n$, $\bar{\omega} \in (L_\infty^n[-H,0])^*$, $(-\omega_o) \in N_1^*$, $\lambda_o \ge 0$, such that

$$\bar{z} = \lambda_o\,D_x\varphi\,[x^o(t_1)] \quad , \quad \bar{\omega} = \omega_o \circ D_\Phi\bar{\mathbf{Q}}\,(x^o_{t_1}) , \tag{4.3}$$

$$\bar{z}^T\,\ell_{\theta v}(t_1) + \bar{\omega}\left[D_\Phi\bar{\mathbf{Q}}\,(x^o_{t_1})[(\ell_{\theta v})_{t_1}]\right] \ge 0 , \quad \theta \in [t_o,t_1), \quad v \in \mathbf{U} . \tag{4.4}$$

This inequality is already a necessary condition for optimality.

5. REGULARIZATION AND MAXIMUM PRINCIPLE

Each functional $\omega \in (L_\infty^n[-H,0])^*$ of the form:

$$\omega\,[\ell] = \int_{-H}^{0} w_a(s)\,\ell(s)\,ds , \quad \text{where } w_a \in L_1[-H,0],$$

will be called *absolutely continuous.*

It is known that any linear and continuous functional ω over L_∞ can be written in the form $\omega = \omega_a + \omega_s$, where ω_a is an absolutely continuous functional and ω_s is a singular functional. This representation is unique. With this notation, the second equality in (4.3) can be written in the following form:

$$\bar{\omega}_a + \bar{\omega}_s = (\omega^o_a + \omega^o_s) \circ D_{\Phi}\bar{\mathbf{Q}}(x^o_{t_1}) .$$

A functional which is simultaneously singular and absolutely continuous is equal to zero, then

$$\bar{\omega}_a = \omega^o_a \circ D_{\Phi}\bar{\mathbf{Q}}(x^o_{t_1}) \quad , \quad \bar{\omega}_s = \omega^o_s \circ D_{\Phi}\bar{\mathbf{Q}}(x^o_{t_1})$$

and the relation (4.4) addopts the form:

$$\lambda_o D_x\varphi\, [x^o(t_1)]\, \ell_{\theta v}(t_1) + \int_{-H}^{\theta} w^o_a(s)\, D_x\mathbf{Q}\, [x^o(t_1+s), t_1+s]\, \ell_{\theta v}(t_1+s)\, ds + \omega^o_s\Big[D_{\Phi}\bar{\mathbf{Q}}\, (x^o_{t_1})\Big((\ell_{\theta v})_{t_1}\Big)\Big] \geq \emptyset,\ \theta \in [t_o,t_1) ,\quad v \in \mathbf{U} . \tag{5.2}$$

Define the conjugate function:

$$\Psi^o(t) = \Psi^o_a(t) + \Psi^o_s(t), \quad \text{where}$$

$$\Psi^o_a(t) = -\Lambda_o\, \mathcal{F}\, (t_1,t) - \int_{t-t_1}^{\theta} \Lambda(s)\, \mathcal{F}\, (t_1+s,\, t)\, ds \ ,\ t \in [t_o,t_1], \tag{5.3}$$

and

$$\Lambda_o := \lambda_o\, D_x\varphi\, [x^o(t_1)]\ ,\ \Lambda(s) := \omega^o_a(s)\, D_x\mathbf{Q}\, [x^o(t_1+s), t_1+s\,],$$

$$s \in [-H,\emptyset\,] .$$

$$\Psi^o_s(t) = \int_{t-t_1}^{\emptyset} D_x\mathbf{Q}\, [x^o(t_1+\sigma), t_1+\sigma]\, \mathcal{F}\, (t_1+\sigma,t)\, d\omega^o_s(\sigma) . \tag{5.4}$$

From (5.3) and (2.2)-(2.3), it's not difficult to prove that $\Psi^o_a(t)$ is the solution of the following differential equation:

$$\dot{\Psi}(t) = -\Psi(t)\, D_x f^o(u^o(t),t) - \mathbf{1}_{[t_o,\nu(t_1))}(t)\quad \Psi(\kappa(t))\, D_y f^o(u^o(\kappa(t)), \kappa(t))\, \kappa'(t) + \Lambda(t-t_1) \ ,\quad t \in [t_o,t_1]. \tag{5.5}$$

$$\Psi^o(t_1) = -\Lambda_o \tag{5.6}$$

With this notation, (5.2) addopts the following classical form:

$$\mathcal{H}(x^o(t),x^o(\nu(t)),\Psi^o(t),u^o(t),t) = \max_{u \in \mathbf{U}} \mathcal{H}\, (x^o(t),x^o(\nu(t)),\Psi^o(t),u,t)$$

$$\text{a.e. } t \in [t_o,t_1) , \tag{5.7}$$

where $\mathcal{H}\, (x,y,\Psi,u,t) = \Psi^T f(x,y,u,t)$.

Note also that λ_o and $\Psi^o(t)|_{[\nu(t_1),t_1]}$ can't be simultaneously zeros. We have then, the following:

Theorem: Let be u^o, x^o the optimal control and corresponding trajectory of problem (1.1)-(1.5), with assumptions (A.1)-(A.8). Then there exist $\lambda_o \geq 0$, $\Psi^o_a(t)$, $\Psi^o_s(t)$ non simultaneously zeros, such that $\Psi^o_a(t)$ is the solution of (5.5)-(5.6) and the maximum condition (5.7) holds.

Final remarks:

1) The result can be compared with those of Banks-Kent(1972), Bien-Chyung(1980) ,Colonius(1982) and Ledzewicz-Kowalewska(1985). The first three papers deals with more general problems than (1.1)-(1.5), but for equality <u>end</u> phase conditions. The form of the results are similar and only the methods to obtain them are different from that we use. The regularization of the L_∞ multiplier is largely based on Ledzewicz-Kowalewska(1985), where inequality and equality phase constraints are considered, but only local maximum principle is obtained .

2) A similar sufficient condition to that given in Colonius(1982),can be given for $\lambda^o \neq 0$, following the classical remark of Girsanov(1972).

6. REFERENCES.

1) H.T. Banks, G.A. Kent (1972). *Control of functional differential equations of retarded and neutral type with target sets in function space.* SIAM Journal of Control, 10, pp. 567-593.

2) Z. Bien, D.H. Chyung (1980). *Optimal control of delay systems with a final function condition.* Int. Journal of Control, 32, pp. 539-560.

3) F. Colonius (1982). *The maximum principle for relaxed hereditary differential system with function space end condition.* SIAM Journal of Control, Vol. 20, No.5, pp. 695-712.

4) R.F. Gabasov, F.M. Kirillova (1974). *The maximum principle in the theory of optimal control problems* (Russian). Ed. Nauka y Tejnika. Minsk.

5) I.V. Girsanov (1972). *Lectures on Mathematical Theory of Extremum problems.* Springer Verlag. New York.

6) J. Hale (1977). *Theory of functional differential equations.* 2nd. edition. Springer Verlag. New York.

7) U. Ledzewicz-Kowalewska (1985). *A necessary condition for a problem of optimal control with equality and inequality constraints .* Control and Cybernetics, Vol. 14, No.4, pp. 351-360.

LIST OF CONTRIBUTORS AND PARTICIPANTS

AGUILERA, J.; Universidad Central de Venezuela, Caracas, VENEZUELA.

ALLENDE S.; Facultad de Matemática-Cibernética, U. de La Habana, La Habana, CUBA.

ALVAREZ, L.; ICIMAF, Cuban Acad. of Sc., O #8, La Habana, Cuba.

ALVAREZ, R.M.; ICIMAF, Cuban Academy of Sciences, O #8, La Habana, CUBA.

ALVAREZ, V.; Facultad de Matemática-Cibernética, U. de La Habana, La Habana, CUBA.

BANK, B.; Humboldt University, Dept. of Mathematics, PSF 1297, Berlin, GDR.

BARONTI, M.; Dipartamento dei Matematica, University of Parma, I-43100 Parma, ITALY.

BARRIENTOS, O.; Universidad Autonoma Tomas Frias, Potosi, BOLIVIA.

BLESS, I.; ICIMAF, Cuban Academy of Sciences, O #8, La Habana, CUBA.

BOUZA, C.; Facultad de Matemática-Cibernética, U. de La Habana, La Habana, Cuba.

BROKATE, M.: Math. Dept., Augsburg Univ., Augsburg, FRG.

BROSOWSKI, B.; Math. Dept., Frankfurt Univ., Frankfurt, FRG.

BUNKE, O., Humboldt University, Dept. of Mathematics, PSF 1297, Berlin, GDR.

BUSTAMANTE, J.; Facultad de Matemática-Cibernética, U. de La Habana, La Habana, CUBA.

CACHAFEIRO, M.A.; Dept. Matematica, E.T.S. Ingenieros Industriales, 36280 Vigo, ESPAÑA.

CALA, F.; Facultad de Matemática-Cibernética, U. de La Habana, La Habana, CUBA.

CALVILLO, G.; IPN, Mexico D.F., MEXICO.

CASAS, O.; ISPETP, La Habana, CUBA.

CASTELLANOS, L.; ICIMAF, Cuban Academy of Sciences, O #8, La Habana, CUBA.

CASTRO, A. ; Facultad de Matemática-Cibernética, U. de La Habana, La Habana, CUBA.

CHEN, SHUPIN; Dept. of Mathematics, Zhejiang Univ., Hangzhou, CHINA.

CIESIELSKI, Z.; Mathematical Institute, PAN, ul Abrahama 18, 81-825 Sopot, POLAND.

CRIBEIRO, J.; Facultad de Matemática-Cibernética, U. de La

Habana, La Habana, CUBA.

CSERNYAK, L.; Buzogani ul. 10, Budapest Pf. 35. 1426, HUNGARY.

CUERVO ALVAREZ, M; Inst. Invest. Económicas JUCEPLAN, Plaza de la Revolución, La Habana, CUBA.

CUESTA, L.E.; Facultad de Matemática-Cibernética, U. de La Habana, La Habana, CUBA.

DAHMEN, W.; University of Bielefeld, Fakul. für Mathematik, 48 Bielefeld, FRG.

DRAUX, A.; Université de Lille Flandres Artois, Laboratoire d'Analyse Numérique-Bat M3, 59665 VILLENEUVE D'ASCQ-CEDEX, FRANCE.

DUNHAM, Ch.; Dept. of Computer Science, University of Western Ontario, London, Ontario N6A 5B7, CANADA.

ECHEVERRY CANO, A.; Universidad Nacional de Colombia, Medellin, COLOMBIA

ERDELYI, T.; Mathematical Institute AC-Hungary, Budapest, HUNGARY.

FALCON C. ; Facultad de Matemática-Cibernética, U. de La Habana, La Habana, CUBA.

FEICHTINGER, G.; Inst.Econometry and Op. Res., University of Wien, AUSTRIA.

FERNANDEZ, J.L.; Facultad de Matemática-Cibernética, U. de La Habana, La Habana, CUBA.

FUCHS, W.H.; Math. Dept., Cornell University, Ithaca, New York, USA

GARCIA OLIVEROS, S.; Fac. de Mat., Universidad de Puebla, Puebla, MEXICO.

GARCIA, L.; ICIMAF, Cuban Academy of Sciences, O #8, La Habana, CUBA.

GARCIA, M.A.; Facultad de Matemática-Cibernética, U. de La Habana, La Habana, CUBA.

GNEDENKO, B.V.; Fac. of Math. and Mech., Moscow State Univ., Moscow, USSR.

GOMEZ, J.A.; ICIMAF, Cuban Academy of Sciences, O #8, La Habana, CUBA.

GONCHAR, A.A.; Institute of Mathematics, USSR Acad. of Sc., ul. Vavilova 42, Moscow, USSR.

GOODMAN, T.N.T. ; Dept. of Mathematics, University of Dundee, Dundee, Scotland.

GRAGG, W.; Math. Dept., University of Kentucky, Lexington, USA.

GRAVE de PERALTA, R.; Editorial Pueblo y Educación, 3ra y 60,

Playa, La Habana, CUBA.

GUADALUPE HERNANDEZ, J.; Fac. de Matemática, Universidad de Zaragoza, Zaragoza, ESPAÑA.

GUDDAT, J.;Humboldt University, Dept. of Mathematics, PSF 1297, Berlin, GDR.

GUERRA. F. ; Facultad de Matemática-Cibernética, U. de La Habana, La Habana, CUBA.

GUERRA,V.; ICIMAF, Cuban Academy of Sciences, O #8, La Habana, CUBA.

GUZZARDI R.; Dip. di Mat., Universitá della Calabria, Italy.

HERNANDEZ, R.; Fac. de Matemática, Inst.Pedag. E.J.Varona, La Habana, CUBA.

HERNANDEZ, V.; ICIMAF, Cuban Academy of Sciences, O #8, La Habana, CUBA.

HERRERA, J.M.; Instituto de Hidroeconomia, Calle E #152, Stgo.Cuba, CUBA

HILGERS, J.W.; Michigan Technological University, Michigan 49931, USA.

HING CORTON, R.; Fac. de Cib., Universidad Central, Santa Clara, CUBA.

HINRICHSEN, D.; Dynamical System Centre, University of Bremen, FRG.

ILLAN, J. ; Facultad de Matemática-Cibernética, U. de La Habana, La Habana, CUBA.

IVANOV, K.; Inst. Mathematics, Bulgarian Acad. of Sc., 1090 Sofia, BULGARIA.

JIMENEZ, M.; Facultad de Matemática-Cibernética, U. de La Habana, La Habana, CUBA.

JONGEN, H.T.; Twente University of Technology, Faculty of Applied Mathematics, P.O.Box 217, 7500 AE Enschede, The NETHERLANDS.

KAKES A.; Facultad de Matemática-Cibernética, U. de La Habana, La Habana, CUBA.

KOSLOVAR, E.; ICIMAF, Cuban Academy of Sciences, O #8, La Habana, CUBA.

KUFNER, A.; Mathematical Institute, Chec. Acad. of Sc., Prague, CHECOSLOVAQUIA.

LEMAGNE, J.; Facultad de Matemática-Cibernética, U. de La Habana, La Habana, CUBA.

LOBO HIDALGO, M.; Fac. de Mat., Universidad de Santander, Santander, ESPAÑA.

LOPEZ, G.; Facultad de Matemática-Cibernética, U. de La Habana, La Habana, CUBA.

LOPEZ, G.; Fac. de Mat., Universidad Autonoma de Puebla, Puebla, MEXICO.

LOPEZ, M.; ICIMAF, Cuban Academy of Sciences, O #8, La Habana, CUBA.

LOPEZ, E.F.; Fac. de Mat., Inst. Pedag. E.J. Varona, La Habana, CUBA.

LORCH, L.A.; York University, 4700 Keele st., North York, Ontario M3J 1P3, CANADA.

MARCELLAN, F.; Dept. de Matemática Aplicada, E.T.S. Ingenieros Industriales, 28006 Madrid, ESPAÑA.

MARRERO, A.; Facultad de Matemática-Cibernética, U. de La Habana, La Habana, CUBA.

MARTINEZ, A.; Facultad de Matemática-Cibernética, U. de La Habana, CUBA.

MEDEROS, M.V.; Facultad de Matemática-Cibernética, U. de La Habana, La Habana, CUBA.

MESA, A.; Facultad de Matemática-Cibernética, U. de La Habana, La Habana, CUBA.

MICCHELLI, Ch.A.; Mathematical Science Dpt. IBM Thomas J. Watson Research Center, Yorktown Heights, New York 10598, USA.

MOHEDANO, M.V.; Universidad de Alcahalá de Henares, ESPAÑA.

MOTSCHA, M. ;Regionales Rechenzentrum, U. of Bremen, Bremen, FRG.

NEVAI, P.; Math. Dept., Ohio State Univ., Colombus, Ohio 43210, USA.

NIKOLSKI, S.M.; Mathematical Inst. USSR Acad. of Sc., ul Vavilova 42, Moscow, USSR.

NOWAK, D. ; Humboldt University, Dept. of Mathematics, PSF 1297, Berlin, GDR.

OETTLI, W. ; Dept. of Mathematics, Manheim Univ., Manheim, FRG.

PAPINI, P.L.; Dipartamento di Matematica, University of Bologna, I-40126, ITALY.

de PEDRO, A.;Facultad de Matemática-Cibernética, U. de La Habana, La Habana, CUBA.

PEREZ, I.; Facultad de Matemática-Cibernética, U. de La Habana, La Habana, CUBA.

PEREZ, M.C.; Facultad de Matemática-Cibernética, U. de La Habana, La Habana, CUBA.

PETRUSHEV, P.; Institute of Mathematics, Bulgarian Acad. of Sc., 1090 Sofia, BULGARIA.

PIEDRA, R.; Facultad de Matemática-Cibernética, U. de La Habana, La Habana, CUBA.

PIJEIRA, H. Centro Universitario de Matanzas, Matanzas, CUBA.

PIÑEIRO DIAZ, L.R.; Facultad de Matemática-Cibernética, U. de La Habana, La Habana, CUBA.

POBEDRIA, B.E., Fac. of Math. and Mech., Moscow State Univ., Moscow, USSR.

QUIÑONES, H.; Dept. de Mat., Instituto Técnico Militar, La Habana, CUBA.

RAKHMANOV, E.A.; Mathematical Inst. USSR Acad. of Sc., ul. Vavilova 42, Moscow, USSR.

RODRIGUEZ, C.; Instituto de Investigaciones del Transporte, La Habana, CUBA.

RODRIGUEZ, M.G.; Fac. de Mat., UAM, Mexico D.F., MEXICO.

RODRIGUEZ, R.; Facultad de Matemática-Cibernética, U. de La Habana, La Habana, CUBA.

ROMERO, D.; IIMAS-UNAM, Mexico D.F., MEXICO.

RONVEAUX, A.; Mathematical Physics, Facultes Universitares N-D de la Paix, 5000 Namur, BELGIUM.

ROVNA, E.A.; Fac. of Math., Univ. of Grodno, Grodno 230009, USSR.

RUIZ, A.; Facultad de Matemática-Cibernética, U. de La Habana, La Habana, CUBA.

RUSSELL, D.; York University, 4700 Keele st., North York, Ontario, M3J 1P3, CANADA.

SAFF, E.B.; Math. Dept, University of South Florida, , Tampa, FL 33620, USA.

SANDIORE, G.; Dept. Matematica Aplicada, E.T.S. Ingenieros Industriales, 28006 Madrid, ESPAÑA.

SASTRE, L.; CENIC, aptdo. 6990, Havana, CUBA.

SCHLICHTING, G.; Math. Inst. TUM, Munich. FRG.

SILVA, C.; Facultad de Matemática-Cibernética, U. de La Habana, La Habana, CUBA.

STECHKIN, S.B.; Mathematical Inst. USSR Acad. of Sc., ul. Vavilova 42, Moscow, USSR.

SUAREZ, M.; MES, 23 y F, Vedado, La Habana, CUBA.

SZABADOS, J.; Inst. Mathematics, Hungarian Acad. of Sc., Budapest, HUNGARY.

TEMLYAKOV, V.N., Mathematical Inst. USSR Acad. of Sc., ul. Vavilova 42, Moscow. USSR.

TIKHOMIROV, V.M.; Fac. of Math. and Mech., Moscow State Univ., Moscow, USSR.

TIMMERMANS, C.A.; University of Delft, Koevorde 9, 8446 NB, Heerenveen, The NETHERLANDS.

TRIGGIANI, R.; Dept. of Math., University of Florida, Gainesville, Florida 36211, USA.

VALDES, C.; Facultad de Matemática-Cibernética, U. de La Habana, La Habana, CUBA.

VARDANIAN, T.A.; CINAG, Universidad de La Habana, La Habana, CUBA.

VERDAGUER, R.; Facultad de Matemática-Cibernética, U. de La Habana, La Habana, CUBA.

VERTESI, P.; Mathematical Inst. Hungarian Academy of Sciences, Budapest, P.O.Box 127, HUNGARY.

VIDAL, C.M.; Facultad de Matemática-Cibernética, U. de La Habana, La Habana, CUBA.

VILARIÑO, D.; Facultad de Matemática-Cibernética, U. de La Habana, La Habana, CUBA.

VINUESA, J.; Facultad de Ciencias, Univ. de Santander, Santander, ESPAÑA.

van WICKEREN, E.; Math. Dept., Technische Hochschule Aachen, Aachen, FRG.

Vol. 1259: F. Cano Torres, Desingularization Strategies for Three-Dimensional Vector Fields. IX, 189 pages. 1987.

Vol. 1260: N.H. Pavel, Nonlinear Evolution Operators and Semigroups. VI, 285 pages. 1987.

Vol. 1261: H. Abels, Finite Presentability of S-Arithmetic Groups. Compact Presentability of Solvable Groups. VI, 178 pages. 1987.

Vol. 1262: E. Hlawka (Hrsg.), Zahlentheoretische Analysis II. Seminar, 1984–86. V, 158 Seiten. 1987.

Vol. 1263: V.L. Hansen (Ed.), Differential Geometry. Proceedings, 1985. XI, 288 pages. 1987.

Vol. 1264: Wu Wen-tsün, Rational Homotopy Type. VIII, 219 pages. 1987.

Vol. 1265: W. Van Assche, Asymptotics for Orthogonal Polynomials. VI, 201 pages. 1987.

Vol. 1266: F. Ghione, C. Peskine, E. Sernesi (Eds.), Space Curves. Proceedings, 1985. VI, 272 pages. 1987.

Vol. 1267: J. Lindenstrauss, V.D. Milman (Eds.), Geometrical Aspects of Functional Analysis. Seminar. VII, 212 pages. 1987.

Vol. 1268: S.G. Krantz (Ed.), Complex Analysis. Seminar, 1986. VII, 195 pages. 1987.

Vol. 1269: M. Shiota, Nash Manifolds. VI, 223 pages. 1987.

Vol. 1270: C. Carasso, P.-A. Raviart, D. Serre (Eds.), Nonlinear Hyperbolic Problems. Proceedings, 1986. XV, 341 pages. 1987.

Vol. 1271: A.M. Cohen, W.H. Hesselink, W.L.J. van der Kallen, J.R. Strooker (Eds.), Algebraic Groups Utrecht 1986. Proceedings. XII, 284 pages. 1987.

Vol. 1272: M.S. Livšic, L.L. Waksman, Commuting Nonselfadjoint Operators in Hilbert Space. III, 115 pages. 1987.

Vol. 1273: G.-M. Greuel, G. Trautmann (Eds.), Singularities, Representation of Algebras, and Vector Bundles. Proceedings, 1985. XIV, 383 pages. 1987.

Vol. 1274: N. C. Phillips, Equivariant K-Theory and Freeness of Group Actions on C*-Algebras. VIII, 371 pages. 1987.

Vol. 1275: C.A. Berenstein (Ed.), Complex Analysis I. Proceedings, 1985–86. XV, 331 pages. 1987.

Vol. 1276: C.A. Berenstein (Ed.), Complex Analysis II. Proceedings, 1985–86. IX, 320 pages. 1987.

Vol. 1277: C.A. Berenstein (Ed.), Complex Analysis III. Proceedings, 1985–86. X, 350 pages. 1987.

Vol. 1278: S.S. Koh (Ed.), Invariant Theory. Proceedings, 1985. V, 102 pages. 1987.

Vol. 1279: D. Ieşan, Saint-Venant's Problem. VIII, 162 Seiten. 1987.

Vol. 1280: E. Neher, Jordan Triple Systems by the Grid Approach. XII, 193 pages. 1987.

Vol. 1281: O.H. Kegel, F. Menegazzo, G. Zacher (Eds.), Group Theory. Proceedings, 1986. VII, 179 pages. 1987.

Vol. 1282: D.E. Handelman, Positive Polynomials, Convex Integral Polytopes, and a Random Walk Problem. XI, 136 pages. 1987.

Vol. 1283: S. Mardešić, J. Segal (Eds.), Geometric Topology and Shape Theory. Proceedings, 1986. V, 261 pages. 1987.

Vol. 1284: B.H. Matzat, Konstruktive Galoistheorie. X, 286 pages. 1987.

Vol. 1285: I.W. Knowles, Y. Saitō (Eds.), Differential Equations and Mathematical Physics. Proceedings, 1986. XVI, 499 pages. 1987.

Vol. 1286: H.R. Miller, D.C. Ravenel (Eds.), Algebraic Topology. Proceedings, 1986. VII, 341 pages. 1987.

Vol. 1287: E.B. Saff (Ed.), Approximation Theory, Tampa. Proceedings, 1985–1986. V, 228 pages. 1987.

Vol. 1288: Yu. L. Rodin, Generalized Analytic Functions on Riemann Surfaces. V, 128 pages, 1987.

Vol. 1289: Yu. I. Manin (Ed.), K-Theory, Arithmetic and Geometry. Seminar, 1984–1986. V, 399 pages. 1987.

Vol. 1290: G. Wüstholz (Ed.), Diophantine Approximation and T scendence Theory. Seminar, 1985. V, 243 pages. 1987.

Vol. 1291: C. Mœglin, M.-F. Vignéras, J.-L. Waldspurger, Corres dances de Howe sur un Corps p-adique. VII, 163 pages. 1987

Vol. 1292: J.T. Baldwin (Ed.), Classification Theory. Proceedi 1985. VI, 500 pages. 1987.

Vol. 1293: W. Ebeling, The Monodromy Groups of Isolated Singu ties of Complete Intersections. XIV, 153 pages. 1987.

Vol. 1294: M. Queffélec, Substitution Dynamical Systems – Spe Analysis. XIII, 240 pages. 1987.

Vol. 1295: P. Lelong, P. Dolbeault, H. Skoda (Réd.), Sémin d'Analyse P. Lelong – P. Dolbeault – H. Skoda. Seminar, 1985/1 VII, 283 pages. 1987.

Vol. 1296: M.-P. Malliavin (Ed.), Séminaire d'Algèbre Paul Dubre Marie-Paule Malliavin. Proceedings, 1986. IV, 324 pages. 1987.

Vol. 1297. Zhu Y.-l., Guo B.-y. (Eds.), Numerical Methods for Pa Differential Equations. Proceedings. XI, 244 pages. 1987.

Vol. 1298: J. Aguadé, R. Kane (Eds.), Algebraic Topology, Barcel 1986. Proceedings. X, 255 pages. 1987.

Vol. 1299: S. Watanabe, Yu.V. Prokhorov (Eds.), Probability The and Mathematical Statistics. Proceedings, 1986. VIII, 589 pa 1988.

Vol. 1300: G.B. Seligman, Constructions of Lie Algebras and t Modules. VI, 190 pages. 1988.

Vol. 1301: N. Schappacher, Periods of Hecke Characters. XV, pages. 1988.

Vol. 1302: M. Cwikel, J. Peetre, Y. Sagher, H. Wallin (Eds.), Func Spaces and Applications. Proceedings, 1986. VI, 445 pages. 1988

Vol. 1303: L. Accardi, W. von Waldenfels (Eds.), Quantum Probab and Applications III. Proceedings, 1987. VI, 373 pages. 1988.

Vol. 1304: F.Q. Gouvêa, Arithmetic of p-adic Modular Forms. VIII, pages. 1988.

Vol. 1305: D.S. Lubinsky, E.B. Saff, Strong Asymptotics for Extre Polynomials Associated with Weights on ℝ. VII, 153 pages. 1988.

Vol. 1306: S.S. Chern (Ed.), Partial Differential Equations. Proce ings, 1986. VI, 294 pages. 1988.

Vol. 1307: T. Murai, A Real Variable Method for the Cauchy Transfo and Analytic Capacity. VIII, 133 pages. 1988.

Vol. 1308: P. Imkeller, Two-Parameter Martingales and Their Quad tic Variation. IV, 177 pages. 1988.

Vol. 1309: B. Fiedler, Global Bifurcation of Periodic Solutions w Symmetry. VIII, 144 pages. 1988.

Vol. 1310: O.A. Laudal, G. Pfister, Local Moduli and Singularities. 117 pages. 1988.

Vol. 1311: A. Holme, R. Speiser (Eds.), Algebraic Geometry, S dance 1986. Proceedings. VI, 320 pages. 1988.

Vol. 1312: N.A. Shirokov, Analytic Functions Smooth up to Boundary. III, 213 pages. 1988.

Vol. 1313: F. Colonius, Optimal Periodic Control. VI, 177 pag 1988.

Vol. 1314: A. Futaki, Kähler-Einstein Metrics and Integral Invariants. 140 pages. 1988.

Vol. 1315: R.A. McCoy, I. Ntantu, Topological Properties of Spaces Continuous Functions. IV, 124 pages. 1988.

Vol. 1316: H. Korezlioglu, A.S. Ustunel (Eds.), Stochastic Analysis a Related Topics. Proceedings, 1986. V, 371 pages. 1988.

Vol. 1317: J. Lindenstrauss, V.D. Milman (Eds.), Geometric Aspects Functional Analysis. Seminar, 1986–87. VII, 289 pages. 1988.

Vol. 1318: Y. Felix (Ed.), Algebraic Topology – Rational Homoto Proceedings, 1986. VIII, 245 pages. 1988

Vol. 1319: M. Vuorinen, Conformal Geometry and Quasiregu Mappings. XIX, 209 pages. 1988.